Low-Temperature Combustion Strategies

Low-Temperature Combustion Strategies: Fundamentals and Applications is a comprehensive guide to one of the most promising approaches in advanced engine technology—low-temperature combustion (LTC). Bridging the gap between fundamental science and engineering practice, this book explores the mechanisms, strategies, and real-world implementations of LTC systems to enable cleaner, more efficient internal combustion engines. Covering a range of cutting-edge topics—including homogeneous charge compression ignition, stratified charge compression ignition, dual-fuel systems, combustion chemistry, emission control, and advanced engine configurations—this book provides in-depth analysis and engineering insights.

Features:

- Presents the LTC techniques, including combustion modeling, combustion modes, optimization techniques, and emission formation control.
- Delves into advanced combustion modes offering detailed insights into their operation, control strategies, and optimization techniques.
- Discusses advanced combustion control and optimization techniques.
- Includes real-world case studies and applications in automotive and power generation sectors.
- Explores the future directions and emerging trends in LTC research, including the integration of renewable fuels and prospects for commercialization.

Practical case studies, current research findings, and emerging trends make this an indispensable reference for students, researchers, and professionals in automotive engineering, combustion science, and sustainable energy systems.

Low-Temperature Combustion Strategies

Fundamentals and Applications

Edited by

Manisha Priyadarshini,
Sumit Kanchan, and Swastik Pradhan

CRC Press
Taylor & Francis Group
Boca Raton London New York

CRC Press is an imprint of the
Taylor & Francis Group, an informa business

Designed cover image: © Sumit Kanchan

First edition published 2026
by CRC Press
2385 NW Executive Center Drive, Suite 320, Boca Raton FL 33431

and by CRC Press
4 Park Square, Milton Park, Abingdon, Oxon, OX14 4RN

CRC Press is an imprint of Taylor & Francis Group, LLC

ISBN: 9781032983240 (hbk)
ISBN: 9781041060802 (pbk)
ISBN: 9781003633631 (ebk)

DOI: 10.1201/9781003633631

Typeset in Times
by codeMantra

Contents

Preface

The urgent global demand for cleaner and more energy-efficient transportation has intensified the focus on advanced combustion technologies. Among them, **low-temperature combustion (LTC)** has emerged as a transformative concept with the potential to drastically reduce emissions while enhancing engine efficiency. This book, *Low-Temperature Combustion Strategies: Fundamentals and Applications*, was conceived to provide a comprehensive and structured understanding of LTC strategies, addressing both foundational principles and their real-world applications in modern internal combustion engines.

This book is motivated by the swift advancement of LTC technologies and the increasing volume of research that crosses various fields, including thermodynamics, fluid mechanics, combustion chemistry, and fuel science. Despite the existence of many studies and technical papers, there is a lack of consolidated resources that align with the curriculum and connect theory to practice. This book aims to be a vital reference for undergraduate and postgraduate students, researchers, practicing engineers, and industry professionals who wish to enhance their comprehension of LTC and its significance for the next generation of engine systems.

This book, structured in chapters that incrementally develop fundamental concepts, addresses the science of combustion at lower temperatures; the functioning principles of LTC modes like homogeneous charge compression ignition, Premixed Charge Compression Ignition (PCCI) and Reactivity Controlled Compression Ignition (RCCI); fuel adaptability; emissions characteristics; and the significance of alternative fuels. Current challenges in LTC, such as ignition control, load range limitations, and integration with hybrid systems, are given special attention. Additionally, this book examines how computational modeling, diagnostics, and control strategies can be employed to optimize LTC engines.

We hope that this work contributes meaningfully to academic discourse, industrial innovation, and sustainable energy education. We believe that as researchers, educators, and engineers engage more deeply with LTC, the path toward environmentally responsible mobility and energy systems will become more attainable.

We express our sincere gratitude to all the researchers, collaborators, reviewers, and institutions whose contributions have informed and inspired the contents of this book.

Dr. Manisha Priyadarshini
Dr. Swastik Pradhan
Mr. Sumit Kanchan

Acknowledgments

We would like to express our deepest gratitude to all those who supported and guided us throughout the development of this book. This work would not have been possible without the contributions, encouragement, and expertise of numerous individuals and institutions. First and foremost, we are thankful to our academic mentors and research supervisors for their invaluable guidance and continuous support. Their insights and constructive feedback greatly shaped the quality and scope of our work.

We extend our appreciation to our colleagues, collaborators, and fellow researchers whose discussions, suggestions, and shared experiences helped refine many of the ideas presented in this book. We are also grateful to the technical staff and research assistants who supported the simulations, experiments, and case study documentation with diligence and precision. Our sincere thanks go to the institutions and funding bodies that provided the necessary resources and infrastructure to carry out this research, as well as to the industry partners who shared practical insights and data to strengthen the real-world relevance of our findings.

Finally, we would like to thank our families for their patience, understanding, and unwavering support throughout this endeavor. Their encouragement kept us motivated and focused, and for that, we are truly grateful.

About the Editors

Manisha Priyadarshini holds a Ph.D. in Mechanical Engineering from the National Institute of Technology, Rourkela. She works as Assistant professor in CGU-Bajaj engineering skills training centre, C V Raman Global University, Bhubaneswar, India. With more than 40 research publications in reputed international journals and conferences, her work focuses on advanced mechanical aspects and manufacturing technologies with energy efficiency. She has collaborated with researchers, academicians, and industry experts on several projects and serves as a peer reviewer for published journals.

Sumit Kanchan is an automobile engineer, researcher, and learning architect with expertise in building industry-aligned ecosystems for EV, R&D, and AI-driven learning. As Head-Projects & Capacity Building at ISIEINDIA, he leads national initiatives including Project Saksham, AI-integrated EV test-bench development, and multi-sector capacity-building programs. He has authored 30+ papers, filed 15 patents, and developed 30+ EV lab and production frameworks. Formerly at LPU, he guided teams to 90+ awards and mentored startups generating ₹6 crore+ annually. He is an Innovation Ambassador and a Ph.D. researcher in low-temperature combustion at NIT Jalandhar.

Swastik Pradhan received his Ph.D. in Production Engineering from NIT Rourkela and currently serves as Assistant Professor in the School of Mechanical Engineering at Lovely Professional University, Punjab, India. With over 7 years of teaching and research experience, he has guided more than 20 research projects and published over 60 research papers. His expertise lies in CAD/CAM, traditional and nontraditional machining, FEA, and industrial safety. Dr. Pradhan is also an active peer reviewer for leading journals.

Contributors

Geeta Arora
Department of Mathematics,
Lovely Professional University
Phagwara, India

Harpreet Singh Bains
Department of Quality and Risk Management
Statnett SF
Oslo, Norway

Manpreet Singh Bains
School of Mechanical Engineering
Lovely Professional University
Phagwara, India

Abhishek Barua
Department of Automobile Engineering
Parala Maharaja Engineering College
Berhampur, India

Zhannat Bekbolatova
Department of Power Engineering
Satbayev University
Almaty, Kazakhstan

Suman Chatterjee
Advanced Processing Technology Research Centre
Dublin City University
Dublin, Ireland

Aditya Chauhan
Department of Mechanical Engineering
Clemson University
Clemson, South Carolina

Abhijit Cholkar
SEAM
South East Technological University
Waterford, Ireland
and
I-Form SFI Research Centre for Advanced Manufacturing
University College Dublin
Dublin, Ireland

Sudhanshu Dogra
School of Mechanical Engineering
Lovely Professional University
Phagwara, India

Masilamani Gowtham
R&D Laboratory, Department of Aeronautical Engineering
Excel Engineering College
Namakkal, India

Nancy Gulati
School of Engineering, Science and Environment
University of Newcastle
Newcastle, Australia

Jaya Gupta
Lovely Professional University
Phagwara, India

Hardik
School of Mechanical Engineering
Lovely Professional University
Phagwara, India

Siddharth Jeet
Department of Pyrometallurgy & Materials Engineering
CSIR-IMMT
Bhubaneswar, India

Sumit Kanchan
Research and Development Cell, ISIEINDIA Private Limited
Noida, India

Amit Kumar
Department of Mechanical Engineering
Dr. B R Ambedkar NIT
Jalandhar, India

Pratibha Mahajan
Department of Artificial Intelligence
Vishwakarma University
Pune, India

Almagul Mergalimova
S. Seifullin Kazakh Agrotechnical Research University
Energy Institute
Astana, Kazakhstan

Saulesh Minazhova
Department of Power Engineering
Satbayev University
Almaty, Kazakhstan

Monalin Mishra
Department of Environment & Sustainability
CSIR-IMMT
Bhubaneswar, India

Shubham Mishra
Lovely Professional University
Phagwara, India

Dillip Kumar Mohanta
Department of Mechanical Engineering
Centurion University
Bhubaneswar, India

Kamalakanta Muduli
School of Mechanical Engineering
Papua New Guinea University of Technology
Lae, Papua New Guinea

Nidhi
Department of Mechanical Engineering
Galgotias University
Greater Noida, India

Bulbul Ongar
Department of Power Engineering
Satbayev University
Almaty, Kazakhstan

Manas Ranjan Padhi
Department of Mechanical Engineering
Centurion University
Bhubaneswar, India

Surya Narayan Panda
Department of Production Engineering
Birsa Institute of Technology, Sindri, Dhanbad, India

Swastik Pradhan
Department of Mechanical Engineering
Lovely Professional University
Phagwara, India

Ramakrishnan Prakash
School of Mechanical Engineering
VIT
Vellore, India

Bhanu Priya
School of Electronics & Electrical Engineering
Lovely Professional University
Jalandhar, India

Manisha Priyadarshini
CGU- Bajaj Engineering Skills Training Centre
C. V. Raman Global University
Bhubaneswar, India

Mahendra Kumar Rath
Department of Mechanical Engineering
Centurion University
Bhubaneswar, India

Sibasis Harihar Sahu
Department of Automobile Engineering
Parala Maharaja Engineering College
Berhampur, India

Sangram Keshari Samal
Mechanical Engineering Department
National Yang Ming Chiao Tung University
Hsinchu, Taiwan

Sarbjot Singh Sandhu
Department of Mechanical Engineering
NIT Jalandhar
Jalandhar, India

Yerlan Sarsenbayev
Department of Power Engineering
Satbayev University
Almaty, Kazakhstan

Ankur Shah
Controls Engineer II
Entek International
Wake Forest, North Carolina

Aashish Sharma
School of Mechanical Engineering
Lovely Professional University
Phagwara, India

Sumit Shoor
Department of Mechanical Engineering
Lovely Professional University
Phagwara, India

Sangam Kumar Singh
Department of Mechanical Engineering
NIT Rourkela
Rourkela, India

Satsimran Singh
Department of Mechanical Engineering
NIT Jalandhar
Jalandhar, India

Santosh Kumar Tamang
Department of Mechanical Engineering
North Eastern Regional Institute of Science & Technology (NERIST)
Nirjuli, India

Amit Thakur
Department of mechanical engineering
School of Mechanical Engineering
Lovely Professional University
Punjab, India

Yashvardhan
School of Mechanical Engineering
Lovely Professional University
Phagwara, India

Anara Yegzekova
Department of Power Engineering
ALT University
Almaty, Kazakhstan

1 Introduction to Low-Temperature Combustion

Manisha Priyadarshini, Sangram Keshari Samal, Abhijit Cholkar, and Swastik Pradhan

1.1 BACKGROUND AND MOTIVATION

The global energy and transportation landscape is undergoing a transformative shift driven by increasingly stringent environmental regulations, concerns over fossil fuel depletion, and the urgent need to combat climate change. Internal combustion engines (ICEs), although mature and widely used, remain significant contributors to greenhouse gas emissions and air pollutants such as nitrogen oxides (NO_x), particulate matter (PM), unburned hydrocarbons (HC), and carbon monoxide (CO). These emissions contribute to severe environmental and health issues, including global warming, urban smog, and respiratory diseases.

To meet these difficulties, contemporary combustion research is concentrating on creating engine technologies that reduce emissions while maintaining performance and fuel efficiency. Of the different strategies examined, low-temperature combustion (LTC) has proven to be a promising method. LTC functions under conditions that simultaneously reduce NO_x and soot formation, unlike traditional spark-ignition (SI) or compression-ignition (CI) systems, providing a cleaner and more efficient combustion route.

The potential of LTC technologies lies not only in their ability to drastically reduce emissions but also in their compatibility with both fossil-derived and renewable fuels. This makes them attractive for a future where energy systems must balance efficiency, sustainability, and environmental compliance. As such, LTC strategies are at the forefront of next-generation engine research, supported by advances in engine design, fuel chemistry, and control methodologies.

1.2 OVERVIEW OF COMBUSTION IN INTERNAL COMBUSTION ENGINES

Understanding the basics of combustion in ICEs and the constraints of conventional systems is crucial to recognizing LTC's development and significance. Generally, ICEs are divided into two main types:

DOI: 10.1201/9781003633631-1

- **SI Engines:** Operate by premixing air and fuel before introducing the mixture into the combustion chamber, where it is ignited by a spark plug. SI engines typically operate on gasoline and are characterized by smooth combustion but relatively high levels of unburned HC and CO.
- **CI Engines:** Commonly known as diesel engines, these rely on the autoignition of fuel injected directly into highly compressed hot air. CI engines exhibit higher thermal efficiency but suffer from increased NO_x and particulate emissions due to heterogeneous fuel–air mixing and high-temperature combustion zones.

1.2.1 Combustion Emissions and Challenges

Thermal NO_x is produced when combustion temperatures in SI and CI engines surpass 2000 K. Furthermore, due to the varied nature of combustion in CI engines, areas with an abundance of fuel are created. These regions are the source of soot particle creation. The main mechanisms for the formation of pollutants include:

- **NO_x:** Formed via the Zeldovich mechanism in high-temperature regions.
- **PM:** Results from incomplete combustion of fuel-rich pockets.
- **CO and HC:** Result from incomplete combustion in low-temperature or quenching zones.

These emissions compromise air quality and restrict compliance with contemporary emission standards without costly aftertreatment systems.

1.3 WHAT IS LOW-TEMPERATURE COMBUSTION (LTC)?

LTC encompasses a range of combustion methods designed to foster autoignition-dominated combustion at reduced in-cylinder temperatures, usually below 1700 K. By steering clear of the high-temperature areas that promote NO_x and soot generation, LTC facilitates a cleaner combustion process with minimal emissions, frequently eliminating the need for catalytic converters or diesel particulate filters (DPFs). The characteristics of LTC are as follows:

- Lower peak combustion temperatures (typically between 1100 and 1600 K)
- Autoignition-based combustion without external ignition (or with partial control)
- Premixed or partially premixed charge to ensure better air–fuel homogeneity
- Fast heat release, often over a narrow crank angle window
- Use of dilute mixtures, either via exhaust gas recirculation (EGR) or lean-burn operation

These traits help to maintain satisfactory thermal efficiency, while NO_x and PM emissions are reduced. Nonetheless, they bring considerable difficulties in managing ignition timing and guaranteeing combustion stability, particularly under high loads or during transient operations.

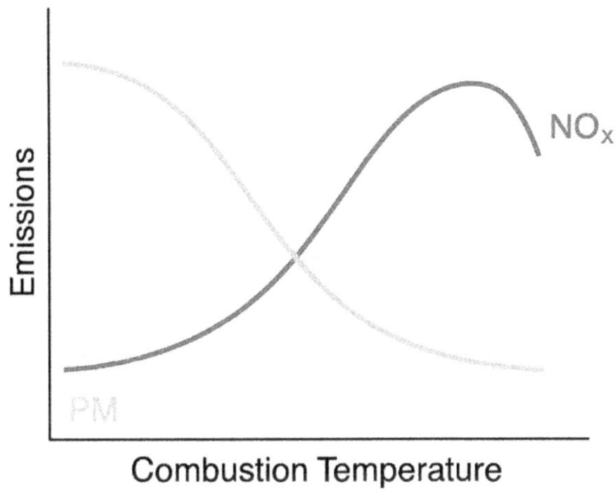

FIGURE 1.1 Variation of combustion temperature and emission for NOx and Particulate Matter.

1.3.1 The Temperature Trade-Off

Temperature has a strong influence on combustion reactions. At elevated temperatures (>2000 K), the generation of NO_x rises owing to thermal mechanisms. Conversely, at low temperatures (<1100 K), the reaction kinetics decelerate, which poses a risk of incomplete combustion and increased HC/CO emissions. The goal of LTC strategies is to achieve a balance by functioning within a moderate temperature range, thereby reducing NO_x and PM while maintaining efficiency. Figure 1.1 shows the emission trends as the function of in-cylinder combustion temperature.

1.3.2 Types of LTC Strategies

Several practical implementations of LTC have emerged, including:

- Homogeneous Charge Compression Ignition (HCCI)
- Reactivity Controlled Compression Ignition (RCCI)
- Partially Premixed Compression Ignition (PPCI or PPC)
- Stratified Charge Compression Ignition (SCCI)

All of these strategies target the optimization of fuel–air mixture preparation, ignition timing, and combustion phasing to ensure a clean and efficient operation. The main differences are in the management of charge stratification, reactivity control, and timing control.

1.3.3 Why LTC Matters

The compelling benefits of LTC include:

- Near-zero NO_x and soot emissions
- High indicated thermal efficiency
- Flexibility with various fuels, including biofuels and alternative fuels
- Potential for simplified engine architecture with reduced aftertreatment requirements

Nevertheless, to put it into practice, one must surmount challenges including control of combustion timing, a limited operational range, cold-start problems, and transient stability. Advanced engine control systems, fuel design, and model-based combustion optimization are actively addressing these challenges, which will be discussed in subsequent chapters.

1.4 KEY LTC STRATEGIES

LTC encompasses a range of innovative combustion strategies aimed at achieving clean and efficient combustion under controlled thermal conditions, rather than being a single process. The methods of introducing, mixing, and igniting fuel in the combustion chamber vary among these strategies. This section will examine the main LTC techniques: HCCI, PPCI, RCCI, SCCI, and dual-fuel LTC, each with its advantages and drawbacks.

1.4.1 Homogeneous Charge Compression Ignition (HCCI)

HCCI incorporates the most advantageous features of both CI and SI engines. In HCCI, an air–fuel mixture with uniform composition is injected into the cylinder, where it ignites spontaneously due to the heat produced by compression. At the time of combustion, there is no direct fuel injection or spark plug; rather, because conditions are uniform throughout, the entire charge ignites almost all at once. The main characteristics of HCCI are:

- **Homogeneous Mixture:** As with SI engines.
- **Compression Autoignition:** Similar to CI engines.
- **Rapid Combustion:** Takes place within a limited crank angle, leading to a shorter combustion duration.
- **Thin Conditions:** Made possible through lean mixtures or EGR

Some of its benefits are:

- Ultra-low NO_x and soot emissions due to the absence of high-temperature zones and fuel-rich pockets
- High thermal efficiency from rapid and complete combustion
- Potential for multi-fuel usage (gasoline, ethanol, and biofuels)

Recent studies have concentrated on employing low-octane fuels, engine mapping, and chemical kinetic modeling to improve HCCI control and broaden its usable range [1].

1.4.2 Partially Premixed Compression Ignition (PPCI/PPC)

A hybrid technique called PPCI involves partially premixing the fuel with air prior to ignition. This strategy strikes a balance between the controllability of CI combustion and the clean emissions of HCCI by providing greater control over the air–fuel equivalency ratio and combustion phasing. Fuel injection usually takes place during the compression stroke in PPCI, enabling partial mixing and stratification. This improves timing control by delaying autoignition. Its capacity to regulate in-cylinder temperatures, which considerably lowers the production of nitrogen oxides (NO_x), is one of its primary characteristics. Partial premixing of fuel and air also reduces soot emissions. Advanced fuel injection techniques, which provide precise combustion process management, are the main means of controlling ignition timing in PPCI.

One benefit of PPCI over HCCI is its increased operational flexibility, which enables it to handle higher loads and a wider range of engine speeds. Its practical viability is further increased by the ease with which it can be integrated into current diesel engine platforms with only minor adjustments. The requirement for complex fuel injection systems, such as high-pressure common rail configurations, is one of the difficulties PPCI faces. Control may become more difficult due to the combustion process's extreme sensitivity to fuel type and injection timing. Furthermore, extra emission control measures are still required to meet strict regulatory standards under specific load conditions. PPCI has gained traction in commercial applications due to its balance of efficiency, emission control, and operational feasibility [2].

1.4.3 Reactivity Controlled Compression Ignition (RCCI)

RCCI is a dual-fuel LTC technology that controls combustion by using two fuels with different reactivities. A low-reactivity fuel, such as gasoline or natural gas, is premixed with air and inducted into the cylinder during the compression stroke, whereas a high-reactivity fuel, such as diesel, is injected immediately. Combustion is the result of spatial and temporal interactions between the two fuels.

Because of its LTC properties, RCCI has ultra-low emissions of soot and nitrogen oxides (NO_x), among other noteworthy benefits. Additionally, it offers fuel flexibility, enabling the use of renewable fuel blends like gasoline and biodiesel, supporting the objectives of sustainable energy. Moreover, RCCI is an energy-efficient substitute for traditional combustion modes since it retains a high thermal efficiency even when operating at partial load.

But RCCI has significant drawbacks. To control the independent delivery of two fuels, sophisticated engine control and fuel injection systems are needed. These dual fuel systems must be precisely coordinated to achieve optimal performance, which increases the complexity of the system. Its broad use in commercial applications may also be constrained by the requirement for complex hardware and control mechanisms, which raises costs and complicates packaging.

Studies by Ref. [3] demonstrated that RCCI engines could achieve 90% reductions in NO_x and 80% reductions in soot compared to conventional diesel engines.

1.4.4 Stratified Charge Compression Ignition (SCCI)

A combustion technique called SCCI uses fuel–air mixtures that are spatially stratified to accomplish controlled ignition sequencing. Staged combustion is made possible by this stratification, which improves combustion stability and moderates peak pressure rise rates. In contrast to HCCI, which uses a completely mixed air–fuel charge, SCCI purposefully leaves some degree of mixture heterogeneity. More controlled combustion phasing is made possible by the strategic management of this nonuniformity. To achieve the desired stratification, SCCI is usually implemented using multiple fuel injection events during the intake and compression strokes.

Improved control over combustion phasing through mixture stratification is one of SCCI's main advantages; this makes it more resilient to a greater range of engine loads and speeds. Furthermore, it requires only minor structural modifications to be converted to traditional diesel engines. SCCI does, however, have some drawbacks. For efficient operation, precise control over air–fuel mixing and injection timing is essential. Because of the localized high-temperature areas, NO_x emissions are typically marginally higher than with HCCI or RCCI. More sophisticated fuel delivery hardware is required because the injection system must be able to manage several fuel injections events per engine cycle.

1.4.5 Dual-Fuel LTC Approaches

Using two distinct fuel types—usually one gaseous and one liquid—to enhance combustion control, fuel flexibility, and engine performance is known as dual-fuel combustion. Diesel and compressed natural gas, diesel and ethanol, and gasoline and biodiesel are examples of common dual-fuel combinations. With this method, a smaller amount of pilot fuel is injected during the compression stroke to start combustion through autoignition, while the primary fuel is premixed with air and introduced into the cylinder during the intake stroke.

There are various benefits to this approach. It lessens dependency on a single fuel source, enabling more flexibility in fuel choice, such as the utilization of renewable or low-carbon fuels. Additionally, dual-fuel combustion offers more control over combustion characteristics and ignition timing. Nonetheless, the approach poses significant difficulties. To efficiently manage both fuel streams, dual fuel injection systems and complex control logic are needed. Cold-start problems and trouble forming the ideal mixture can result from specific fuel combinations. Furthermore, careful calibration and monitoring are required because the resulting emissions profile is susceptible to the physical and chemical characteristics of the chosen fuels (Table 1.1).

1.5 THERMODYNAMIC AND CHEMICAL FUNDAMENTALS OF LTC

Analyzing the performance, efficiency, and emissions characteristics of LTC requires an understanding of the thermodynamic and chemical principles underlying the process. LTC is governed by a complex interaction between in-cylinder mixing,

TABLE 1.1
Summary of LTC Strategies

Strategy	Fuel Type	Control Level	NO_x/Soot	Combustion Phasing	Key Limitation
HCCI	Single	Low	Very low	Difficult	Narrow operating range
PPCI	Single	Moderate	Low	Moderate	Injection timing
RCCI	Dual	High	Ultra-low	Excellent	Complex system
SCCI	Single	Moderate	Low	Moderate	Injection complexity
Dual-Fuel	Dual	Variable	Low	Variable	Fuel management

fuel oxidation chemistry, ignition delay, and heat release rates. The basic ideas that underlie LTC operation are covered in this section. The key thermodynamic distinction between LTC and traditional combustion lies in the temperature and pressure trajectories followed during the combustion process.

a. Reduced Peak Temperatures
   Because of the Zeldovich mechanism, combustion temperatures in traditional diesel or SI engines can rise above 2300 K, which results in a significant production of NO_x. On the other hand, LTC processes suppress the formation of thermal NO_x by operating at lower peak temperatures, usually below 1800 K.
b. Compression and Autoignition
   The primary source of autoignition in LTC systems is isentropic compression of the air–fuel mixture. When the piston gets closer to top dead center, the cylinder's internal temperature and pressure increase, triggering chemical reactions that ignite the engine on their own without the aid of an external ignition source like a glow plug or spark plug. Important elements that influence the exact timing of autoignition include the fuel's reactivity, the charge's initial temperature and pressure, and the amount of dilution delivered through EGR or the use of extra air. Together, these parameters dictate the start and trajectory of combustion in LTC systems.
c. Heat Release Characteristics
   One distinctive feature of LTC is its multistage heat release characteristic. Fuel oxidation at relatively low temperatures, usually below 1000 K, causes the first phase, often referred to as low-temperature heat release (LTHR). This stage serves as a crucial prelude to complete ignition and helps to gradually raise the pressure. The process enters the high-temperature heat release phase as the temperature rises more, during which full combustion occurs more quickly. In LTC engines, the LTHR phase can significantly affect the features of pressure rise and combustion noise, as well as the overall combustion phasing.

1.5.1 Chemical Kinetics of LTC

The chemistry of LTC is highly fuel-dependent and involves complex oxidation pathways that are sensitive to temperature, pressure, and fuel–air mixture characteristics.

a. **Low-Temperature Oxidation**:
   At low temperatures (600–900 K), the following reactions dominate:
   1. Fuel + O_2 → RO_2 (alkylperoxy radicals)
   2. RO_2 → QOOH (hydroperoxide radicals)
   3. QOOH + O_2 → ketohydroperoxides + OH•

   These low-temperature reactions initiate chain branching, accelerating the rate of heat release and eventually leading to autoignition. The cool flame phenomenon, observed in HCCI engines, is attributed to this phase.
b. **High-Temperature Oxidation**:
   Once the temperature exceeds ~1000 K:
   - Thermal decomposition of peroxides occurs
   - Rapid oxidation of intermediates such as CO and H_2
   - Main heat release through complete combustion of hydrocarbon chains

   This phase results in the main ignition event and determines the bulk of energy conversion.
c. **The Behavior of the Negative Temperature Coefficient (NTC)**:
   Raising the temperature paradoxically slows down the rate of reaction under the NTC regime, where LTC operates. This surprising effect results from the struggle between chain-propagating and chain-terminating processes. The NTC area is crucial for phasing control in RCCI and HCCI modes.
d. **Comparing Chemical and Physical Time Scales**:
   The ignition delay in LTC is significantly influenced by the residence and mixing time scales. If chemical processes proceed too slowly in comparison to piston motion, misfire may occur. Conversely, an overly brief delay leads to high pressure rise rates.

1.5.2 Role of Exhaust Gas Recirculation (EGR)

EGR is a commonly employed technique in LTC systems to regulate combustion characteristics and reduce emissions. It functions by lowering the in-cylinder oxygen concentration and absorbing heat, which helps to reduce peak combustion temperatures. Additionally, EGR extends the ignition delay, allowing for more precise control over the timing of combustion. Beyond these effects, EGR also influences fuel reactivity and enhances pre-ignition chemical activity by altering the thermodynamic conditions within the cylinder. For example, in gasoline-fueled HCCI engines, EGR rates between 30% and 50% are typically used to effectively suppress NO_x emissions while maintaining stable combustion performance.

1.5.3 Fuel Properties and Reactivity Index

The success of LTC strategies is tightly coupled with fuel characteristics, particularly:

- Octane number (ON) for gasoline-type fuels
- Cetane number (CN) for diesel-type fuels
- Ignition delay sensitivity
- Volatility and latent heat of vaporization

In RCCI and dual-fuel modes, pairing fuels with contrasting reactivity (e.g., gasoline + diesel) enables stratified ignition control and spatial heat release tailoring.

For dual-fuel or RCCI modeling, a normalized or empirical reactivity index may be constructed based on ignition delay or fuel composition. One example of a derived, simplified reactivity index formula is:

$$RI = \frac{1}{ID}$$

where, RI = Reactivity index of the blended fuel, ID = ignition delay (in crank angle degrees or milliseconds).

1.5.4 Combustion Stability and Knock

Controlling combustion stability is an important issue because LTC has a small window of stable combustion. Unpredictable engine performance can arise from significant cycle-to-cycle variation in in-cylinder pressure and heat release, which are crucial indicators of instability in LTC. The tendency for knock, especially at higher loads in modes like RCCI and HCCI, further limits engine operation, and rapid pressure increase rates can also result in excessive combustion noise.

To overcome these issues, researchers and engine engineers have implemented several control systems aimed at stabilizing combustion. One common tactic is to use in-cylinder pressure feedback devices, which enable real-time monitoring and combustion parameter change. Variable valve actuation is another effective method that allows for dynamic control of combustion phasing to maintain stability in a range of operating conditions. Furthermore, strategies such as fuel stratification and dual injection techniques are employed to tailor the air–fuel mixture's ignition timing and distribution, reducing sensitivity to operating fluctuations and extending the stable operating range of LTC systems.

1.5.5 Heat Transfer and Efficiency Considerations

LTC offers several mechanical and thermodynamic advantages because of its inherent ability to function at lower in-cylinder temperatures. Lower combustion temperatures increase total thermal efficiency by lowering thermal losses to the cylinder walls, especially in lean or highly diluted circumstances. Additionally, the reduced temperatures result in less mechanical and thermal stress on engine components, potentially extending engine longevity and durability. This advantage does come with some trade-offs, although. Excessive charge dilution or premature heat loss quenching the combustion process can cause incomplete fuel oxidation. As a result, emissions of carbon monoxide (CO) and unburned HC may increase, making it challenging to regulate emissions and guarantee full combustion in LTC engines.

1.5.6 Modeling and Simulation Tools

Advanced LTC strategies require precise modeling of in-cylinder processes. Popular tools include:

- CHEMKIN and Cantera for detailed chemical kinetics
- GT-Power, AVL FIRE, and CONVERGE computational fluid dynamics (CFD) for engine cycle simulation
- Zero-D and 3D models for combustion chamber dynamics

Kinetic models include multi-zone HCCI models, which segment the cylinder into regions with varying thermal and chemical states, capturing real-world combustion more accurately.

1.6 APPLICATIONS OF LTC IN ENGINES AND POWER SYSTEMS

The potential of LTC techniques to lower emissions and boost efficiency across a range of internal combustion (IC) engine platforms has garnered a lot of interest. Although automobile engines have been the subject of the majority of study, LTC technologies are also being employed more and more in military engines, hybrid cars, and stationary power production systems. This section looks at particular applications, advantages, and obstacles while highlighting the primary areas where LTC is having an influence.

1.6.1 LTC in Automotive Engines

Stricter emissions regulations and the global push for increased fuel efficiency have led to the automobile industry continuing to be the most active developer of LTC. To reduce NO_x and PM emissions in passenger cars without the need for expensive aftertreatment systems, automakers are investigating LTC. Additionally, by allowing lean or diluted combustion, LTC can increase part-load efficiency and promote gasoline compression ignition, a competitive alternative to traditional SI. Under a range of load and speed situations, Honda's Advanced Valve Timing and Lift Electronic Control (VTEC) HCCI prototype, for instance, showed reliable LTC performance with exhaust rebreathing and variable valve timing [4].

In commercial vehicles, particularly heavy-duty diesel applications, LTC modes such as dual-fuel RCCI and PPCI offer significant advantages. These include reduced reliance on aftertreatment systems such as selective catalytic reduction and DPF; reduced fuel expenses by partially replacing alternative fuels like natural gas or alcohols; and stable combustion even at high torque outputs, provided that combustion phasing is appropriately managed. One prominent example is the dual-fuel RCCI engine, which was developed by Navistar and the University of Wisconsin–Madison and met EPA 2010 pollution regulations without requiring any form of aftertreatment [5].

1.6.2 LTC in Range-Extended and Hybrid Electric Cars

LTC engines also work well with hybrid electric vehicle (HEV) systems because of their constant combustion behavior and outstanding part-load efficiency. Their ability to operate within precise, well-defined regimes makes them ideal for constant-speed applications, particularly in series hybrid systems. In plug-in hybrids or electric vehicles, LTC engines can be run at preset high-efficiency points, which reduces pollutants while the engine is operating and, by increasing fuel economy, lowers the cost of charging batteries.

1.6.3 LTC in the Production of Stationary Power

Stationary engines, such as those used in power plants and microgrid systems, provide another possible application for LTC technology. In these circumstances, LTC can significantly reduce fuel usage and pollution emissions, which is particularly important in urban or environmentally sensitive areas. The lower combustion temperatures associated with LTC also extend the engine components' operational life. Fuels like natural gas, syngas, and biogas are frequently used in LTC-capable stationary engines. For example, it has been shown that burning simulated syngas in HCCI mode reduces NO_x emissions by more than 90% when compared to conventional SI engines.

1.6.4 LTC in Military and Off-Highway Engines

Off-road and military applications like construction equipment and defense vehicles can also benefit greatly from LTC technology. Some of the primary advantages include reduced infrared signature from lower exhaust temperatures, quieter operation for stealth operations, and less heat stress on engine parts for increased dependability in difficult situations. Although these engines often operate at extremely changing loads and speeds, which can jeopardize LTC stability, advancements in flexible fuel strategies and adaptive control systems are helping to overcome these challenges.

1.6.5 LTC Engines with Two Fuels

Dual-fuel engines that adopt LTC methods like RCCI or multi-fuel PPCI can use a low-reactivity primary fuel (like gasoline or ethanol) and a high-reactivity pilot fuel (like diesel or dimethyl ether (DME)). This configuration allows for fine control of combustion phasing through discrete injection events, aids decarbonization activities, and provides fuel sourcing flexibility. Additionally, by improving charge uniformity, dual-fuel LTC lowers the soot-NO_x trade-off. In a case study from the

Wisconsin Engine Research Center, it was demonstrated that a 1.9L RCCI engine running on gasoline and diesel could achieve ultra-low emissions and over 50% indicated thermal efficiency [6].

1.6.6 LTC Using Renewable and Alternative Fuels

LTC's broad fuel compatibility makes it especially attractive for emerging sustainable energy systems. Among the fuels studied in LTC contexts are alcohols like ethanol, methanol, and butanol, which offer high octane ratings and effective evaporative cooling; biodiesel, which promotes low-temperature ignition due to its high CN; and DME, which is valued for its high volatility and reactivity. Research is also being done on hydrogen and ammonia blends that could allow for carbon-free combustion with characteristics akin to LTC. These fuels are commonly used in dual-fuel or RCCI systems to maximize performance and benefit from their complementing properties.

1.6.7 LTC in the Aerospace and Marine Industries

Although it is still in its infancy, LTC is being researched for particular applications in the maritime and aerospace sectors. Marine engines operating in emission control areas in particular may benefit from LTC's assistance in meeting stringent environmental restrictions. The aerospace industry is looking into LTC concepts for unmanned aerial vehicles to increase fuel efficiency and lower acoustic signatures, as well as for auxiliary power units in tiny aircraft to support low-noise operation. However, adoption in these areas is challenging due to complex thermal management needs and control systems.

1.6.8 Real-World Difficulties in LTC Applications

Before LTC can be put into effect, a number of practical challenges must be resolved, despite its great advantages. Because of LTC's short operating range, combustion stability is quite sensitive to changes in engine load and speed. The complexity of overall management is further increased by the requirement for intricate control over fuel injection, EGR, and combustion phasing in LTC systems. Low in-cylinder temperatures and high dilution levels make reliable ignition difficult in cold-start situations. Additionally, LTC performance is sometimes significantly impacted by fuel properties, necessitating dual-fuel setups or specially blended fuels. Therefore, if LTC is to be feasible in widely used applications, engine developers must carefully balance emissions reduction, thermal efficiency, control complexity, and total cost.

1.7 HISTORICAL DEVELOPMENT AND RESEARCH TRENDS IN LOW-TEMPERATURE COMBUSTION

The concept of LTC has evolved significantly over the last century, driven by the need for cleaner, more efficient IC engines. This section provides a chronological overview of key developments in LTC technologies and discusses recent research trends shaping the future of combustion science.

1.7.1 The HCCI Concept

The foundation of LTC is the early study of HCCI. In the 1970s and 1980s, researchers began studying the autoignition of uniform air–fuel mixes under compression. Stanglmaier and Roberts (1999) [7] conducted significant research on HCCI, demonstrating that it was feasible to simultaneously reduce NO_x and soot. Early studies in natural gas and *n*-heptane autoignition also laid the theoretical groundwork for LTC operation in reciprocating engines. These experiments suggest that compression ignition of a premixed charge could avoid soot, prevent NO_x formation, and drastically lower peak temperatures—all of which are important markers of LTC combustion.

In the 1990s and 2000s, governments and businesses started to fund research into advanced combustion modes as environmental concerns grew. Several LTC tactics were developed at this time:

- Premixed Charge Compression Ignition (PCCI), is a partial premixing technique that bridges the gap between HCCI and conventional diesel.
  SCCI is a technique used to control ignition by stratifying fuel.
- RCCI is a dual-fuel strategy that gives you more control over the reactivity and phasing of combustion.

These techniques sought to address the drawbacks of pure HCCI, specifically its limited working range and problems with combustion phasing.

1.7.2 Advances in Fuel Chemistry and Surrogate Modeling

A key factor in LTC is fuel behavior. Important work concentrated on the following, starting in the late 2000s:

- Chemical kinetic modeling for LTC fuels.
- Surrogate fuels like primary reference fuels, toluene reference fuels, and fuels for advanced combustion engines.
- Using sensitivity analysis to find species that regulate ignition.

High-fidelity chemical kinetic models, such as AramcoMech and LLNL, aided in simulating LTC combustion under various circumstances. This made it possible for researchers to comprehend how emissions, LTHR, and ignition delay are impacted by fuel structure (chain length, branching, and oxygenates).

1.7.3 Role of Optical Diagnostics and CFD Tools

Laser-based diagnostic methods and optical engines have emerged as essential resources for examining the intricate processes related to LTC. Critical combustion events, such as fuel stratification in dual-fuel LTC systems, ignition kernel formation, and flame propagation under extremely diluted conditions, can be directly visualized thanks to these sophisticated diagnostics. Among the most popular methods are chemiluminescence imaging, which records the locations and intensities of reaction zones; laser-induced incandescence, which is used to detect soot formation; and planar laser-induced fluorescence, which offers spatially resolved information on the distributions of fuel and intermediate species.

CFD simulations have been essential in comprehending and improving LTC processes in tandem with experimental data. Comprehensive three-dimensional modeling of spray dynamics, mixture formation, autoignition, and heat release is made possible by software systems like CONVERGE, KIVA, and AVL FIRE. By providing more in-depth understanding of combustion phasing, pollutant generation, and the consequences of changing engine operating conditions, these simulations enhance experimental research and ultimately direct the creation of LTC engine designs that are cleaner and more efficient.

1.7.4 LTC in the Era of Electrification and Decarbonization

LTC research has shifted more and more toward investigating sustainable and future-ready technologies as a result of the increased focus on electrification and global carbon neutrality goals. The utilization of alternative fuels including ethanol, methanol, ammonia, and hydrogen, as well as the creation of carbon-neutral dual-fuel systems that combine conventional and renewable fuels, are currently major areas of focus. Furthermore, there is growing interest in integrating LTC with hybrid powertrains, which offers the possibility of combining low emissions and high thermal efficiency while the engine is running.

Current LTC research addresses issues such as engine start–stop cycles and transient load circumstances, emphasizing real-world applicability. Another big worry is making sure it works with a variety of fuels, many of which have different physical characteristics and levels of reactivity. Researchers are using sophisticated digital control algorithms to govern fuel injection, EGR, and combustion phasing in real-time to satisfy these needs. As a result, LTC is becoming more widely acknowledged as a transitional technology that can facilitate carbon-neutral combustion by efficiently utilizing synthetic and renewable fuels, in addition to being a means of achieving ultra-low emissions.

Current research focuses on the following areas:

- **AI and Machine Learning in LTC:** For the prediction of combustion phasing, knock management, and reduction of cycle-to-cycle variability.
- **Fuel Reactivity Tailoring:** Using fuel blends and additives to improve ignition timing and LTHR.
- **EGR Techniques:** Dynamic EGR maintains combustion stability under sporadic loads.
- **Advanced Sensors and Closed-Loop Control:** Real-time feedback systems for adaptive LTC operation.

1.8 CHALLENGES AND FUTURE OUTLOOK OF LOW-TEMPERATURE COMBUSTION

Although LTC has great potential for reducing emissions and improving thermal efficiency, there are a number of operational and technological obstacles to overcome before it can be used in commercial ICEs. This section discusses LTC's main drawbacks and suggests future areas of inquiry that could fully realize its promise in the worldwide movement toward sustainable, effective, and clean energy and transportation systems.

1.8.1 Major Challenges in Implementing LTC

The lack of direct control in LTC systems, particularly HCCI, makes combustion phasing control difficult. Temperature, pressure, and chemical kinetics all influence autoignition, which can result in knock, misfire, and incomplete combustion. Phasing requires sophisticated control mechanisms, which add complexity to the system. LTC modes function well at restricted engine speeds and loads.

LTC systems have difficulties starting engines because of sensitivity of autoignition temperature, especially in cold areas. Under transient operations, it is challenging to maintain a regulated combustion profile. Since not all commercial fuels are suitable for LTC operation, fuel availability and sensitivity are crucial. Dual-fuel systems are necessary for advanced LTC methods, which raises infrastructure and costs. Biofuels, hydrogen, and ammonia are examples of alternative fuels that have problems with handling, storage, and supply chains. Optimized exhaust aftertreatment systems are required because nonideal combustion can result in CO and unburned hydrocarbons (UHC) emissions.

1.8.2 Future Outlook and Research Directions

Due to autoignition temperature sensitivity, LTC systems have trouble starting engines, particularly in cold climates. Maintaining a controlled combustion profile during transient operations is difficult. Fuel availability and sensitivity are critical because not all commercial fuels are appropriate for LTC operation. Advanced LTC techniques require dual-fuel systems, which increases infrastructure and expenses. Alternative fuels with handling, storage, and supply chain issues include biofuels, hydrogen, and ammonia. Due to the possibility of CO and UHC emissions from less-than-ideal combustion, optimized exhaust aftertreatment systems are necessary.

1.9 RESEARCH AND APPLICATION SIGNIFICANCE

LTC, has garnered a lot of interest recently as a possible way to reduce harmful emissions and increase engine efficiency. LTC offers a unique platform for study that integrates enhanced fuel chemistry, thermodynamic control, and combustion dynamics. Its interdisciplinary character makes it particularly attractive for academic research since it promotes contributions from fields such as materials science, mechanical engineering, chemical engineering, and control systems [8].

LTC approaches such as HCCI, PCCI, RCCI, and Spark Assisted Compression Ignition (SACI) allow combustion to occur at lower peak temperatures, reducing NO_x and PM emissions without requiring complex aftertreatment systems [9,10]. Researchers have developed complex models and diagnostics to examine the emissions behavior, combustion phasing, and ignition delay in LTC systems under various conditions. For better visualization, surrogate fuels and optical diagnostics are commonly employed [11]. A solid framework for the design of practical engines has been formed by these efforts, which have led to a clearer understanding of mixture stratification, autoignition behavior, and chemical heat release features.

From an application perspective, the potential incorporation of LTC into cutting-edge powertrains—particularly for HEVs and low-carbon transportation systems—is noteworthy. LTC modes are ideal for driving cycles in urban areas where traditional engines struggle with pollutants and thermal efficiency because of their high efficiency at partial loads [12]. To meet carbon neutrality targets and adjust to changing fuel conditions, LTC can also be utilized with a range of fuels, including gasoline, diesel, ethers, alcohols, and bioderived substitutes [13].

For military, naval, and stationary power generation applications, LTC technologies offer operational advantages such as reduced infrared signatures, lower combustion temperatures, and smoother combustion. The increasing interest in renewable fuels such as ethanol, methanol, and biodiesel for LTC engines also demonstrates a sustainable transition strategy in line with global climate targets [14].

LTC's research and application significance can be explained by its technical promise and strategic importance as a transitional technology that connects current IC systems with future zero-emission powertrains.

REFERENCES

1. Yao, M., Zheng, Z., & Liu, H. (2009). Progress and recent trends in homogeneous charge compression ignition (HCCI) engines. *Progress in Energy and Combustion Science*, *35*(5), 398–437.
2. Kalghatgi, G. (2013). Developments in internal combustion engines and implications for combustion science and future transport fuels. *Proceedings of the Combustion Institute*, *34*(1), 301–318.
3. Kokjohn, S.L., Hanson, R.M., Splitter, D.A., & Reitz, R.D. (2011). Fuel reactivity-controlled compression ignition (RCCI): A pathway to controlled high-efficiency clean combustion. *International Journal of Engine Research*, *12*(3), 209–226.
4. Zhao, H. (2012). *HCCI and CAI Engines for the Automotive Industry*. Woodhead Publishing, New York.
5. Reitz, R.D., & Duraisamy, G. (2015). Review of high efficiency and clean reactivity-controlled compression ignition (RCCI) combustion in internal combustion engines. *Progress in Energy and Combustion Science*, *46*, 12–71.
6. Hanson, R., & Reitz, R. (2013). Transient RCCI operation in a light-duty multi-cylinder engine. *SAE International Journal of Engines*, *6*(3), 1694–1705.
7. Stanglmaier, R.H., & Roberts, C.E. (1999). Homogeneous charge compression ignition (HCCI): Benefits, compromises, and future engine applications. *SAE Transactions*, *1999*, 2138–2145.
8. Singh, A.P., & Agarwal, A.K. (2017). Low-temperature combustion: An advanced technology for internal combustion engines. In: Agarwal, A.K., Martínez, A.G., Kalwar, A., & Valera, H. (eds) *Advances in Internal Combustion Engine Research*, pp. 9–41. Springer, Singapore.
9. Batool, S., Naber, J.D., & Shahbakhti, M. (2021). Multi-mode low temperature combustion (LTC) and mode switching control. In: Agarwal, A.K., Martínez, A.G., Kalwar, A., & Valera, H. (eds) *Advanced Combustion for Sustainable Transport*, pp. 43–93. Springer, Singapore.

10. Chauhan, B.V., Sayyed, I., Vedrantam, A., Garg, A., Bharti, S., & Shukla, M. (2021). State of the art in low-temperature combustion technologies: HCCI, PCCI, and RCCI. In: Agarwal, A.K., Martínez, A.G., Kalwar, A., & Valera, H. (eds) *Advanced Combustion for Sustainable Transport*, pp. 95–139. Springer, Singapore.
11. Lundgren, M. (2017). *Optical Diagnostics of Gasoline Compression Ignition: HCCI-PPC-Diffusion Combustion* (Doctoral dissertation, Lund University).
12. Krishnasamy, A., Gupta, S.K., & Reitz, R.D. (2021). Prospective fuels for diesel low temperature combustion engine applications: A critical review. *International Journal of Engine Research*, *22*(7), 2071–2106.
13. Olson, A.L. (2024). *Beyond Ethanol and Biodiesel: The Potential of Glycerol Derivatives and C1–C4 Alcohols for Motor Fuel Applications*. Lund University, Lund.
14. Oloyede, C.T., Itabiyi, O.E., Popoola, O.A., Jekayinfa, S.O., Olaniyan, M.A., Adebayo, A.O., Ogunkunle, O., Zamri, M.F.M.A., & Fattah, I.M.R. (2024). Navigating prospects and challenges for green fuels for achieving economical, environmental and ecological resilience: A scientific review. *Biofuels*, *15*(7), 929–941.

2 Fundamentals of Low-Temperature Combustion Chemistry and Kinetics

Abhishek Barua, Suman Chatterjee, Yerlan Sarsenbayev, Sibasis Harihar Sahu, Monalin Mishra, Siddharth Jeet, and Sangam Kumar Singh

2.1 INTRODUCTION

Low-temperature combustion (LTC) has emerged as a transformative strategy for achieving high thermal efficiency and ultra-low emissions in advanced internal combustion engines. Unlike conventional high-temperature combustion (HTC) processes, LTC operates under reduced flame temperatures, which significantly mitigates the formation of nitrogen oxides (NO_x) and particulate matter (PM). The drive for stricter emission norms, improved fuel economy, and compatibility with a wide range of fuels has pushed researchers and engine manufacturers to explore and refine LTC strategies. This chapter aims to lay the groundwork for understanding the fundamental chemistry and kinetics involved in LTC, providing a foundation for subsequent discussions on fuel design, engine architecture, and control strategies.

LTC refers to a family of combustion strategies that aim to initiate and complete combustion at lower temperatures compared to conventional spark-ignition or diesel combustion (See Figure 2.1). These strategies rely primarily on autoignition and delayed heat release to control combustion phasing and improve emissions characteristics [1].

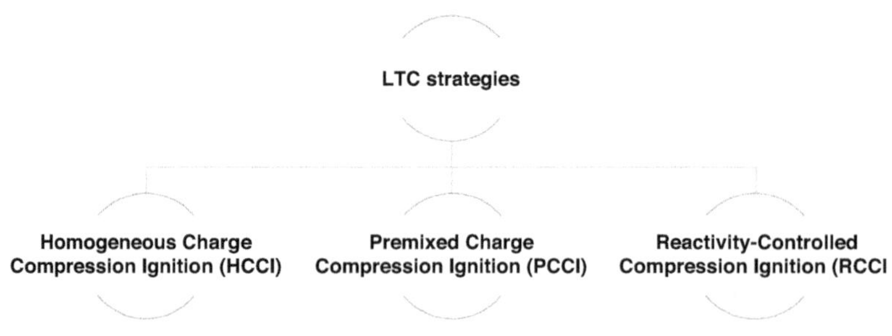

FIGURE 2.1 Different LTC strategies.

DOI: 10.1201/9781003633631-2

LTC relies heavily on the chemical kinetics of fuel oxidation at intermediate temperatures (600–1000 K). This regime is characterized by two-stage ignition phenomena: a cool flame followed by a hot ignition. Unlike in HTC where radical chain reactions dominate, LTC processes involve low-temperature chain branching and complex radical pool interactions (See Table 2.1).

Figure 2.2 represents the ignition delay as a function of temperature, clearly indicating the negative temperature coefficient (NTC) behavior in the 700–900 K range, which is a defining feature of LTC kinetics.

Modern engine technologies demand a fine balance between performance, efficiency, and emission compliance (See Table 2.2). LTC offers a potential pathway to meet these challenges, particularly under the tightening global emissions regulations such as Euro VI, BS VI, and EPA Tier 3 standards. Benefits of LTC in Engine Design:

TABLE 2.1
Key Characteristics of LTC

Feature	LTC	Conventional Combustion
Combustion mode	Autoignition	Flame propagation (SI), diffusion (CI)
Temperature	<1500 K	>2000 K
NO_x emission	Very low	High
Soot emission	Low to moderate	Moderate to high
Combustion control	Challenging	Easier (with spark or injection timing)

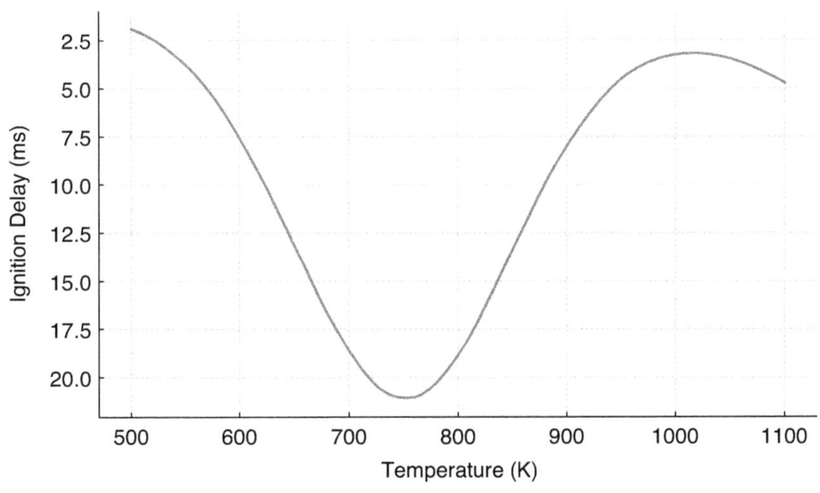

FIGURE 2.2 Temperature vs. ignition delay characteristics of hydrocarbons in LTC region.

TABLE 2.2
Comparison of Engine Emissions under Different Combustion Modes

Engine Mode	NO_x Emissions (g/kWh)	PM Emissions (g/kWh)	Efficiency (%)
SI engine	1.2	0.02	28–32
CI engine	0.6	0.05	38–42
LTC engine (HCCI)	~0.05	~0.01	40–44

- **Ultra-Low Emissions:** Combustion at lower temperatures limits NO_x formation due to reduced thermal dissociation of nitrogen and oxygen.
- **High Thermal Efficiency:** LTC allows for lean and premixed combustion which reduces heat losses and enhances fuel economy.
- **Fuel Flexibility:** LTC can work with a broad range of fuels, including gasoline, diesel, alcohols, and biofuels.
- **Reduced Aftertreatment Needs:** Lower emissions reduce reliance on expensive aftertreatment systems like Selective Catalytic Reduction (SCR) and Diesel Particulate Filters (DPF).

The primary objective of this chapter is to provide a comprehensive account of the chemical kinetics and reaction mechanisms that govern LTC processes. It focuses on understanding the low-temperature oxidation pathways of fuels and analyzing the behavior of critical intermediates such as peroxy radicals, ketohydroperoxides (KHP), and aldehydes. This chapter also explores factors influencing reaction dynamics, including reaction rate constants, pressure effects, and fuel-specific characteristics. Additionally, it delves into modeling approaches—both detailed and reduced kinetic mechanisms—that are essential for simulating LTC behavior. To support these insights, this chapter reviews key experimental techniques used to study LTC chemistry, such as rapid compression machines (RCMs), shock tubes (STs), and flow reactors.

2.2 BASIC CONCEPTS OF COMBUSTION CHEMISTRY

Understanding the foundational chemistry behind combustion is essential to developing LTC technologies. Unlike conventional HTC, LTC aims to minimize emissions and improve thermal efficiency by operating under kinetically controlled conditions. This section discusses the structural aspects of fuels, the complex web of combustion reactions, and how temperature fundamentally alters reaction mechanisms [2].

2.2.1 Fuel Structure and Composition

The structure and composition of fuels significantly influence their combustion behavior, ignition delay, and emission profile (See Table 2.3). Fuels consist mainly of hydrocarbons, with varying molecular structures that affect reactivity:

TABLE 2.3
Common Fuels and Their Structural Characteristics

Fuel Type	Molecular Formula	Structural Characteristics	Autoignition Temp (°C)	Cetane Number
n-Heptane	C_7H_{16}	Straight-chain alkane	~204	56
Iso-octane	C_8H_{18}	Branched alkane	~372	15
Ethanol	C_2H_5OH	Alcohol (oxygenated)	~365	8
Toluene	C_7H_8	Aromatic hydrocarbon	~480	12
Dimethyl ether	CH_3OCH_3	Ether (oxygenated compound)	~235	55

- Alkanes (saturated hydrocarbons) have single bonds, for example, *n*-heptane (C_7H_{16})
- Alkenes (unsaturated hydrocarbons) contain double bonds, for example, ethylene (C_2H_4)
- Aromatics possess ring structures, for example, toluene (C_7H_8)
- Oxygenates, such as ethanol (C_2H_5OH), include oxygen atoms

Fuel structure dictates bond dissociation energies and the formation of radicals, key to combustion chemistry.

Figure 2.3 illustrates how ignition delay varies across different fuel molecules under LTC scenarios. Fuels like dimethyl ether (DME) and *n*-heptane, which possess strong low-temperature reactivity, exhibit shorter ignition delays, making them suitable for LTC strategies. In contrast, toluene, with its aromatic ring structure, shows

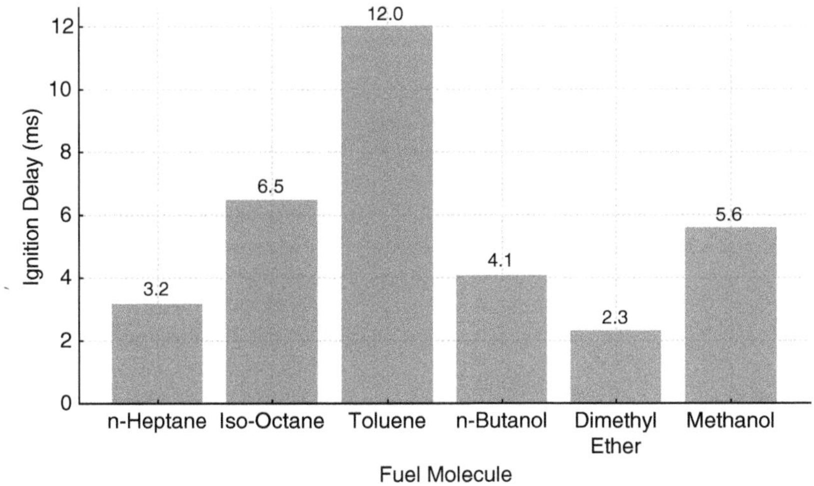

FIGURE 2.3 Molecular structure vs. ignition delay at LTC conditions.

significantly longer delays, reflecting its lower reactivity under LTC conditions. This comparison highlights the critical role of molecular structure in fuel behavior and combustion design. Branched hydrocarbons and aromatics show longer ignition delays at LTC-relevant temperatures due to resonance stabilization and steric effects. Straight-chain alkanes and oxygenates generally favor early-stage low-temperature oxidation.

2.2.2 Combustion Reaction Pathways

Combustion occurs via a series of elementary reactions, initiated by the thermal decomposition of fuel molecules. The pathways differ significantly under low- and high-temperature regimes. At LTC conditions (600–1000 K), low-temperature oxidation becomes significant, involving chain-branching mechanisms initiated by peroxy radicals (RO_2•). Table 2.4 represents the stages of hydrocarbon combustion, combining chemical reactions with explanatory labels:

At lower temperatures, chain propagation via peroxy intermediates (RO_7, QOOH) dominates. As temperature increases, the chemistry transitions to high-temperature chain branching via H• and O• radicals. In LTC, the formation and progression of specific chemical intermediates play a crucial role in dictating ignition behavior and heat release characteristics (See Table 2.5). The RO_2• is formed when fuel radicals (R•) react with oxygen, initiating low-temperature oxidation and setting the stage for multistage ignition. This radical undergoes isomerization to form QOOH, a key species that promotes chain branching through subsequent reactions. QOOH then reacts with additional oxygen to produce KHP, which are thermally unstable and decompose to release heat, contributing significantly to the main ignition event. Meanwhile, hydroxyl radicals (OH•) are generated through hydrogen abstraction and chain propagation reactions, serving to amplify the radical pool and accelerate the overall reaction kinetics. Together, these intermediates form a tightly coupled

TABLE 2.4
Stages of Hydrocarbon Combustion

Stage	Reaction	Description
1. Fuel decomposition	$RH + \cdot OH \rightarrow \cdot R + H_2O$	Hydrocarbon molecules (RH) react with hydroxyl radicals (·OH), forming alkyl radicals (·R) and water
2. Low-temperature oxidation	$\cdot R + O_2 \rightarrow RO_2\cdot \rightarrow QOOH \rightarrow$ Ketohydroperoxides (KHP) + OH	The alkyl radical (·R) reacts with oxygen to form a $RO_2\cdot$, which rearranges to form QOOH. QOOH leads to KHP and new OH radicals, sustaining low-temperature chain branching
3. High-temperature branching	$\cdot H + O_2 \rightarrow \cdot OH + \cdot O$	At elevated temperatures, hydrogen radicals (·H) react with O_2, forming highly reactive ·OH and ·O radicals, initiating fast, exothermic chain-branching

TABLE 2.5
Major Intermediates and Their Roles

Intermediate	Formation Pathway	Role in LTC
RO_2•	R• + O_2	Low-temp oxidation initiator
QOOH	RO_2• isomerization	Promotes chain branching
KHP	QOOH + O_2	Heat release and ignition
OH•	H-abstraction and chain propagation	Radical pool amplifier

TABLE 2.6
Comparison of LTC and HTC

Parameter	Low-Temperature Combustion (LTC)	High-Temperature Combustion (HTC)
Temperature range (K)	600–1000	1300–2500
Dominant mechanism	Peroxy/ketohydroperoxide chain	Thermal chain-branching (O, OH, H)
Ignition delay	Longer and multistage	Short and single-stage
Emissions	Low NO_x, low soot	High NO_x and potential soot
Pressure sensitivity	High	Moderate
Application	HCCI, RCCI, and LTC engines	SI and CI engines

network of reactions that define the LTC regime's characteristic two-stage ignition behavior and enable efficient, low-emission combustion.

2.2.3 Differences between High- and Low-Temperature Combustion

LTC fundamentally differs from HTC in mechanisms, temperature regimes, emission products, and kinetic sensitivity (See Table 2.6).

Figure 2.4 compares the heat release profiles for LTC and HTC. The LTC curve shows an earlier, lower-temperature peak (around 750 K), characteristic of cool flame heat release, while HTC peaks sharply at higher temperatures (around 1200 K), representing conventional rapid combustion. This illustrates the staged ignition behavior and emission-control potential of LTC regimes.

2.3 LOW-TEMPERATURE COMBUSTION MECHANISMS

LTC is a promising strategy for simultaneously reducing NO_x and PM emissions in internal combustion engines. Unlike conventional combustion, LTC operates at lower temperatures (700–1200 K) and exploits complex chemical kinetics. The underlying mechanisms involve pre-ignition reactions, cool flame behavior, and two-stage ignition, all of which contribute to unique thermal and chemical behaviors like the NTC phenomenon [3].

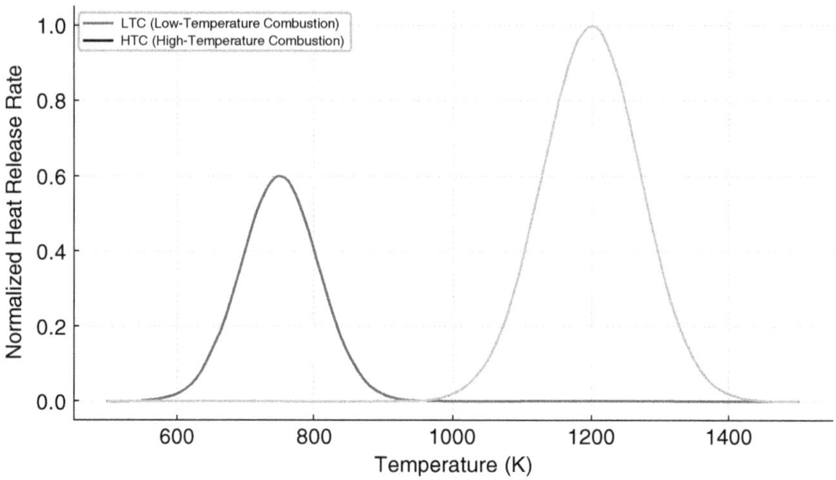

FIGURE 2.4 Heat release rate vs. temperature for LTC and HTC.

2.3.1 Pre-Ignition Chemistry

Pre-ignition chemistry represents the initial chemical reactions that occur at relatively low temperatures (500–700 K) before autoignition. These reactions primarily involve the oxidation of fuel molecules by oxygen, forming intermediate species such as RO_2 and hydroperoxides (ROOH). In this regime, alkyl radicals react with O_2, forming RO_2, which further undergoes isomerization to QOOH (a hydroperoxyalkyl radical), a key intermediate in the LTC pathway.

These reactions (See Table 2.7) are critical for heat release and influence the subsequent stages of LTC by promoting low-temperature chain-branching pathways, thereby increasing reactivity at modest temperatures without flame propagation.

2.3.2 Cool Flame Phenomenon

The cool flame is a luminous, weakly exothermic reaction stage observed during the initial phase of LTC, typically occurring between 600 and 800 K. It is characterized by a noticeable temperature rise (20–100 K) due to low-temperature exothermic reactions, without achieving full ignition. Characteristics of cool flame are:

TABLE 2.7
Key Reactions in Pre-Ignition Phase

Reaction	Description
$R + O_2 \rightarrow RO_2$	Formation of peroxy radicals
$RO_2 \rightarrow QOOH$	Intramolecular isomerization
$QOOH + O_2 \rightarrow O_2QOOH \rightarrow ROOH + OH$	Chain branching via hydroperoxides

- Visible bluish glow under dark conditions.
- Accompanied by pressure and temperature oscillations.
- Mainly results from exothermic decomposition of hydroperoxides (ROOH → OH + products).

Cool flames are critical in delaying autoignition and contribute to staged energy release, a fundamental feature exploited in advanced LTC modes like Homogeneous Charge Compression Ignition (HCCI).

2.3.3 Negative Temperature Coefficient Behavior

The NTC region is a unique feature of LTC where the overall reactivity decreases with increasing temperature, contrary to Arrhenius behavior. This typically occurs between 700 and 900 K and is associated with a shift from chain-branching to chain-propagating reactions. NTC Behavior Dynamics are:

- **At Low Temperatures:** QOOH formation dominates → chain branching → increased reactivity.
- **At NTC Temperatures:** β-scission reactions compete with isomerization → decreased QOOH → reduced reactivity.
- **At Higher Temperatures:** Traditional high-temperature chain branching resumes.

The NTC behavior is crucial for multistage ignition and is often exploited in engine strategies for controlled combustion phasing and emissions reduction. Figure 2.5 shows how reactivity initially increases with temperature (600–700 K), dips in the

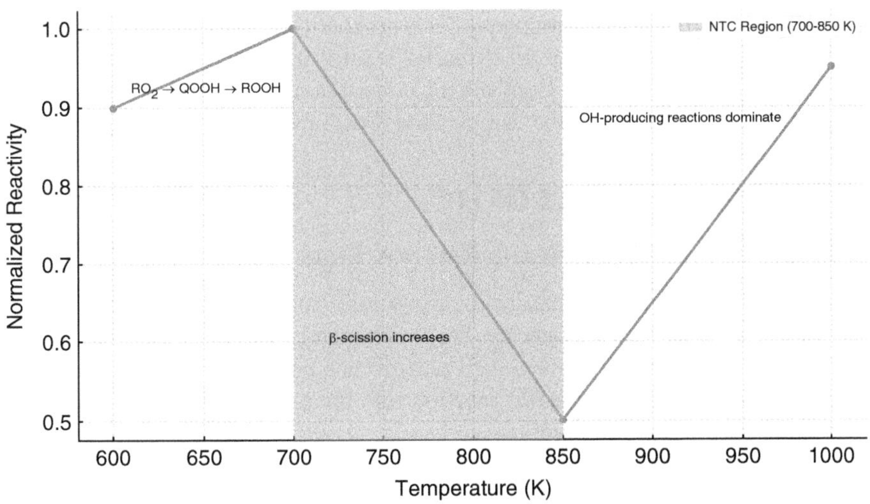

FIGURE 2.5 NTC behavior: temperature vs. reactivity.

TABLE 2.8
Key Characteristics of LTC Mechanisms

Phenomenon	Temperature Range (K)	Key Intermediates	Impact on Combustion
Pre-Ignition	500–700	RO_2, QOOH	Initiates slow oxidation
Cool Flame	600–800	ROOH, H_2O_2	Partial heat release and extends ignition delay
NTC Behavior	700–900	Competing pathways	Slows reactivity and controls phasing
Two-Stage Ignition	600–1000+	QOOH, OH, H_2O_2	Enables stratified combustion

NTC region (700–850 K) due to enhanced β-scission reactions, and then rises again as high-temperature OH-producing reactions dominate (850–1000 K). This characteristic dip is critical for multistage ignition in LTC regimes and enables better control over combustion phasing and emissions.

2.3.4 Two-Stage Ignition Phenomenon

Two-stage ignition is a defining characteristic of LTC chemistry, marked by two distinct rises in temperature during the ignition process. The first stage, known as low-temperature ignition, is initiated by cool flame reactions involving intermediates like QOOH and ROOH, leading to a modest temperature increase of around 20–100 K. This phase is relatively slow and controlled [4]. The second stage occurs after the buildup of highly reactive species such as OH radicals and H_2O_2, triggering a rapid and intense exothermic reaction that causes a sharp temperature rise, often exceeding 500 K. This sequential ignition process is crucial in enabling precise control over combustion phasing, making it an essential feature in advanced engine technologies like HCCI and Reactivity-Controlled Compression Ignition (RCCI), where high efficiency and ultra-low emissions are desired (See Table 2.8).

2.4 CHEMICAL KINETICS OF LTC

2.4.1 Elementary Reactions and Reaction Rates

LTC involves complex chemical mechanisms that can be understood by breaking them down into elementary reactions. These elementary steps represent the simplest possible chemical transformations, such as hydrogen abstraction, isomerization, and radical recombination. The overall reaction rate for an elementary process is governed by the Arrhenius equation:

$$k(T) = A \cdot T^n \cdot \exp\left(-\frac{E_a}{RT}\right) \tag{2.1}$$

TABLE 2.9
Influence of Pressure and Temperature on Reaction Kinetics

Temperature (K)	Pressure (bar)	Dominant Reaction Pathway	Observed Effect
600	1	RO_2 formation	Slow reaction rate
700	10	$QOOH + O_2 \rightarrow OOQOOH$	Accelerated chain branching
800	20	Decomposition of OOQOOH	Increased reactivity
900	1	Thermal cracking	Transition to high-T kinetics

Where:

- $k(T)$ = reaction rate constant,
- A = pre-exponential factor,
- Ea = activation energy,
- R = universal gas constant,
- T = temperature,
- n = temperature exponent.

In LTC, radical chain propagation and branching reactions play a pivotal role. Important species such as OH, HO_2, and RO_2 govern the kinetics, especially under lean and moderate temperature conditions.

2.4.2 Pressure and Temperature Effects

The reactivity of LTC systems is highly sensitive to both pressure and temperature. At lower temperatures (600–900 K), the oxidation chemistry is dominated by low-temperature chain branching via RO_2. Pressure influences the stabilization of intermediate species like QOOH, impacting ignition delay and fuel conversion (See Table 2.9).

As the pressure increases, the thermal stabilization of key intermediates enhances, which lowers the required temperature for ignition and promotes efficient fuel utilization under LTC conditions.

2.4.3 Autoignition Delay and Chemical Timescales

Autoignition delay is a critical indicator of fuel reactivity and is a function of the overall kinetics. It represents the time from the start of compression until the onset of ignition. In LTC, due to the complexity of reaction pathways, ignition delay is significantly affected by chain branching and radical pool buildup. Figure 2.6 shows a typical relationship between temperature and autoignition delay at a constant pressure. As the temperature increases, the delay decreases exponentially due to faster reaction kinetics and greater radical production. This behavior is essential for understanding the timing and control of ignition in LTC engines [5].

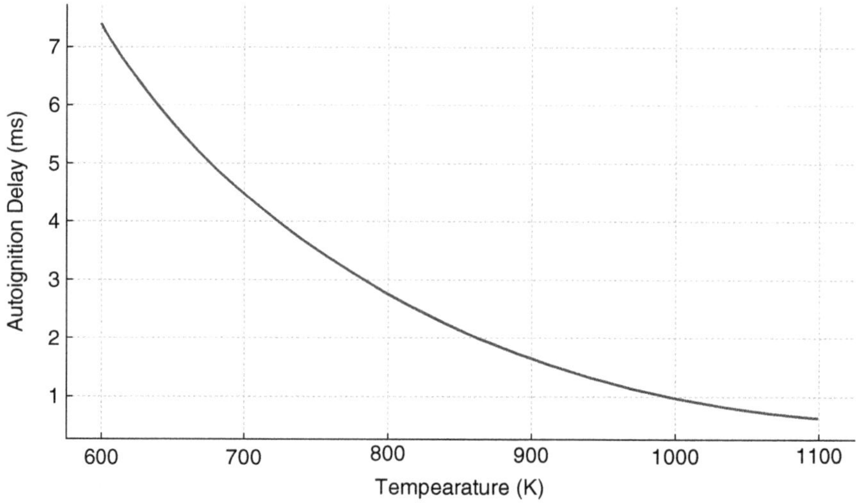

FIGURE 2.6 Temperature and autoignition delay at constant pressure.

TABLE 2.10
Comparison of Fuel Structures on LTC Reactivity

Fuel Type	Structure Type	Autoignition Delay (ms at 700 K)	Reactivity	Dominant Pathway
n-heptane	Linear alkane	12	High	$RO_2 \rightarrow QOOH \rightarrow$ Chain branching
Iso-octane	Branched alkane	38	Moderate	RO_2 stabilization
Toluene	Aromatic	120	Low	Slow oxidation
Cyclohexane	Cycloalkane	90	Moderate	Ring-opening pathways

2.4.4 Influence of Fuel Structure on Kinetics

Fuel molecular structure plays a vital role in determining the reactivity and chemical pathways in LTC. Straight-chain alkanes tend to have higher reactivity due to easier hydrogen abstraction and formation of reactive intermediates like RO_2 and QOOH. Branched alkanes, aromatics, and cyclic compounds often show slower low-temperature chemistry due to steric hindrance and stable ring structures [6]. Understanding fuel-specific kinetics is crucial for LTC engine designers aiming to tailor ignition characteristics for improved efficiency and reduced emissions (See Table 2.10).

2.5 EXPERIMENTAL AND COMPUTATIONAL TOOLS

Understanding the complex behavior of LTC requires a synergy between experimental diagnostics and computational modeling. This section explores key methodologies used to study LTC phenomena, focusing on STs, RCMs, flow reactors and jet-stirred reactors (JSRs), and computational approaches such as kinetics modeling and sensitivity analysis [7].

2.5.1 Shock Tube and Rapid Compression Machine Studies

STs and RCMs are indispensable in capturing ignition delay times, species profiles, and reaction rates under controlled conditions relevant to LTC [8].

- STs provide nearly instantaneous high-temperature and high-pressure environments, ideal for investigating ignition delays at constant volume.
- RCMs simulate engine-like compression processes at lower temperatures and higher pressures, closely mimicking LTC conditions in real engines.

These devices enable precise control over initial conditions and yield reproducible results, crucial for validating kinetic models. Typically, the ignition delay is measured as the time between the arrival of the reflected shock/compression and a rapid rise in pressure or radical species (e.g., OH*) emission (See Table 2.11).

Figure 2.7, of ignition delay time (μs) against temperature (800–1400 K), typically reveals:

TABLE 2.11
Comparison between Shock Tubes and Rapid Compression Machines

Parameter	Shock Tube	Rapid Compression Machine
Temperature range (K)	900–2000	600–1100
Pressure range (atm)	1–100	5–50
Time resolution (μs)	High	Moderate
End gas temperature control	Limited	Accurate
Relevance to LTC	Moderate to high	High

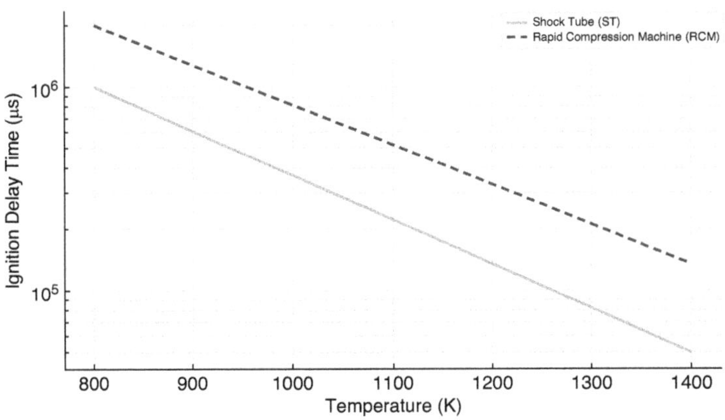

FIGURE 2.7 Ignition delay time vs. temperature.

TABLE 2.12
Operating Conditions and Features of Reactors

Reactor Type	Temp. Range (K)	Pressure (atm)	Key Features
Flow reactor	300 –1500	1–30	Plug flow, axial gradients
Jet-stirred reactor	500 –1300	1–20	Ideal mixing, uniform conditions

- Shorter delays at higher temperatures for both ST and RCM.
- RCM delays are longer at the same temperature due to compression heat losses and lower post-compression temps.

2.5.2 Flow Reactors and Jet-Stirred Reactors

Flow reactors and JSRs are employed to measure species concentrations at steady-state conditions over a range of residence times, temperatures, and pressures [9].

- Flow Reactors (laminar or turbulent) simulate plug-flow conditions to observe intermediate species and end products.
- JSRs, due to their near-perfect mixing, provide uniform temperature and composition throughout the reactor, making them ideal for kinetic studies.

These reactors are typically coupled with diagnostics such as gas chromatography, Fourier-transform infrared spectroscopy, and mass spectrometry to quantify combustion intermediates and validate reaction mechanisms (See Table 2.12).

2.5.3 Computational Kinetics and Mechanism Development

Computational modeling plays a vital role in understanding LTC by simulating thousands of reactions among hundreds of species [10].

- Mechanism generation tools such as Reaction Mechanism Generator and EXGAS automate the creation of detailed kinetic mechanisms.
- Simulation platforms such as CHEMKIN, Cantera, and OpenSMOKE++ are used to simulate ignition delays, species evolution, flame speeds, and reactor behavior.

These simulations guide experimental efforts by predicting key reaction pathways and intermediate species. Reduced or skeletal mechanisms are often derived to enable faster simulations in engine-like computational fluid dynamics (CFD) environments (See Table 2.13).

2.5.4 Sensitivity and Uncertainty Analysis

To prioritize reactions for experimental verification and mechanism refinement, sensitivity and uncertainty analyses are essential.

TABLE 2.13
Common Computational Tools

Software	Purpose	Strengths
CHEMKIN	Reactor and flame modeling	Extensive database, robust solvers
Cantera	Multi-phase and flame simulations	Python integration, open-source
RMG	Automated mechanism generation	Rapid kinetic modeling

TABLE 2.14
Sensitivity Coefficients Example (for Ignition Delay at 1000 K)

Reaction	Sensitivity Coefficient	Interpretation
$H + O_2 \rightleftharpoons OH + O$	+0.45	Strong promoter of ignition
$CH_3 + O_2 \rightleftharpoons CH_3O + O$	+0.20	Moderate influence
$CH_3 + H \rightleftharpoons CH_4$	−0.10	Inhibits ignition

- Sensitivity analysis determines the impact of individual reaction rate constants or thermodynamic parameters on global observables (e.g., ignition delay).
- Uncertainty quantification assesses how parameter uncertainties propagate through a kinetic model to affect predictions.

These analyses help identify rate-limiting steps, highlight poorly characterized reactions, and guide mechanism reduction (See Table 2.14).

2.6 FUEL EFFECTS ON LTC CHEMISTRY

The chemical nature of fuels significantly influences LTC behavior, including autoignition delay, heat release rate, intermediate species formation, and emissions. LTC processes rely heavily on low-temperature chain-branching reactions, and different classes of fuels—alkanes, alkenes, aromatics, oxygenates, and surrogate blends—exhibit varied reactivity due to molecular structure and functional groups. Understanding these differences is crucial for optimizing fuel design and engine calibration for advanced LTC modes like HCCI and RCCI [11].

2.6.1 Alkanes, Alkenes, and Aromatic Fuels

a. Alkanes (saturated hydrocarbons) exhibit strong low-temperature reactivity due to favorable pathways for RO_2 and KHP formation, leading to efficient chain branching. *n*-heptane and *n*-decane are commonly used alkanes in LTC studies and demonstrate short ignition delays (See Table 2.15).
b. Alkenes, with at least one double bond, show moderate LTC reactivity. The presence of unsaturation introduces resonance-stabilized radicals that slow

TABLE 2.15
Ignition Delay and Heat Release Characteristics

Fuel Type	Example	Ignition Delay (ms @ 850 K)	Heat Release Rate	LTC Reactivity
Alkane	*n*-Heptane	7	High	High
Alkene	1-Octene	10	Moderate	Medium
Aromatic	Toluene	18	Low	Low

TABLE 2.16
Oxygenated Fuel Characteristics

Fuel Type	Example	Ignition Delay (ms @ 850 K)	NO_x Emission Tendency	Soot Formation
Alcohol	Ethanol	11	Low	Very low
Ether	Dimethyl ether	5	Moderate	Negligible
Ester	Methyl butanoate	9	Moderate	Low

down chain branching. They tend to produce more unsaturated intermediates and soot precursors.

c. Aromatic fuels like toluene or xylene have the lowest reactivity under LTC conditions. Aromatic rings stabilize radicals and interrupt chain branching. However, they are key in forming polycyclic aromatic hydrocarbons, contributing to soot emissions [12–14].

2.6.2 Oxygenated Fuels (e.g., Alcohols, Ethers, and Esters)

Oxygenated fuels, such as ethanol, DME, and methyl esters, have gained attention due to their clean-burning nature and favorable LTC properties. Their built-in oxygen content promotes more complete combustion and reduces soot formation.

- **Alcohols:** Have moderate reactivity. Ethanol, for example, shows longer ignition delays than alkanes but burns cleanly (See Table 2.16).
- **Ethers (e.g., DME):** Extremely reactive at low temperatures due to weak C–H bonds adjacent to oxygen atoms.
- **Esters (e.g., Methyl Butanoate):** Represent biodiesel surrogates and exhibit reactivity depending on chain length and oxygenation level.

2.6.3 Surrogate Fuels and Real Fuel Blends

To study real-world fuels in a controlled environment, surrogate fuels are employed—combinations of pure components that mimic physical and chemical properties of complex commercial fuels [15].

TABLE 2.17
Comparison Between Surrogate and Real Fuels

Fuel Type	Composition Example	Target Fuel	LTC Model Compatibility	Cost
PRF	*n*-Heptane + iso-Octane	Gasoline	High	Low
TRF	Toluene + PRF	Aromatic gasoline	Medium	Medium
Biodiesel Surrogate	Methyl decanoate + *n*-heptane	Biodiesel	Medium	Medium
Real Fuel Blend	Ethanol-Gasoline (E10)	Commercial blend	Low (complexity)	High

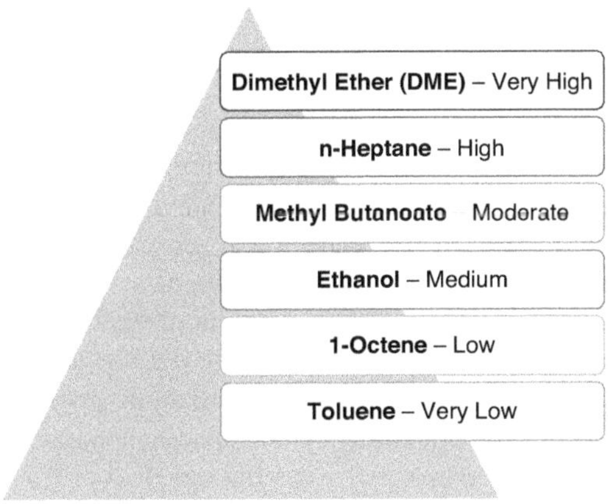

FIGURE 2.8 Reactivity ranking (typical order under LTC).

- Primary Reference Fuels like *n*-heptane and iso-octane are used for gasoline.
- Toluene Reference Fuels include aromatics to better represent commercial fuels.
- Real fuel blends, especially those containing ethanol or biodiesel, present complex interactions between components influencing ignition delay and emissions.

Surrogate fuels are essential for developing chemical kinetic models, especially for LTC regimes where ignition timing and intermediate species formation are sensitive to molecular structure (See Table 2.17).

2.6.4 Reactivity Ranking and Ignition Behavior

Understanding the relative reactivity of fuels under LTC is essential for fuel selection and engine control. Reactivity is often ranked using ignition delay time, cool flame behavior, and peak heat release characteristics [11]. This ranking helps in formulating dual-fuel strategies (e.g., RCCI), where a highly reactive fuel like DME is paired with a low-reactivity fuel like gasoline (See Figure 2.8).

TABLE 2.18
Comparison of Detailed vs. Reduced Mechanisms in Combustion Modeling

Aspect	Detailed Mechanism	Reduced Mechanism
Number of species	>1000 species	50–300 species
Number of reactions	>5000 reactions	200–1000 reactions
Computational cost	Very high	Moderate to low
Accuracy	High (close to experiments)	Moderate (depends on reduction method)
Suitability for CFD integration	Low (computationally prohibitive)	High (more feasible)

2.7 KINETIC MODELING AND MECHANISM REDUCTION

2.7.1 Detailed vs. Reduced Kinetic Mechanisms

In combustion modeling, particularly under low-temperature regimes, detailed kinetic mechanisms provide a comprehensive representation of the chemical processes. These mechanisms often involve thousands of elementary reactions among hundreds or even thousands of species. They are essential for gaining accurate insights into reaction pathways, intermediate species formation, and autoignition behavior. However, such mechanisms are computationally expensive and impractical for use in multidimensional CFD simulations [16,17]. Reduced mechanisms, on the other hand, are derived by eliminating redundant species and reactions without significantly compromising accuracy for specific operating conditions. These mechanisms can be skeletal (retaining essential species) or further simplified using quasi-steady-state or lumping approaches (See Table 2.18).

2.7.2 Skeletal Mechanism Generation

Skeletal mechanisms serve as an intermediate step between detailed and fully reduced mechanisms. They are generated by systematically eliminating species and reactions that have negligible influence on the desired output (e.g., ignition delay and flame speed). This can be achieved using:

- **Reaction Path Analysis:** Identifies major and minor pathways contributing to key outputs.
- **Sensitivity Analysis:** Ranks species and reactions based on their influence on observables.
- **Directed Relation Graphs:** Determines species interconnectivity and truncates the graph below a certain threshold.
- **Principal Component Analysis:** Helps identify redundant dimensions in the reaction manifold.

Skeletal mechanisms preserve the fundamental kinetics while significantly reducing computational demands, making them suitable for engine-relevant conditions like low temperature and high pressure.

2.7.3 Quasi-Steady-State and Lumped Species Approaches

Further reduction is achieved by assuming that certain intermediate species quickly reach a steady-state (quasi-steady-state approximation [QSSA]) or by lumping chemically similar species together.

- QSSA simplifies the differential equations by setting the time derivatives of fast-reacting intermediates to zero.
- Lumping groups multiple isomers or minor pathways into a single representative species to reduce the dimensionality of the system.

These methods introduce approximations but are powerful in reducing stiffness in the ordinary differential equations (ODEs), thereby accelerating simulations. They are widely used in real-time engine simulations and control-oriented modeling.

Reduced chemical kinetic mechanisms offer several benefits, particularly in CFD applications. By simplifying reaction networks, they significantly reduce the stiffness of chemical ordinary differential equations (ODEs), which in turn accelerates convergence in CFD solvers and decreases the overall computational time. Additionally, these mechanisms require less memory, making them suitable for large-scale or real-time simulations. However, the simplification comes with trade-offs. A major challenge is the loss of detailed chemical fidelity, as key intermediates—especially those crucial in LTC processes—may be omitted. This can lead to inaccuracies in predicting ignition delays, pollutant formation, and overall combustion behavior under certain operating conditions [18].

2.7.4 Integration with CFD and Engine Models

For real-world applications, kinetic models must be integrated with CFD and engine simulation tools such as CONVERGE, OpenFOAM, and ANSYS Fluent. This integration involves:

- **Tabulation Methods (e.g., Flamelet and ILDM):** Precompute reaction rates and store them in lookup tables to avoid on-the-fly calculations.
- **In-situ Adaptive Tabulation:** Stores previously computed reaction states and reuses them when similar conditions are encountered.
- **Dynamic Adaptive Chemistry:** Activates subsets of the mechanism dynamically based on local thermochemical conditions.

Some of the key considerations are:

- Accurate reproduction of ignition delays, flame structure, and pollutant formation.
- Compatibility with turbulence–chemistry interaction models (e.g., RANS and LES).
- Real-time performance in control applications (e.g., for HCCI or RCCI engines).

The integration of reduced mechanisms with CFD enables predictive simulations of engine combustion phenomena like knock, cool flame formation, and autoignition under LTC strategies. However, model fidelity must be rigorously validated against experimental data [19,20].

2.8 CONCLUSIONS

LTC is a transformative combustion regime that occurs at relatively low temperatures, typically between 600 and 1000 K. This mode promotes high thermal efficiency while drastically reducing emissions such as nitrogen oxides (NO_x) and PM. The chemical kinetics of LTC differ fundamentally from traditional HTC. A defining feature is its multistage ignition behavior, comprising a low-temperature "cool flame" phase (first-stage ignition) followed by a high-temperature "main ignition" phase (second-stage ignition). These stages are governed by intricate low-temperature oxidation pathways involving RO_2 and QOOH intermediates, which contribute to chain-branching reactions essential for ignition. A key kinetic phenomenon in LTC is the NTC behavior, where reaction rates decrease with increasing temperature within a certain range. This counterintuitive trend results from a competition between chain-propagation and chain-branching mechanisms. Fuel molecular structure, pressure, and dilution significantly impact ignition delay and heat release profiles, underlining the sensitivity of LTC to operating conditions. For example, RO_2 and QOOH dominate in LTC regimes, while radicals like OH and fast reactions like NO_x formation are more prominent at higher temperatures. Understanding the relative roles of these species enables more precise control of LTC ignition timing and heat release characteristics. The kinetic principles of LTC inform the development of advanced combustion strategies such as HCCI, Premixed Charge Compression Ignition (PCCI), and RCCI. These techniques exploit low-temperature reactivity to minimize soot and NO_x emissions while maximizing thermal efficiency. However, precise control over autoignition timing remains a major design challenge. Fuel chemistry plays a central role in LTC. Fuels like *n*-heptane and DME, which exhibit strong low-temperature reactivity, are ideal for stable LTC operation. Alternatively, fuel blending strategies using combinations of high- and low-reactivity fuels (e.g., diesel with alcohols or gasoline with ethers) offer tunability in combustion phasing. Exhaust gas recirculation is also commonly employed to enhance low-temperature pathways while suppressing undesirable high-temperature reactions. Each combustion strategy has its own design trade-offs. HCCI provides ultra-low emissions but suffers from limited load control. RCCI, through dual-fuel injection, enables better ignition phasing but adds complexity to fuel delivery. PCCI, based on timing-controlled pre-injection, achieves a homogeneous air-fuel mixture but faces challenges with emissions under high load conditions. Designers must balance fuel selection, injection strategies, and in-cylinder conditions to realize the full benefits of LTC-based operation.

The future of LTC lies in advancing both fundamental understanding and practical implementation. A primary research focus is the refinement of detailed chemical kinetic models that accurately capture pressure-sensitive and multistage ignition behavior. Mechanisms such as HyChem and LLNL surrogates are increasingly used to simulate real fuel blends with better predictive capability.

Machine learning is beginning to play a transformative role in combustion research. AI models can rapidly predict ignition delay, optimize combustion phasing, and assist in the reduction of complex chemical mechanisms for real-time engine control. Simultaneously, low-carbon fuels such as alcohols, ethers, and bio-derived compounds are being evaluated for LTC compatibility, aiming to couple clean combustion with reduced greenhouse gas emissions. Advanced diagnostic tools, such as Planar Laser-Induced Fluorescence, Infrared imaging, and Laser-Induced Fluorescence, are improving our ability to visualize intermediates like RO_2 and QOOH in real time. These diagnostics validate and refine chemical models, closing the loop between theory and experiment. LTC principles also hold significant promise in hybrid and range-extended engines, where part-load conditions dominate and efficient, clean combustion is essential. Emerging research areas include catalyst-assisted LTC, high-pressure LTC operation (>30 bar), and electrification-integrated LTC platforms.

REFERENCES

1. Pilling, M. J. *Low-Temperature Combustion and Autoignition*. Amsterdam: Elsevier, 1997.
2. Gardiner, W. C., Jr., and D. B. Olson. Chemical kinetics of high temperature combustion. *Annual Review of Physical Chemistry* 31, no. 1 (1980): 377–99.
3. Walker, R. W., and C. Morley. Basic chemistry of combustion. *Comprehensive Chemical Kinetics* 35 (1997): 1–124.
4. Mishra, D. P. *Fundamentals of Combustion*. New Delhi: PHI Learning Pvt. Ltd., 2007.
5. Maurya, R. K. *Characteristics and Control of Low Temperature Combustion Engines*. Cham, Switzerland: Springer, 2018.
6. El-Mahallawy, F., and S. D. Habik. *Fundamentals and Technology of Combustion*. Amsterdam: Elsevier, 2002.
7. McAllister, S., J. Y. Chen, and A. C. Fernandez-Pello. *Fundamentals of Combustion Processes*. New York: Springer, 2011.
8. Cavaliere, A., and M. De Joannon. Mild combustion. *Progress in Energy and Combustion Science* 30, no. 4 (2004): 329–66.
9. Lackner, M., Á. Palotás, and F. Winter. *Combustion: From Basics to Applications*. Hoboken, NJ: John Wiley & Sons, 2013.
10. Hosseini, S. E., ed. *Fundamentals of Low Emission Flameless Combustion and Its Applications*. Cambridge, MA: Academic Press, 2022.
11. Undavalli, V. K., P. Berwal, and B. Khandelwal. Chemical kinetics of hydrogen combustion. In: Kumar, S., Agarwal, A.K., Khandelwal, B., and Singh, P. (eds) *Ammonia and Hydrogen for Green Energy Transition*, pp. 239–63. Singapore: Springer Nature Singapore, 2024.
12. Kirkpatrick, A. T., and K. K. Kuo. *Principles of Combustion*. Hoboken, NJ: John Wiley & Sons, 2024.
13. Nemitallah, M. A., M. A. Habib, and A. Abdelhafez. *Hydrogen for Clean Energy Production: Combustion Fundamentals and Applications*. Cham: Springer Nature, 2024.
14. Zhan, H., S. Li, G. Yin, E. Hu, and Z. Huang. Experimental and kinetic study of ammonia oxidation and NO_x emissions at elevated pressures. *Combustion and Flame* 263 (2024): 113129.

15. Huang, Q., R. Yang, J. Liu, T. Xie, and J. Liu. Investigation of the mechanism behind the surge in nitrogen dioxide emissions in engines transitioning from pure diesel operation to methanol/diesel dual-fuel operation. *Fuel Processing Technology* 264 (2024): 108131.
16. Zhao, Z., Y. Wang, J. Zhang, J. Liang, Y. Zhang, F. Zhao, and Q.-D. Wang. A shock-tube experimental and kinetic simulation study on the autoignition of methane at ultra-lean and lean conditions. *Heliyon* 10, no. 14 (2024): e34204.
17. Sharipov, A. S., B. I. Loukhovitski, A. V. Pelevkin, and M. R. Korshunova. A detailed kinetic submechanism for OH* chemiluminescence in hydrogen combustion revisited. Part 1. *Combustion and Flame* 263 (2024): 113407.
18. de Uña, A. P.. Experimental and kinetic analysis of acetaldehyde low-temperature oxidation using a rapid compression machine. PhD diss., RWTH Aachen, 2024.
19. Virt, M., and M. Zöldy. Realization of low temperature combustion in an unmodified diesel engine. *Cognitive Sustainability* 3 (2024): 2.
20. Liu, Z., X. He, C. Zhao, and X. Zhou. Experimental and kinetic of hydrogen blending effects on lean flammability limit of gasoline fuels. *Fuel* 384 (2025): 133936.

3 Thermodynamics of Low-Temperature Combustion

Sumit Shoor, Harpreet Singh Bains, Manpreet Singh Bains, and Swastik Pradhan

3.1 INTRODUCTION

3.1.1 IMPORTANCE OF COMBUSTION IN ENERGY SYSTEMS

As the primary means of transforming the chemical energy contained in fuels into thermal, mechanical, or electrical energy, combustion is a basic process in energy systems. It has supported worldwide industrialization and economic growth for more than a century by supplying process heat, producing electricity, and powering engines in almost every sphere of civilization. Because of its high energy density, controllability, and compatibility with established infrastructure, combustion is still necessary even in the face of growing emphasis on renewable energy technology. When burned, fossil fuels like coal, oil, and natural gas supply more than three-quarters of the world's primary energy, allowing for anything from internal combustion engine propulsion in transportation to large-scale electricity generation in thermal power plants. Particularly in applications that need continuous power or high-temperature heat, like chemical processing, cement manufacturing, and metal refining, combustion systems provide scalability and reliability that many renewable alternatives cannot match. Modern combustion research also keeps pushing the limits of efficiency and environmental sustainability with the development of low-emission burners, advanced fuel injection systems, lean-burn combustion, and low-temperature combustion (LTC) techniques. By lowering dangerous emissions including carbon dioxide, unburned hydrocarbons (UHCs), and nitrogen oxides (NO_x), these developments seek to lessen the negative environmental effects of combustion-based technologies. Additionally, the flexibility of combustion systems makes it possible to gradually incorporate alternative fuels like hydrogen, biofuels, and synthetic hydrocarbons, offering a decarbonization transition route without requiring the abandonment of current infrastructure. In hybrid energy systems, combustion also serves as a supporting component, frequently serving as a load-balancing or backup mechanism to offset the unpredictable nature of wind and solar energy. Combustion is therefore not only a fundamental part of the energy systems that are in place now, but it is also a vital part of the worldwide shift to cleaner, more sustainable energy futures. Flow diagram of the thermodynamic of LTC is shown in Figure 3.1.

DOI: 10.1201/9781003633631-3

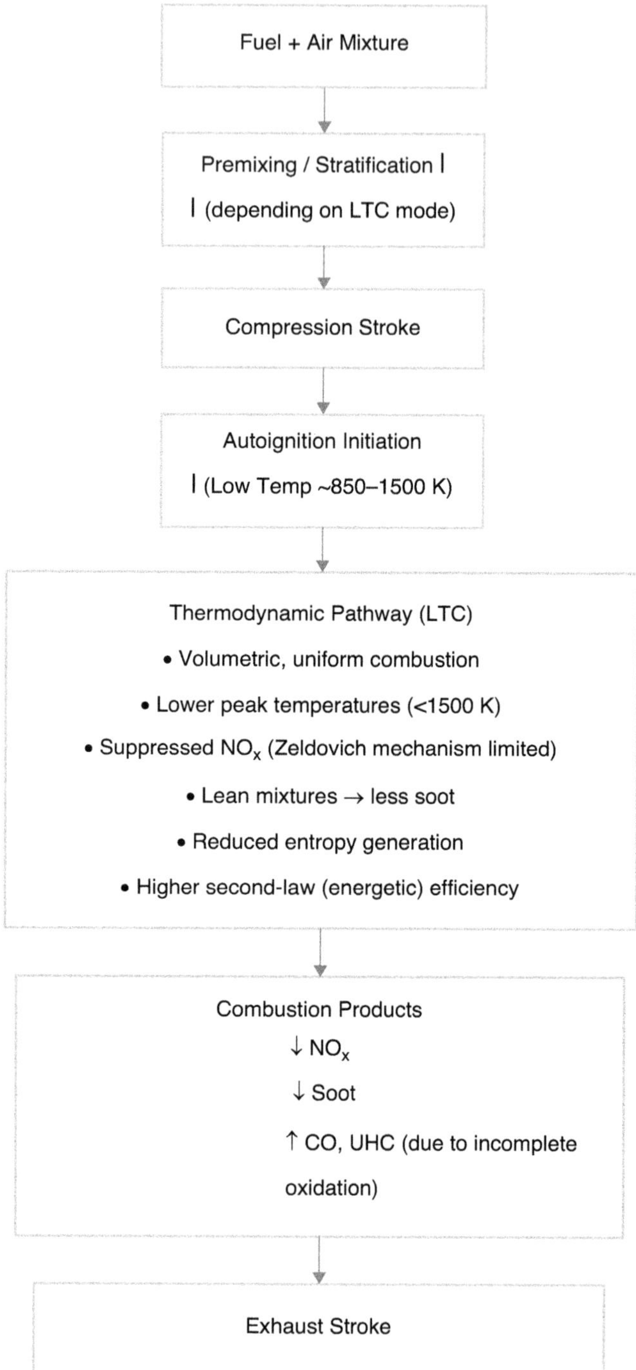

FIGURE 3.1 Flow diagram of thermodynamic of low-temperature combustion.

LTC modes—such as homogeneous charge compression ignition (HCCI), reactivity-controlled compression ignition (RCCI), and premixed charge compression ignition (PCCI)—have garnered significant attention for their thermodynamic benefits in internal combustion engines. These strategies reduce peak combustion temperatures, leading to lower NO_x and soot emissions while enhancing thermal efficiency through improved combustion phasing and reduced exergy destruction [1,2]. The distributed nature of LTC allows more uniform heat release, thereby lowering entropy generation and increasing second-law efficiency [3]. Additionally, studies by Refs. [4] and [5] emphasize the importance of low-temperature chemical kinetics, such as the role of alkylperoxy (RO_2) and hydroperoxy (QOOH) radicals in shaping ignition delay and heat release. Engine experiments by Refs. [6,7] and further demonstrate that careful thermodynamic control—via exhaust gas recirculation (EGR), intake boost, or fuel reactivity management—can push LTC engines toward brake thermal efficiencies above 45%. Though a modern engineering phenomenon, LTC reflects a timeless human endeavor to master heat with purpose and precision, an effort subtly echoed in Herodotus's accounts of ancient societies adapting fire for constructive use.

3.1.2 Traditional Combustion vs. Low-Temperature Combustion (LTC)

Conventional combustion uses high-temperature flame-based processes that provide quick heat release and distinct flame fronts [8–14]. In these engines, fuel and air combine and ignite either by compression (in diesel engines) or a spark (in spark-ignition (SI) engines). Because of their stability, controllability, and maturity over a variety of fuels and operating situations, these traditional technologies are frequently employed. However, they are naturally linked to high peak temperatures, which encourage the production of soot and nitrogen oxides (NO_x), two important pollutants that are governed by contemporary emission rules. LTC, on the other hand, is a sophisticated technique that seeks to reduce these pollutants by reducing the combustion temperature and postponing the rate of heat release. To do this, LTC encourages extremely diluted or premixed fuel–air combinations to autoignite at lower peak temperatures without a flame front. The LTC category includes methods including RCCI, PCCI, and HCCI. These methods produce slower, volumetric combustion and a more even temperature distribution, which greatly lowers NO_x and particulate matter (PM). Narrow working ranges, limited controllability of ignition timing, and increased emissions of UHCs and carbon monoxide due to partial oxidation at lower temperatures are some of the technical difficulties associated with LTC. In contrast to conventional combustion, LTC necessitates exact regulation of in-cylinder pressure, intake temperature, and mixture formation to guarantee steady and effective operation. Even though conventional combustion is still popular because it is straightforward and dependable, LTC offers a viable way to achieve extremely low emissions and high thermal efficiency, particularly in applications for advanced internal combustion engines that aim to comply with more stringent environmental regulations.

3.1.3 Motivation for LTC: Efficiency and Emissions

The desire to improve thermal efficiency and lower hazardous emissions—two essential objectives in the thermodynamics associated with modern energy conversion—is the main driving force behind the development of LTC systems. Despite being reliable and well-established, traditional combustion systems lose a lot of energy because of high-peak temperatures, insufficient heat recovery, and large irreversibilities in the combustion process. In addition, high combustion temperatures cause localized fuel-rich zones that produce soot and encourage the thermal generation of nitrogen oxides (NO_x), both of which are serious health and environmental hazards. By purposefully reducing combustion temperatures using lean mixes, EGR, and premixed fuel–air methods, LTC overcomes these thermodynamic and environmental constraints. According to thermodynamic theory, LTC promotes more even heat release and improved temperature distribution in the combustion chamber, which can lower entropy production and raise the engine's overall second-law efficiency. The premixed nature of the charge reduces soot because there are no locally rich combustion zones, and the lower peak temperatures in LTC greatly reduce thermal NO_x generation in terms of emissions. LTC is a desirable approach for next-generation internal combustion engines because of its combined advantages of high efficiency and ultra-low NO_x and particle emissions, especially in light of stricter regulations and global decarbonization initiatives. Even though there are still issues with managing cold-start emissions, stabilizing operation at part load, and regulating ignition timing, the thermodynamic benefits of LTC offer compelling reasons for its further advancement and incorporation into high-efficiency, low-emission energy systems.

3.1.4 Scope and Objectives of this Chapter

1. To offer a thorough thermodynamic framework for understanding systems involving LTC.
2. To evaluate the efficiency, emission characteristics, and heat release of LTC in comparison to conventional combustion.
3. To investigate the main LTC tactics from a thermodynamic standpoint, including RCCI, PCCI, and HCCI.
4. To determine the sources of efficiency improvements by analyzing the energy and exergy flows in LTC systems.
5. To investigate LTC's potential for reducing emissions, with a focus on suppressing NO_x and PM.
6. To discuss the challenges and limitations of LTC from a thermodynamic control and stability viewpoint.
7. To introduce simulation techniques and modeling tools for the thermodynamic assessment of LTC systems.

3.2 FUNDAMENTALS OF THERMODYNAMICS IN COMBUSTION

The fundamental ideas of thermodynamics control combustion, a complicated exothermic chemical reaction involving a fuel and an oxidant [14–19]. Analyzing, designing, and optimizing combustion-based energy systems requires an understanding of these foundational concepts. Fundamentally, combustion is the process of turning fuels' chemical energy into thermal energy, which can then be used to produce mechanical or electrical work. Heat release, work output, and system efficiency during combustion can all be calculated using the first law of thermodynamics, or the conservation of energy. By assessing entropy generation and exergy destruction, the second law of thermodynamics offers information about the quality of energy conversion. These principles are essential for figuring out variables like flame temperature, thermal efficiency, and the gas turbine and internal combustion engine performance limits. To assess combustion characteristics under different circumstances, chemical thermodynamics also offers tools for calculating the heat of reaction, equilibrium product composition, and adiabatic flame temperatures. Thermodynamic analysis is even more important in contemporary energy systems, particularly those aiming for ultra-low emissions and high efficiency, like LTC, to comprehend how to regulate combustion phasing, minimize irreversibilities, and maximize fuel–air interactions. This section provides the framework for a more thorough examination of advanced combustion regimes like LTC by introducing the basic thermodynamic rules as they relate to combustion systems.

3.2.1 First and Second Laws of Thermodynamics in Low-Temperature Combustion

To comprehend the efficiency and emission benefits of LTC systems in comparison to conventional combustion processes, it is necessary to examine them via a view of the basic laws of thermodynamics. By taking into consideration the internal energy changes, work interactions, and heat transfer related to the combustion process, the first law of thermodynamics—also referred to as the energy conservation principle—manages the energy balance in combustion systems. Chemical reactions in LTC release energy from the fuel, which is then transformed into residual heat and productive work, such as engine piston movement. LTC distributes the energy release more evenly during the expansion stroke, minimizing heat losses and enhancing thermal utilization, in contrast to conventional combustion, which frequently displays high peak temperatures and abrupt heat-release rates. As a result of improved work extraction and lower cooling needs, first-law efficiency increases. The quality and irreversibility of energy conversion in combustion processes are better understood by applying the second law of thermodynamics, which addresses entropy and the directionality of energy transformations. Significant entropy creation, or lost work potential or exergy destruction, results from steep thermal gradients and rapid reaction rates in conventional high-temperature combustion. In contrast, LTC minimizes entropy formation and mitigates isolated hotspots by operating at lower combustion temperatures and more homogenous mixture conditions. Because more of the fuel's chemical energy may be transformed into meaningful work with fewer irreversibilities, second-law (exergy) efficiency may increase as a result. Additionally, LTC's lower peak temperatures automatically restrict the production of thermal NO_x emissions, which is consistent with the second law's focus on reducing waste and inefficiencies. But LTC systems also have thermodynamic issues like decreased reaction completeness and sensitivity to mixture preparation, which, if left unchecked, can lead to higher CO and UHC emissions. Therefore, to assess and optimize LTC solutions, a comprehensive thermodynamic study of both the first and second laws is necessary. These laws offer recommendations for cutting emissions and enhancing sustainability in advanced combustion systems in addition to aiding in the quantification of energy conversion performance.

3.2.2 Enthalpy and Internal Energy in Reacting Systems

During combustion, the internal energy of a reacting system is altered by the rearrangement of chemical bonds [19–23]. The internal energy (*u*) of a reacting system is the sum of the microscopic energy in a control volume, including the electronic, vibrational, rotational, and translational modes of the individual species as well as the chemical energy associated with molecular bonds. Whether in traditional or LTC systems, the thermodynamic analysis of combustion depends on two fundamental state functions: internal energy (*u*) and enthalpy (h). These quantities, which represent the energy stored in a system and its capacity for nonmechanical work, must be understood to apply the first law of thermodynamics to reacting systems.

The change in internal energy ΔU for a closed system going through a constant-volume combustion process is directly related to the heat emitted by the chemical reaction less any energy losses, as shown in equation (3.1)

$$\Delta U = Q_v = \sum n_p u_p(T) - \sum n_r u_r(T) \tag{3.1}$$

where u_p, u_r are the molar internal energies of reactants and products, respectively.

However, in open systems with continuous flow, such as gas turbines and the intake-exhaust cycles of reciprocating engines, the enthalpy (h) is more significant. Enthalpy takes into account both flow work and internal energy is shown in equation 3.2.

$$h = u + pv \tag{3.2}$$

It is especially helpful for examining constant-pressure processes, which are typical of combustion devices found in the real world. Using enthalpy, the first law may be applied to a steady-flow reactor, such as the combustor in a gas turbine is shown in equation 3.3

$$\sum \dot{m}_{\text{in}} h_{\text{in}} + \dot{Q} - \dot{W} = \sum m \tag{3.3}$$

Thermodynamically defining LTC hence requires precise assessment of internal energy and enthalpy using temperature-dependent specific heat models and actual chemical equilibrium data. In current combustion systems, these characteristics serve as the foundation for cycle analysis, heat-release modeling, and efficiency optimization.

3.2.3 Gibbs Free Energy and Spontaneity of Combustion

The Gibbs free energy (G) is a key factor in assessing the spontaneity and chemical equilibrium of a reacting system in the thermodynamic analysis of combustion. The definition of Gibbs free energy is shown in equation 3.4.

$$G = H - TS \tag{3.4}$$

where S is the system's entropy, T is the absolute temperature, and H is the enthalpy. It is especially important to real-world combustion processes, many of which take place at circumstances of nearly constant pressure, since it indicates the greatest non-expansion work that can be obtained from a closed system at constant temperature and pressure.

The spontaneous nature of a chemical reaction is determined by the change in Gibbs free energy (ΔG). When ΔG is negative, the reaction is thermodynamically favorable and can continue without the need for additional energy. The Gibbs free energy change is nearly always significantly negative in combustion reactions, which are generally very exothermic and entropy-increasing, especially under normal conditions (298 K, 1 atm). For instance, methane combustion is shown in equation 3.5.

$$CH_4 + 2O_2 \rightarrow CO_2 + 2H_2O \tag{3.5}$$

indicates that the reaction is spontaneous and releases useful energy, as evidenced by its high negative standard Gibbs free energy shift ($\Delta G \approx -818$ kJ/mol).

3.2.4 Equilibrium vs. Actual Combustion Systems

Differentiating between idealized equilibrium combustion and real combustion systems is crucial in the thermodynamic analysis of combustion. Because of kinetic, transport, and system limits, actual combustion processes frequently differ greatly from equilibrium simulations, which offer important insights into the maximum attainable performance and product composition. According to the theory of equilibrium combustion, the reacting mixture has had enough time and favorable circumstances (temperature, pressure, and mixing) to reach chemical equilibrium, when the system's Gibbs free energy is at its lowest.

At equilibrium: Only thermodynamic characteristics such as temperature, pressure, and fuel-to-air ratio—determine the makeup of the products. Assuming complete combustion, no intermediary species (such as CO or unburned hydrocarbons (UHCs)) are left.

3.3 CHARACTERISTICS OF LOW-TEMPERATURE COMBUSTION

LTC is an advanced combustion strategy designed to achieve higher thermal efficiency and ultra-low emissions by controlling the combustion process at relatively lower peak temperatures compared to conventional SI and compression-ignition (CI) engines. RCCI, PCCI, and HCCI are all included in the LTC family of ideas.

Despite operating differently, these techniques have similar thermodynamic and combustion-related traits that characterize the LTC regime.

3.3.1 Low-Temperature Combustion (LTC)

It relates to a group of advanced combustion techniques designed to bring together the low emissions of SI engines and the great efficiency of diesel engines. At fairly low temperatures and in highly diluted conditions, like with excess air or EGR, LTC works by encouraging the autoignition of a well-mixed or stratified fuel–air mixture. The controlled decrease in combustion temperature that helps prevent the emission of thermal nitrogen oxides (NO_x) without sacrificing thermal efficiency is what distinguishes LTC. In contrast to typical engines, which use a propagating flame front, combustion happens by spontaneous ignition at several sites in the chamber. Traditional diesel combustion.

3.3.2 Autoignition Behavior and Chemical Kinetics in Low-Temperature Combustion

Autoignition, or the spontaneous ignition of a fuel–air combination under the influence of high temperature and pressure, is the fundamental component of LTC systems. This process occurs without the need for an external spark or flame.

Complex chemical changes control the onset and properties of autoignition, which are very different from conventional high-temperature combustion processes. When a fuel–air mixture's local temperature and pressure surpass the fuel's critical ignition conditions, autoignition takes place. Lower local flame temperatures and quick heat release are the results of this volumetric combustion. Numerous important parameters impact the autoignition behavior in LTC, including: EGR levels and dilution, mixture homogeneity, compression ratio, heat transfer rates, in-cylinder temperature and pressure histories, and fuel reactivity (cetane or octane number). To prevent premature ignition or misfire, the timing of autoignition in LTC modes like HCCI or RCCI is crucial and needs to be carefully managed. This presents a serious thermodynamic problem because the ignition is dependent on time rather than being deliberately activated.

3.3.3 Compression Ignition Modes in Low-Temperature Combustion: HCCI, RCCI, and PPCI

LTC refers to a number of cutting-edge compression ignition techniques that use autoignition at relatively low flame temperatures to produce high thermal efficiency with ultra-low emissions. Partially Premixed Compression Ignition (PPCI), RCCI, and HCCI are the three LTC techniques that are most frequently researched and used. A different strategy for managing mixture formation, ignition timing, and combustion phasing is represented by each of these combustion modes. By using compression heating instead of a spark or fuel injection event at ignition timing, HCCI entails the autoignition of a homogeneous, premixed fuel–air mixture throughout the combustion chamber.

TABLE 3.1
Comparison of LTC Combustion Modes

Feature	HCCI	RCCI	PPCI
Charge preparation	Fully premixed	Dual-fuel, stratified reactivity	Partially premixed
Ignition control	Compression temperature	Fuel reactivity and injection timing	Injection timing
NO_x and soot emissions	Very low	Very low	Low
Load range	Narrow	Wide	Moderate
Fuel requirement	Single fuel	Dual fuels (high and low reactivity)	Single fuel (diesel)
Implementation complexity	Moderate	High	Low to moderate

3.3.4 Advantages of Low-Temperature Combustion (LTC) over High-Temperature Regimes

A crucial advancement in internal combustion engine technology, LTC aims to maximize thermodynamic efficiency while lowering flame temperatures. In comparison to traditional high-temperature combustion regimes like CI and SI diesel engines, LTC provides a number of significant benefits in terms of fuel efficiency, pollution control, and thermal performance. Using the Zeldovich mechanism, high-temperature combustion zones typically above 2000 K promote the generation of thermal NO_x. On the other hand, LTC virtually completely prevents NO_x generation by limiting in-cylinder temperatures to below 1700 K. This lowers system complexity and cost by enabling LTC engines to run with little to no aftertreatment for NO_x management. Local fuel-rich zones cause traditional diffusion flames to produce a lot of soot, particularly in diesel engines. Premixed or partially premixed combustion is encouraged by LTC modes like HCCI and RCCI, which prevent fuel-rich areas and result in combustion that is almost soot-free. Comparison of LTC Combustion Modes is shown in Table 3.1.

3.4 THERMODYNAMIC ANALYSIS OF LOW-TEMPERATURE COMBUSTION (LTC) PROCESSES

Understanding LTC's efficiency potential and emission behavior in comparison to conventional combustion modes requires a thermodynamic study. Heat release, entropy generation, energy conservation, and exergy balance are all aspects of classical thermodynamics that may be used to properly examine LTC processes, despite their chemical and kinetic complexity.

3.4.1 Heat Release and Combustion Phasing

Heat release in traditional high-temperature combustion usually occurs in two stages: premixed and diffusion-controlled. On the other hand, autoignition-driven heat release and a more uniform or stratified-premixed charge are characteristics of LTC modes including HCCI, RCCI, and PPCI, which frequently produce a single-stage or weakly staged heat release.

For a closed system, the first law can be written in equation (3.6)

$$dU = \delta Q - \delta W \tag{3.6}$$

In an engine cylinder:

$$\delta Q = dU + PdV \tag{3.7}$$

Where:

- dU is the change in internal energy,
- PdV is the boundary work done by the gas,
- δQ is the heat added due to fuel combustion.

3.4.2 Exergy Analysis of LTC Combustion Cycles

An essential thermodynamic method for assessing the caliber of energy transitions taking place in LTC systems is energy analysis. Exergy takes irreversibility into consideration and determines the greatest amount of beneficial work that may be done once a system approaches equilibrium with its surroundings, in contrast to energy, which is conserved.

Energy analysis shows how much of the fuel's chemical potential is transformed into mechanical work and how much is lost to inefficiencies like incomplete combustion, heat loss, and exhaust emissions in LTC engines, where combustion occurs at lower temperatures and frequently involves homogeneous or partially premixed mixtures.

3.4.3 Entropy Generation and Irreversibilities in Thermodynamic Analysis of LTC Processes

When assessing the performance of LTC engines, entropy generation is a critical determinant of irreversibility in thermodynamic systems. Entropy, which represents missed opportunities to transform fuel energy into useful work, is produced in any real combustion process by departures from ideal behavior, including heat losses, friction, mixing, and finite-rate chemical reactions [23–27]. In contrast to traditional high-temperature combustion regimes, LTC systems are deliberately made to limit entropy formation. They are distinguished by their lean mixes, low combustion temperatures, and frequently homogeneous charge preparation. The primary sources of irreversibility in LTC include thermal gradients during combustion, incomplete combustion of fuel (leading to UHCs and CO emissions), and heat transfer to engine walls. Unlike conventional SI or CI engines where flame fronts and steep temperature gradients dominate, LTC modes such as HCCI, RCCI, and PPCI promote a more uniform heat release, thereby reducing local entropy generation.

3.5 CHEMICAL KINETICS AND REACTION MECHANISMS

The foundation for comprehending combustion behavior in LTC systems is thermal kinetics and reaction mechanisms. LTC mainly depends on low-temperature oxidation pathways and autoignition chemistry, in contrast to traditional high-temperature combustion, which usually includes rapid flame propagation controlled by thermal ignition. The kinetics of fuel oxidation, including the production and consumption of intermediate species such as hydroperoxides (ROOH), alkylperoxy (RO_2), and keto-hydroperoxides (KHP), have a significant impact on the combustion process in LTC. The kinetics of fuel oxidation, including the production and consumption of intermediate species such as hydroperoxides (ROOH), alkylperoxy (RO_2), and ketohydroperoxides (KHP), have a significant impact on the combustion process in LTC [28].

3.5.1 Low-Temperature Oxidation Chemistry

Since it controls the ignition behavior, heat release properties, and emissions profile under sub-flame propagation conditions, low-temperature oxidation chemistry is essential to the functioning of LTC systems. In contrast to traditional high-temperature combustion, which is primarily characterized by thermal NO_x generation and flame propagation driven by radicals, LTC processes depend on a series of slower, temperature-sensitive reactions that take place between roughly 600 and 1000 K.

The chemistry starts when oxygen or radical species remove a hydrogen atom from a fuel molecule (RH), creating an alkyl radical (R·). Predicting ignition delay, maximizing combustion phasing, and reducing CO, HC, and PM emissions in LTC systems all depend on an understanding of the ability to simulate low-temperature oxidation chemistry.

3.5.2 The Negative Temperature Coefficient (NTC)

In LTC chemistry, behavior is a special and important phenomenon that is especially pertinent to LTC modes like RCCI and HCCI. The reaction rate rises with temperature in most chemical processes. But in the NTC zone, which is typically between 700 and 900 K, the autoignition delay inexplicably rises as the temperature rises.

The intricate interaction between different chemical pathways is the cause of this surprising behavior. Chain-branching processes involving hydroperoxyalkyl (QOOH) and ketohydroperoxide (KHP) intermediates predominate at lower temperatures, which speeds up the accumulation of radicals and encourages ignition. Thermal breakdown of these intermediates is preferable to chain branching when the temperature rises into the NTC range, which lowers the rate of radical production and delays ignition. Reactivity decreases as a result, which is typical of the NTC regime [29].

3.5.3 Coupling of Chemical Kinetics with Thermodynamic Analysis

For LTC systems to be effectively modeled and optimized, thermodynamic analysis and chemical kinetics must be combined. Chemical kinetics explains the microscopic reaction paths and velocity that dictate how and when combustion takes place, whereas thermodynamics supplies the macroscopic limitations, such as energy conservation, entropy generation, and equilibrium limits. The timing and advancement of heat release in LTC modes, where autoignition takes the place of flame propagation, are controlled by the precise kinetics of intermediary species such as RO_2, QOOH, and ketohydroperoxides in addition to the overall energy balance. These species specify the features of pollutant generation, combustion phasing, and ignition delay at low temperatures.

3.6 EXPERIMENTAL STUDIES AND THERMODYNAMIC MEASUREMENTS

Understanding and verifying the thermodynamic and chemical properties of LTC systems requires experimental research. To create precise kinetic mechanisms and thermodynamic models, these studies offer fundamental data on ignition delay, pressure rise, temperature distribution, and emissions. Techniques like shock tubes, optical engines, and rapid compression machines are frequently used to examine combustion processes in controlled environments and model situations relevant to LTC. These platforms enable researchers to use sophisticated diagnostics like gas chromatography-mass spectrometry, laser-induced fluorescence (LIF), and Fourier-transform infrared spectroscopy to measure intermediate species like formaldehyde, hydrogen peroxide, and ketohydroperoxides and isolate low-temperature oxidation processes.

3.6.1 Techniques for Temperature and Pressure Measurement

Analyzing thermodynamic behavior in LTC systems requires precise temperature and pressure readings. High-frequency piezoelectric transducers are frequently employed to record in-cylinder pressure with fine time resolution, allowing for heat release and pressure–volume analysis, because of the fast and spatially variable combustion events. Low flame luminosity and dispersed reaction zones make it more difficult to determine temperature in LTC. Particularly in HCCI engines, optical methods such as Coherent Anti-Stokes Raman Spectroscopy, infrared thermography, two-line atomic fluorescence, and LIF provide high-resolution, non-intrusive temperature data [30]. Even though they respond more slowly, thermocouples are used for average measurements in the intake and exhaust zones. By combining these methods, it is possible to identify exergy losses, analyze ignition timing, and validate the model.

3.6.2 Thermodynamic Property Estimation during LTC

Estimating thermodynamic properties during LTC is vital for analyzing combustion performance and efficiency. LTC occurs between 600 and 1200 K, where chemical kinetics and thermodynamics strongly interact. Key properties like enthalpy, internal energy, specific heats, and entropy are determined using experimental data, empirical correlations, and tools such as CHEMKIN, Cantera, and NASA CEA, which use NASA polynomials. Complex intermediates like peroxides require detailed data. In-cylinder pressure combined with the first law of thermodynamics helps estimate heat release and work output. Accurate property estimation enhances modeling, aiding cleaner and more efficient LTC system development.

3.6.3 Heat Release Analysis from Experimental Data

Understanding LTC features, such as ignition time, combustion duration, and energy efficiency, requires an understanding of heat release analysis. It employs piezoelectric sensors to measure in-cylinder pressure and applies the first law of thermodynamics to pressure-volume data. For simplicity, the popular single-zone heat release model disregards heat transmission and crevice effects and assumes a uniform charge [31]. This technique, which is specific to LTC, separates the low-temperature heat release and high-temperature heat release phases. Accurate heat release analysis is crucial for monitoring fuel behavior, optimizing injection tactics, and enhancing engine thermal efficiency and emissions.

3.6.4 Case Studies on LTC Engines

Practical insights into techniques such as PPCI, RCCI, and HCCI are provided via case studies on LTC engines. Although ignition timing regulation is difficult, HCCI engines running on gasoline, diesel, DME, and ethanol exhibit reduced NO_x and soot emissions as well as great thermal efficiency. By optimizing combustion phasing through the use of dual fuels, RCCI increases efficiency and load flexibility. PPCI maintains low emissions while extending LTC to greater loads. These investigations validate LTC principles and direct the design of future low-emission engines using in-cylinder pressure, emissions data, and thermodynamic analysis.

3.7 LTC AND EMISSIONS THERMODYNAMICS

By operating below 1500 K and minimizing the generation of NO_x caused by high-temperature processes, LTC lowers engine emissions. Soot and hot patches are reduced by its uniform heat release and lean, homogeneous air–fuel combinations. By turning more chemical energy into productive labor, LTC improves second-law (exergetic) efficiency thermodynamically by reducing entropy production and pressure rise rates. However, because of slower oxidation kinetics, incomplete combustion at low temperatures can result in an increase in UHCs and carbon monoxide (CO). Although LTC provides cleaner combustion, efficiency and emissions must be balanced through careful management of fuel reactivity and combustion phasing.

3.7.1 NO_x Formation and Suppression at Low Temperatures

The Zeldovich mechanism, which is active above around 1800 K, is the main source of nitrogen oxide (NO_x) emissions from high-temperature combustion. Conventional engines frequently produce far more NO_x than this. By keeping peak temperatures below 1500 K, LTC techniques including HCCI, RCCI, and PCCI inhibit the generation of NO_x. Lean, uniform combustion with progressive heat release is made possible by LTC, which reduces hot patches and sharp thermal gradients that cause NO_x to be produced. NO_x is further constrained by the lower amounts of nitrogen and oxygen radicals. LTC provides a cleaner combustion route without the need for aftertreatment systems by thermodynamically avoiding high-temperature zones, which makes it perfect for satisfying stringent emissions regulations.

3.7.2 Unburned Hydrocarbons and CO Emissions

LTC lowers soot and NO_x, but because of incomplete fuel oxidation, it frequently raises UHC and carbon monoxide (CO). Full fuel breakdown is hampered by lower in-cylinder temperatures and slower reaction kinetics in LTC modes like HCCI and RCCI, particularly below ~850–1000 K. Fuel may remain unburned in chilly border layers and fissures in the absence of a flame front [32]. As thermodynamic sinks, these areas raise entropy and decrease efficiency. When oxygen is present but temperatures are too low for complete CO-to-CO_2 conversion, CO emissions are particularly significant under lean, low-load circumstances, necessitating better in-cylinder management or aftertreatment.

3.7.3 Soot Behavior under LTC Regimes

In traditional combustion, fuel-rich zones and high temperatures that facilitate PAH chemistry are the sources of soot production. LTC techniques, such as HCCI, RCCI, and PPCI, on the other hand, significantly minimize soot because they use lean, premixed mixtures that avoid rich pockets and lower combustion temperatures (<1600 K). Soot nucleation and growth are suppressed by LTC's consistent, volumetric heat emission. In an oxygen-rich environment, even little soot precursors usually undergo oxidation. Diesel particulate filters are frequently no longer necessary as a result of the considerable reduction in soot emissions [33]. However, in cooler climates, incomplete oxidation may result in more CO and UHC, offsetting this benefit.

3.8 CHALLENGES AND FUTURE DIRECTIONS

Although LTC lowers NO_x and soot, it has drawbacks, such as challenging combustion phasing control, particularly in HCCI, which depends on the kinetics of chemicals. It is difficult to operate steadily with different loads and speeds, and incomplete combustion increases emissions of CO and UHC. Calibration is made more difficult by LTC's sensitivity to fuel characteristics and intake conditions. Predictive combustion models, enhanced fuel reactivity control, alternate fuels, and tactics like RCCI and advanced EGR are examples of future solutions. LTC's full thermodynamic and environmental potential may be unlocked by integrating it with AI-based control and hybrid systems, which could improve performance and stability.

3.8.1 Cold Start and Combustion Stability

Two of the main issues with LTC systems are cold start and combustion stability. Low in-cylinder and wall temperatures during cold starts prevent autoignition in modes such as RCCI and HCCI, leading to misfires, delayed ignition, and elevated emissions of CO and UHC. Control is made more difficult by LTC's sensitivity to temperature, pressure, and fuel reactivity, particularly during transients or load changes. EGR, exact control of intake conditions, and occasionally external heating or fuel reformulation are necessary for stable combustion. Commercial feasibility depends on resolving these problems, and research is being done to enhance ignition performance and reliability using improved controllers, fuel additives, and hybrid thermal systems.

3.8.2 Research Gaps and Emerging Trends

LTC has advanced; however, adoption is hampered by significant gaps. Insufficient comprehension of low-temperature kinetics, particularly in the NTC region, makes combustion control and predictive modeling more difficult. Because LTC is sensitive to temperature, pressure, and fuel chemistry, real-time regulation under varying loads is still challenging. Development is further limited by the lack of thermodynamic data for alternative fuels and intermediates. AI-based models, sophisticated diagnostics, and interest in low-carbon fuels like hydrogen and ammonia are examples of emerging solutions. Unlocking LTC's full efficiency and emissions-reduction potential requires interdisciplinary study combining thermodynamics, chemistry, and control engineering as emission limits tighten.

REFERENCES

1. Shirvani, S., et al. (2021). Thermodynamic energy and exergy analysis of low-temperature combustion strategies. *SAE International Journal of Engines*, 14(3): 345–368.
2. Reader, G. T.; Asad, U.; Zheng, M. (2013). Energy efficiency trade-off with phasing of HCCI combustion. *International Journal of Energy Research*, 37(3): 200–210.
3. Feng, B., et al. (2023). A comprehensive review of energy regeneration and conversion technologies based on mechanical–electric–hydraulic hybrid energy storage systems in vehicles. *Applied Sciences*, 13(7): 4152.
4. Hughes, K. J., et al. (2006). Evaluation of models for the low temperature combustion of alkanes through interpretation of pressure–temperature ignition diagrams. *Physical Chemistry Chemical Physics*, 8(27): 3197–3210.
5. Dewey, N. S.; Rotavera, B. (2023). Reaction mechanisms of alkyloxiranes for combustion modeling. *Combustion and Flame*, 252: 112753.
6. Asad, U.; Han, X.; Zheng, M. (2012). An empirical study to extend engine load in diesel low temperature combustion. *SAE International Journal of Engines*, 5(3): 709–717.
7. Battin-Leclerc, F. (2008). Detailed chemical kinetic models for the low-temperature combustion of hydrocarbons with application to gasoline and diesel fuel surrogates. *Progress in Energy and Combustion Science*, 34(4): 440–498.
8. Ju, Y.; Lefkowitz, J. K.; Reuter, C. B.; Won, S. H.; Yang, X.; Yang, S.; Sun, W.; Jiang, Z.; Chen, Q. (2016). Plasma assisted low temperature combustion. *Plasma Chemistry and Plasma Processing*, 36(1): 85–105.

9. Rotavera, B.; Taatjes, C. A. (2021). Influence of functional groups on low-temperature combustion chemistry of biofuels. *Progress in Energy and Combustion Science*, 86: 100925.
10. Wang, M.; Kukkadapu, G.; Fang, R.; Pitz, W. J.; Sung, C. J. (2021). Autoignition study of iso-cetane/tetralin blends at low temperature. *Combustion and Flame*, 228: 415–429.
11. Nadiri, S.; Agarwal, S.; He, X.; Kühne, U.; Fernandes, R.; Shu, B. (2021). Ignition delay of LNG: kinetics and modeling at intermediate to high temperatures. *Fuel*, 302: 121137.
12. Dold, J. W.; Thatcher, R. W.; Omon-Arancibia, A.; Redman, J. (2002). From one-step to chain-branching premixed flame asymptotics. *Proceedings of the Combustion Institute*, 29(2): 1519–1526.
13. Dold, J. W. (2007). Premixed flames modelled with thermally sensitive intermediate branching kinetics. *Combustion Theory and Modelling*, 11(6): 909–948.
14. Im, H. G.; Gupta, S.; Keum, S.; Pal, P. (2014). Modeling of scalar dissipation rates in flamelet models for low temperature combustion engines. arXiv:1412.4285.
15. Anderlohr, J.; Bounaceur, R.; Da Cruz, A. P.; Battin-Leclerc, F. (2009). Modeling of autoignition and NO sensitization for IC engine surrogate fuels. *Combustion and Flame*, 156(2): 505–521.
16. Raman, V.; Hassanaly, M. (2018). Emerging trends in numerical simulations of combustion systems. arXiv:1808.07478.
17. Rameez, P. V.; Ibrahim, M. M. (2024). A comprehensive review on the utilization of hydrogen in low temperature combustion strategies: Combustion, performance and emission attributes. *Journal of the Energy Institute*, 113, 101511.
18. Asad, U.; Divekar, P.; Zheng, X.; et al. (2012). Mode switching control for diesel LTC with fast feedback algorithms. *SAE International Journal of Engines*, 5(3): 850–863.
19. De Ojeda, W.; Bulicz, T.; Han, X.; Zheng, M.; Cornforth, F. (2011). Impact of fuel properties on diesel low temperature combustion. *SAE International Journal of Engines*, 4(1): 188–201.
20. Han, X.; Zheng, M.; Wang, J. (2013). Fuel suitability for low temperature combustion in compression ignition engines. *Fuel*, 109: 336–349.
21. Epping, K., Aceves, S., Bechtold, R., & Dec, J. E. (2002). *The potential of HCCI combustion for high efficiency and low emissions* (No. 2002-01-1923). SAE technical paper.
22. Asad, U.; Han, X.; Zheng, M. (2012). Tightened intake oxygen control for improving diesel low temperature combustion. *SAE International Journal of Engines*, 3(1): 788–800.
23. Han, X.; Tjong, J.; Zheng, M.; Xie, K. (2012). Simultaneous low NO_x and soot combustion with diesel and ethanol in diesel engines. *Journal of Engineering for Gas Turbines and Power*, 134: 11.
24. Jhalani, A.; Sharma, D.; Sharma, P. K.; Singh, D.; Dubey, S. J. P. (2021). A comprehensive review on low-temperature combustion technologies for emission reduction in diesel engines. *International Journal of Automotive & Mechanical Engineering*, 18(4): 9230–9243.
25. Osintsev, K. (2012). Low-temperature combustion technology. *International Scholarly Research Notices*, 2012: 637545.
26. Zhang, Y.; Zhang, Y.; Li, Y.; Xue, J.; Shi, X.; Yang, C. (2022). Low-temperature oxidation characteristics of coals with methane-containing atmosphere. *Journal of Thermal Analysis and Calorimetry*, 147: 3379–3389.
27. Battin-Leclerc, F.; Biet, J.; Hakka, M. H.; Warth, V.; Glaude, P.-A. (2008). Experimental and modeling study of the low-temperature oxidation of large alkanes. *Combustion Kinetics*, 22(4): 2258–2269.
28. Chiavazzo, E.; Asinari, P.; Visconti, F.; Boulouchos, K.; Karlin, I. (2011–2014). The global relaxation redistribution method for reduction of combustion kinetics. *Journal of Chemical Physics*, 141(4): 044102.

29. Bykov, V.; Maas, U. (2007). Extension of intrinsic low-dimensional manifolds concepts. *Combustion Theory and Modeling*, 11(6): 839–862.
30. Ren, Z.; Pope, S. B.; Vladimirsky, A.; Guckenheimer, J. M. (2009). The invariant constrained equilibrium edge preimage curve method for the dimension reduction of chemical kinetics. *Journal of Chemical Physics*, 124(11): 114111.
31. Lebiedz, D.; Reinhardt, V.; Winckler, M. (2010). Entropy-related extremum principles for model reduction of dissipative dynamical systems. *Entropy*, 12(4): 706–719.
32. Chiavazzo, E.; Karlin, I.; Boulouchos, K. (2010). The role of thermodynamics in model reduction when using invariant grids. *Communications in Computational Physics*, 8(4): 701–734.
33. Maas, U.; Pope, S. B. (1992). Simplifying chemical kinetics: intrinsic low-dimensional manifolds in composition space. *Combustion and Flame*, 88(3–4): 239–264.

4 Combustion Modeling and Simulation Techniques

Aditya Chauhan, Ankur Shah, and Santosh Kumar Tamang

4.1 INTRODUCTION

Low-temperature combustion (LTC) is an advanced combustion concept in internal combustion engines that has attracted significant attention for its potential to simultaneously reduce pollutant emissions and improve efficiency [1]. Unlike conventional spark ignition (SI) or compression ignition (CI) combustion, LTC operates with lean or dilute mixtures and suppressed flame temperatures, which can drastically cut emissions of nitrogen oxides (NO_x) and particulate matter (PM) [2]. In fact, LTC has been shown to achieve ultra-low engine-out NO_x and soot while maintaining high thermal efficiency [3,4]. This makes LTC a promising pathway to meet stringent emissions regulations and reduce reliance on aftertreatment systems. At the same time, LTC facilitates the use of alternative fuels and novel engine strategies in pursuit of cleaner and more efficient engines [1].

However, realizing LTC in practical engines poses serious challenges. Chief among these is the difficulty of controlling the ignition timing and combustion rate, since LTC often relies on autoignition chemistry rather than direct ignition sources [2]. The combustion phasing in LTC (e.g., Homogeneous Charge Compression Ignition (HCCI)) is predominantly kinetically controlled, making it sensitive to fuel properties, mixture stratification, and operating conditions. This sensitivity leads to a narrow operating range, with issues like excessive pressure rise rates (causing engine knock or noise) and combustion instability at low loads [5,6]. Moreover, while NO_x and soot are low, the lower combustion temperatures can result in incomplete oxidation, thus higher carbon monoxide (CO) and unburned hydrocarbon (UHC) emissions if the process is not well-managed [7]. In summary, LTC provides a compelling trade-off of emissions and efficiency benefits, but it demands new strategies for combustion control and optimization.

Modeling and simulation have a critical role in advancing LTC research and addressing these challenges. Because LTC regimes involve complex interactions of turbulence, chemistry, and heat transfer, they are difficult to study purely by experiments [8]. Detailed computational fluid dynamics (CFD) modeling allows researchers to "better understand the origin of experimentally observed trends" and to virtually prototype engine designs or operating strategies optimized for LTC [9].

 DOI: 10.1201/9781003633631-4

In the past, engine CFD models were primarily used for qualitative insights—explaining why a certain design or condition yielded better emissions, for example [9]. But modern demands are higher: "providing accurate trends and qualitatively explaining consequences is not enough anymore" [10]. Today, predictive simulation is required, meaning models must quantitatively capture combustion phasing and pollutant formation to be truly useful in designing LTC engines [11]. As noted by Knop et al., implementing accurate combustion and emissions sub-models in 3D CFD codes is "a prerequisite" for using simulation as a reliable design tool for new combustion concepts like LTC [11]. An inadequate understanding or modeling capability for these new combustion regimes has been identified as a barrier to bringing LTC concepts to production [8].

This chapter provides a comprehensive overview of the fundamental principles and techniques for modeling and simulating combustion, with a focus on their application to LTC strategies. We will begin by outlining the governing equations and physical models that describe reacting flows, including how turbulence and combustion are modelled. Then we will discuss chemical kinetics modeling for LTC, comparing detailed and reduced reaction mechanisms and how these are integrated into simulation tools. Next, major combustion modeling approaches (such as flamelet models, the Eddy Dissipation Concept (EDC), PDF methods, and G-equation flame tracking) are reviewed in the context of capturing turbulence–chemistry interactions (TCIs). We then address the modeling of emissions (NO_x, soot, CO, and UHC) and how simulations predict these in LTC conditions. The crucial issue of combustion–turbulence interaction is examined, particularly its role in autoignition and stratified mixtures relevant to LTC. To illustrate the application of these modeling techniques, case studies for gasoline and diesel LTC concepts (e.g., HCCI and reactivity-controlled compression ignition (RCCI)) are presented, including typical setup considerations in CFD (boundary conditions, fuel injection, etc.). We also discuss the challenges and limitations faced (such as chemical solver stiffness and computational cost) and how model validation is performed. Finally, emerging trends and future directions are explored—including the use of machine learning (ML), advanced adaptive mesh and solver strategies, and the impact of high-performance computing on combustion simulations. Throughout this chapter, references to peer-reviewed studies (SAE, ASME, and others) are provided to ground the discussion in established research. By the end, the reader should have a solid understanding of the tools and methods available for modeling combustion in LTC regimes, the strengths and weaknesses of these approaches, and the outlook for continued advancements in this field (Figure 4.1).

4.2 GOVERNING EQUATIONS AND PHYSICAL MODELS

Accurate combustion simulations rest on the foundation of fundamental governing equations that describe conservation of mass, momentum, energy, and species in a reacting flow. These conservation equations are typically written in differential form and can be solved in CFD codes to predict the flow field and flame behavior. For a multicomponent reacting gas mixture, the governing equations can be summarized as follows:

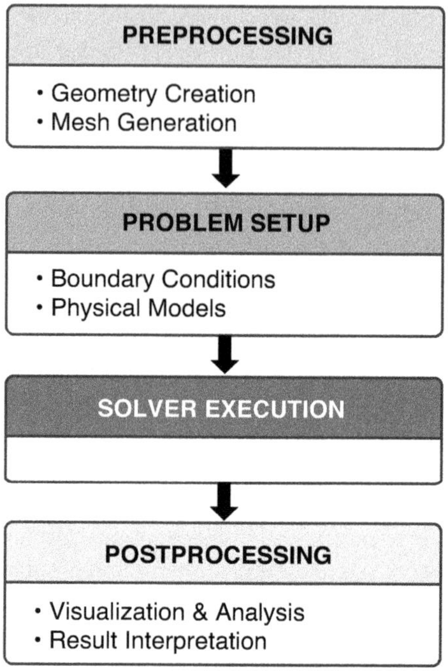

FIGURE 4.1 1CFD workflow.

- **Mass Conservation (Continuity Equation):** This ensures conservation of mass for the fluid. In differential form for an unsteady compressible flow:

$$\frac{\partial \rho}{\partial t} + \nabla \cdot (\rho \mathrm{u}) = 0,$$

where ρ is the density and u the velocity vector. This equation states that the rate of increase of mass in a fluid element plus the net mass flux out is zero, implying mass is neither created nor destroyed [12].
- **Momentum Conservation (Navier–Stokes Equations):** These represent Newton's second law for the fluid, accounting for pressure forces, viscous stresses, and body forces. For each momentum component (in vector form):

$$\frac{\partial (\rho \mathrm{u})}{\partial t} + \nabla \cdot (\rho \mathrm{uu}) = -\nabla p + \nabla \cdot \tau + \rho \mathrm{g},$$

where p is pressure, τ is the viscous stress tensor (related to velocity gradients and fluid viscosity), and g represents body forces (like gravity). In reacting flows, the equation of state (often the ideal gas law) closes the relation between$p$, g, and temperature.

- **Energy Conservation (Energy/Enthalpy Equation):** Energy equations can be written in terms of total energy or sensible enthalpy. A common form (total energy form) is:

$$\frac{\partial(\rho E)}{\partial t}+\nabla\cdot\left[\mathrm{u}(\rho E+p)\right]=\nabla\cdot(k\nabla T)+\rho\mathrm{u}\cdot\mathrm{g}+\dot{q}_{\mathrm{chem}},$$

  where E is total energy per unit mass, T temperature, and k thermal conductivity. The term on the right includes heat conduction ($k\nabla T$), work by body forces, and $\dot{q}_{\mathrm{chem}}$ is the volumetric heat release (or absorption) due to chemical reactions. In combustion, $\dot{q}$couples the energy equation with species equations through exothermic or endothermic reaction heat.
- **Species conservation (Species Transport Equations):** For each chemical species Y_J (mass fraction of species i), a transport equation accounts for convection, diffusion, and reaction source terms:

$$\frac{\partial(\rho Y_i)}{\partial t}+\nabla\cdot(\rho\mathrm{u}Y_i)=-\nabla\cdot J_i+\dot{\omega}_i,$$

  where $J_i$is the diffusive flux of species i (often modeled by Fick's law or more elaborate multicomponent diffusion laws) and $\dot{\omega}_i$ is the net production rate of species \$i\$ from chemical reactions (mass per volume per time). The summation of $\dot{\omega}_i$over all species is zero (ensuring overall mass conservation). These species equations, combined with a reaction mechanism (a set of chemical reactions and associated kinetics), determine how fuel and oxidizer are consumed and products (including pollutants) are formed.

When solving these equations for turbulent reacting flows, additional modeling is required to address the effect of turbulence on transport and mixing. Most engineering combustion simulations use either the Reynolds-Averaged Navier–Stokes (RANS) approach or higher-fidelity approaches like Large Eddy Simulation (LES) or hybrid methods (Detached Eddy Simulation [DES], etc.) to model turbulence. In RANS, the flow fields are decomposed into mean and fluctuating components, and the equations are time- or ensemble-averaged. This averaging introduces unknown correlations (Reynolds stresses and turbulent scalar fluxes) that must be closed with turbulence models [13]. Common two-equation turbulence closures like the k, E or $\dot{\omega}_i$models add transport equations for turbulent kinetic energy k and its dissipation or specific dissipation E or $\dot{\omega}_i$ respectively, to model the effect of turbulent eddies on momentum and scalar transport. RANS models are computationally efficient and have been widely used in combustion CFD (including many engine simulations) to capture the general flow patterns and mixing.

In contrast, LES explicitly resolves the large-scale turbulent structures and only models the small subgrid scales (SGSs). The governing equations in LES are spatially filtered rather than averaged; large eddies that carry most of the energy and do significant mixing are directly simulated, while a SGS model (e.g., Smagorinsky model or one-equation SGS kinetic energy model) accounts for the unresolved smaller eddies [14]. LES can provide a more faithful representation of transient mixing and flame dynamics in combustion, at a higher computational cost than RANS. It has been successfully applied in research to study in-cylinder flows, sprays, and combustion with greater detail [15]. DES and other hybrid RANS/LES approaches attempt to combine the strengths of both: typically using RANS near walls and in steady regions, and LES in core flow regions or unsteady separated zones, to achieve higher accuracy than RANS without the full cost of LES.

The choice of turbulence model has a significant impact on combustion modeling accuracy. For instance, capturing turbulent mixing is crucial for non-premixed or partially premixed flames, and it is equally important for LTC regimes (which often involve highly diluted or stratified mixtures). RANS models, while robust, may miss cycle-to-cycle variation and fine-scale mixing details in engines, whereas LES can capture these but requires massively fine grids and time steps [15]. In simulations of LTC engines, researchers have employed both approaches: RANS (including unsteady RANS for cyclic variations) for parametric studies, and LES for detailed investigation of TCI and cycle-resolved phenomena [16]. A review demonstrated that LES was able to reproduce in-cylinder flow statistics and even capture combustion cyclic variability in ways RANS could not [17]. The computational cost trade-off is often a deciding factor: for design cycles, RANS may be preferred, while LES is used in targeted studies or when sufficient HPC resources are available.

Another crucial aspect of physical modeling is the treatment of boundary conditions and mesh in combustion simulations. Accurate boundary conditions are needed for inflows (e.g., specifying temperature, pressure, and species concentrations or equivalence ratio for reactants), outflows (pressure outlets, etc.), and walls (which require thermal conditions such as wall temperatures or heat transfer coefficients). In engine simulations, boundary conditions can be complex: moving boundaries (piston motion and valve motion) are present, and one often uses time-varying profiles or dynamic mesh approaches to simulate the compression and expansion strokes. High-fidelity simulations may include detailed wall heat transfer models since heat loss strongly affects LTC autoignition and emissions (e.g., colder walls promote wall quenching of reactions leading to UHC emissions).

The mesh (grid) resolution and quality must be sufficient to resolve important flow and flame features. In premixed flame simulations, one might target resolving the front thickness or use combustion models to avoid resolving it fully. In LTC cases like HCCI, where combustion happens volumetrically, there is no thin front to resolve, but the mesh still needs to capture temperature and concentration gradients that lead to ignition. For spray-based LTC (e.g., diesel PPCI), mesh refinement in the fuel spray region and near ignition sites is important. Adaptive mesh refinement (AMR) techniques are emerging as useful for combustion: refining the grid on the fly in regions of sharp gradients or high reaction rates, which can significantly improve accuracy without a uniformly fine mesh everywhere. Good mesh design also helps reduce numerical diffusion, which can otherwise smear out stratification that is critical in LTC simulations (e.g., temperature stratification in HCCI that lengthens combustion duration). Convergence practices in combustion CFD involve gradually refining the mesh and time step until key results (such as ignition timing, peak pressure, and emissions indices) change acceptably little. Solver selection (pressure-based vs. density-based, explicit vs. implicit time stepping) and convergence criteria (monitoring residuals and integrated quantities) must be chosen to handle the stiffness introduced by fast chemistry and the strong nonlinear coupling between flow and reactions.

In summary, the governing equations provide the backbone of combustion models, and turbulence models (RANS/LES/DES), along with careful boundary and mesh considerations, form the physical modeling framework. These allow us to simulate reacting flows by capturing the essential physics: flow advection, diffusion, and chemical source terms. The next sections will build on this by introducing how chemical kinetics are modeled and then how combustion–turbulence interaction is handled through various modeling approaches.

4.3 CHEMICAL KINETICS IN LTC

Chemical kinetics—the detailed description of how fuel and oxidizer molecules react to form products—is at the heart of combustion modeling, especially for LTC modes. LTC strategies like HCCI or RCCI rely heavily on autoignition chemistry; thus, an accurate kinetics model is essential to predict ignition timing, heat release rates, and pollutant formation. There are two broad approaches to incorporating kinetics: detailed chemical mechanisms versus reduced (skeletal or simplified) mechanisms, each with its own role in simulations.

Detailed chemical mechanisms include large numbers of species and reactions to faithfully represent the combustion chemistry of real fuels. For example, a detailed mechanism for a gasoline surrogate (mixtures of iso-octane, *n*-heptane, toluene, etc.) or diesel surrogate (n-heptane, methyl decanoate, etc.) might involve hundreds of species and thousands of elementary reactions. These mechanisms are usually developed and validated against fundamental experiments (ignition delay times from shock tubes, laminar flame speeds, jet-stirred reactor data, etc.) to ensure they reproduce key combustion characteristics [18]. Detailed mechanisms provide high accuracy and can capture subtle kinetic phenomena (like cool flame chemistry, which is important in LTC for controlling ignition phasing). However, including a full detailed mechanism in a 3D CFD simulation is often computationally intractable. The reaction rate equations form a stiff system of Ordinary Differential Equations (ODEs) that must be integrated at each cell and time step, which can dominate CPU time. The stiffness arises from the wide range of timescales in combustion chemistry (e.g., rapid radical recombination versus slow fuel breakdown) and can force extremely small-time steps or advanced implicit ODE solvers.

Because of these challenges, reduced chemical mechanisms are commonly employed for engine CFD. A reduced mechanism aims to capture the most important reaction pathways of the fuel with far fewer species and reactions, thus easing computational load. The process of mechanism reduction is a major topic in combustion research [19]. Techniques for reduction include identifying redundant species or reactions that have minimal impact (using methods like directed relation graphs, sensitivity analysis, or flux analysis), time-scale analysis (quasi-steady-state assumptions for short-lived intermediates), lumping species into groups, and optimization against targets. For instance, Patel et al. [20] developed a reduced reaction mechanism for diesel-fueled HCCI combustion, which retained enough fidelity to predict ignition timing and emissions while being small enough for multidimensional CFD [20,21]. Their reduced mechanism was validated by comparing CFD results to experimental data, ensuring it still captured critical behavior [20]. Similarly, researchers have produced reduced mechanisms for gasoline LTC (primary reference fuels like PRF mechanism for iso-octane/n-heptane) and alternative fuels like ethanol or ammonia for LTC conditions [22,23]. These reduced mechanisms might involve on the order of 20–100 species instead of several hundred, making 3D simulations feasible.

Creating a reduced mechanism often involves mechanism-generation and validation tools. Automated mechanism generators like Reaction Mechanism Generator can propose reaction networks for a given fuel, though typically used for detailed mechanisms. For reduction, software tools (e.g., CHEMKIN-based utilities and COSILab, or features within CONVERGE Studio [24]) allow users to trim detailed mechanisms. Approaches include specifying a target set of conditions (temperature, equivalence ratio range relevant to LTC) and eliminating species that do not appreciably affect ignition or flame propagation in that domain. One must validate the reduced mechanism by comparing its predictions of ignition delay or flame speed against the detailed mechanism (or against experiments). For example, mechanism reduction tools in Converge CFD allow targeting ignition delay and laminar flame speed during reduction, to ensure the skeletal mechanism maintains the same key combustion characteristics as the full mechanism [24,25]. Validation might involve 0D simulations (constant-volume reactors for HCCI ignition timing), 1D flame simulations, and eventually, 3D engine sector simulations to see that pressure curves and emissions match experimental trends.

Integrating the chemical kinetics into CFD solvers requires robust coupling. Software like ANSYS Fluent and CONVERGE offer multiple ways to handle chemistry: Fluent allows importing detailed or reduced mechanisms (often via CHEMKIN format). One can use the built-in species transport with finite-rate chemistry model, where the reaction rates $\dot{\omega}_i$ are computed from Arrhenius kinetics for each reaction in the mechanism. Fluent also provides the EDC option (see next section) to integrate turbulence effects with detailed chemistry, effectively running a reduced stirred reactor in each cell [26]. Another approach in Fluent is the Flamelet Generated Manifold (FGM) or PDF methods where chemistry is pre-solved in tables—but those still rely on having a detailed mechanism solved offline. CONVERGE is widely used for engine combustion, including LTC, and it includes a chemistry solver (typically using the SAGE detailed chemistry solver). CONVERGE can directly use detailed mechanisms and has optimized stiff chemistry integrators for efficiency. It also offers 0D/1D preprocessing tools to run homogeneous reactor simulations (for mechanism tuning or to generate tabulated combustion models) [24]. Notably, CONVERGE's chemistry tools can do mechanism reduction and mechanism merging, which is convenient for users to manage large fuel mechanisms and trim them for CFD use [24].

The governing equations with chemistry become a tightly coupled set: the species equations feed into the energy equation via heat release, and temperature influences reaction rates exponentially (per Arrhenius law). In LTC simulations, it is common to use detailed chemistry solvers that integrate the species and energy equations together implicitly at each time step (because explicit time-stepping would require prohibitively small steps to resolve chemical timescales). This implicit integration is a significant part of simulation cost. Advanced solver techniques like operator splitting can be employed—solving convective transport separately from reaction kinetics in each time step. For example, one may convect species with a larger time step, but sub-step the reaction ODE integration with an adaptive solver. The stiff chemistry ODE solvers (e.g., CVODE or LSODE libraries) use backward differentiation and Jacobian evaluation to handle the rapid reaction transitions, such as the cool-flame and main ignition in HCCI.

In LTC, low-temperature chemistry (e.g., the negative temperature coefficient behavior, peroxide formation, etc.) is particularly important. Detailed mechanisms often include low-temperature oxidation pathways (for large hydrocarbons, via peroxy radicals and so on) that cause the two-stage ignition behavior. Capturing this in reduced mechanisms can be challenging but is necessary to predict the phasing of the first-stage (cool flame) and second-stage ignition in HCCI. Mechanism validation for LTC conditions might involve ensuring the mechanism reproduces the two-stage ignition delay curve and the temperature sensitivity seen in experiments [22].

In summary, chemical kinetics modeling in LTC requires balancing fidelity and computational feasibility. Detailed mechanisms provide the highest accuracy, but reduced mechanisms are generally used in 3D engine simulations to make the problem tractable [20]. The process of reducing and validating mechanisms ensures that the essential kinetic behavior is retained. Integration of these mechanisms into CFD codes like Fluent and CONVERGE enables predictive simulation of LTC phenomena—from the timing of autoignition to the formation of emissions. As we shall see in later sections, some combustion modeling approaches (flamelet models, PDF methods, etc.) offer alternate ways to include kinetics without solving every reaction in every cell, which can be advantageous for complex fuels. Regardless of the approach, a sound kinetic model is indispensable for any LTC simulation, since autoignition and emissions are fundamentally chemistry-driven.

4.4 COMBUSTION SIMULATION TECHNIQUES

Modeling combustion, especially in turbulent flows, requires bridging the gap between the fine-scale chemistry and the larger-scale transport phenomena. Over the years, several combustion modeling frameworks have been developed to handle this TCI, each with its own assumptions and applicability. "Different scientists adhere to different concepts like the laminar flamelet approach, the PDF approach, or the Eddy Dissipation Concept," as Magnussen noted in a review of turbulence–chemistry modeling [27]. In this section, we discuss the key techniques relevant to LTC modeling: flamelet models, the EDC, PDF methods, and the G-equation for flame tracking. These methods were largely developed for high-temperature combustion, but many can be adapted or have variants for low-temperature or kinetically controlled combustion regimes.

4.4.1 Laminar Flamelet Models

The laminar flamelet concept is a powerful approach for modeling non-premixed (and partially premixed) turbulent flames. Originally formulated by Peters and others, it assumes that within a turbulent flame, one can identify thin reaction zones (flamelets) that resemble laminar flames locally, even though they are wrinkled and distributed by turbulence. The idea is to decouple the combustion chemistry from turbulence by precomputing a library of laminar flame solutions under various conditions and then use that library during the CFD simulation. In practice, one often uses mixture fraction Z (a conserved scalar representing fuel-oxidizer blending) as the coordinate for the flamelet library, along with perhaps a scalar dissipation rate (which indicates how much the flamelet is strained by mixing).

In a steady flamelet model, one assumes a flamelet quickly adjusts to local conditions and solves the 1D flamelet equations (parametrized by Z) in advance. The solutions give relationships between Z, temperature, and species mass fractions for a range of conditions (e.g., different strain rates). During the CFD run, each cell doesn't solve chemistry directly; instead, the solver looks up the state (species and temperature) from the flamelet library based on the local mixture fraction and a measure of strain (or progress variable). This dramatically reduces computational cost and allows use of very detailed chemistry in the precomputation. Flamelet models have been very successful for diesel diffusion flames and other non-premixed combustion, capturing phenomena like localized extinction and ignition if unsteady flamelet libraries are used.

For LTC applications: in diesel-type LTC (like partially premixed compression ignition (PCCI)), flamelet models can be applied because combustion may occur as a mix of premixed and diffusion burn. FGM methods, an extension of flamelets, are used in some HCCI/RCCI modeling as well. In FGM, one might tabulate homogeneous autoignition in a mixture fraction-progress space. Essentially, it's like treating an HCCI autoignition as a "flamelet" occurring in homogeneous reactors. This approach can capture the two-stage ignition in LTC by including those regimes in the manifold.

It's worth noting that flamelet models assume the existence of a dominant low-dimensional manifold in composition space (like most variance explained by one or two scalars). In highly diluted LTC combustion, especially homogeneous modes, the concept of a thin flame front is less clear, but manifold methods can still be applied by defining appropriate coordinates (e.g., using reaction progress or a timescale coordinate). For example, tabulated kinetics approaches (also called FTA: flamelet/progress variable approaches) have been developed where autoignition delay or progress variable source terms are tabulated as functions of conditions and then used in CFD [28]. The tabulated kinetics for ignition model mentioned by Knop et al. is an example, wherein ignition delay data from detailed chemistry is stored and used to guide the ignition in the CFD simulation [28]. Such methods help to include detailed chemistry effects without solving full kinetics every time step.

In summary, flamelet and FGM models are valuable for LTC when the combustion can be parameterized by a few variables. They excel at non-premixed LTC (e.g., high-EGR diesel LTC where combustion occurs in diffusion flames at low temperature). They are less straightforward for fully premixed autoignition (pure HCCI) unless modified to tabulate homogeneous reactor outcomes.

4.4.2 Eddy Dissipation Concept (EDC)

The EDC by Magnussen is a widely used turbulent combustion model that attempts to directly couple turbulence with detailed chemical kinetics within the RANS context. It can be seen as an extension of the simpler eddy dissipation model (EDM). The original EDM assumed that reaction rates in a turbulent flame are controlled by the turbulence mixing time—effectively, fuel and oxidizer burn as fast as they mix (suitable for fast chemistry limits). That model is robust but fails when finite-rate chemistry matters (like in NO_x formation, ignition delay, and extinction phenomena) [29].

EDC addresses this by assuming that reactions occur in small turbulent structures (the fine scales) where reactants mix on a microscopic level. It introduces the concept of fine structure regions with a certain volume fraction in each cell and a certain residence time, where the actual chemistry occurs. A key assumption is that within these fine structures, the flow is perfectly mixed (or nearly so) and reaction takes place governed by detailed kinetics (like in a constant-pressure batch reactor). The surrounding fluid is inert and exchanges mass with the reacting regions at some rate.

In practical terms, EDC provides a reaction rate source term to the species equations based on the fraction of the cell that is reacting and the time available for reactions. For example, Magnussen's model defines a *length scale* ξ^* xi *for the fine structures (as some fraction of the Kolmogorov scale) and a* time scale τ for their evolution [26] *the reaction time is* τ. The species conversion in the cell is then modeled by assuming that a certain fraction of the reactants is burned within τ^* according to chemical kinetics. Mathematically, one often writes a source term like:

$$\dot{\omega}_i = \frac{\rho C_\xi \left(Y_i^* - \tilde{Y}_i\right)}{\tau^*}$$

where $\tilde{Y}_i$ is the cell-averaged mass fraction and $\tilde{Y}_i$ is the mass fraction in the reacting fine-structure (obtained by integrating the detailed reaction ODEs over time τ^*, usually assuming constant pressure and using initial conditions from the cell composition) [26]. C is a model constant governing how much of the cell participates in reactions. In essence, EDC runs a small homogeneous reactor calculation inside each CFD cell, over a timescale $\tau^*$to update the species. This couples turbulence and chemistry: faster turbulence (smaller τ^*) gives less time for reactions, emulating quenching, whereas slower mixing (*larger* τ^*) allows reactions to go closer to completion.

EDC has been quite effective for many scenarios and is available in many CFD codes (e.g., ANSYS Fluent's "Finite-Rate/Eddy-Dissipation" model when detailed kinetics are turned on is an EDC-type model). It allows use of detailed reaction mechanisms in RANS simulations without resolving all scales of turbulence, by providing a subgrid model for the reactive fluctuations [27]. For LTC, EDC is attractive because it can handle finite-rate chemistry effects like autoignition. For example, in modeling HCCI combustion with RANS, one might use EDC so that the detailed kinetic mechanism dictates ignition timing, rather than assuming immediate mixing-controlled burn. EDC has been successfully used for partially premixed engine combustion and can predict ignition and burnout phased by chemistry rather than purely by turbulent timescales [30].

One challenge with EDC in LTC is calibrating it for very low-temperature, high-EGR conditions; the standard constants (originally tuned for high-temperature flames) might need adjustment. Researchers have looked at optimizing EDC model constants for mild or flameless combustion regimes [31], which are somewhat analogous to LTC in that chemistry is slower and distributed. An Extended EDC or Modified EDC has been proposed in literature to better fit such conditions [32]. Nevertheless, the core idea remains: EDC provides a "bridge between turbulence and chemistry" [33] by treating the smallest turbulent eddies as the site of reaction and explicitly computing kinetics there. It's a concept that "includes a complete review of the concept and its physical basis" in Magnussen's works [34]. By using EDC, one can incorporate complex fuel chemistry (e.g., a detailed LTC mechanism with low-temperature pathways) in a RANS simulation and still capture the delay and slow-burn characteristics of LTC.

4.4.3 PDF-Based Models

PDF methods form another class of combustion modeling technique, particularly useful when dealing with turbulent fluctuations of composition and temperature that strongly influence reaction rates. In combustion, the joint PDF of species and enthalpy can be used to statistically describe the distribution of thermochemical states in a turbulent flow. Instead of tracking numerous fluctuating species equations and closing their nonlinear reaction source terms with ad hoc models, PDF methods carry the statistical distribution and directly integrate chemistry against this distribution.

In practice, there are prescribed PDF approaches and transported PDF approaches: Prescribed PDF: Often used in RANS context, one assumes a shape (like Beta-function for mixture fraction PDFs in non-premixed flames, or Gaussian for temperature fluctuations) and parameters (mean and variance, etc.) that are obtained from transport equations. Then the reaction rates or species states are computed by integrating the reaction source term over this assumed PDF. The classic example is premixed combustion with an assumed PDF for the progress variable to account for flame wrinkling. Transported PDF: A more rigorous approach where one writes a transport equation for the joint PDF of all reacting scalars. This equation has no chemical source term closure problem—chemistry appears as a conditional term inside it that can be integrated exactly. The difficulty is that the PDF equation is high-dimensional (one dimension per scalar plus physical space and time) and not tractable by finite-volume methods directly. Instead, Monte Carlo methods are often used: a large number of stochastic particles are tracked, each carrying a set of species values, and they move and mix according to models reflecting molecular diffusion and turbulent micro-mixing (e.g., interactions by exchange with the mean, IEM model). Each particle's composition is advanced by the deterministic chemistry ODEs. In effect, a transported PDF simulation is like running many small reactors (particles) that mix and react stochastically.

PDF methods are very powerful for scenarios with strong TCI, such as moderate Damköhler number (Da) flows where fluctuations in composition can lead to distributed reactions, or when modeling emissions that are sensitive to fluctuations (e.g., NO formation which depends on local temperature probabilities). "The PDF approach" was highlighted alongside flamelets and EDC as one of the main concepts for turbulence–chemistry coupling [27].

For LTC, PDF methods can be beneficial especially in capturing ignition scatter and emissions. For instance, in HCCI engines, if there is some inhomogeneity (thermal or mixture stratification), the ignition will not occur uniformly but with some distribution. A PDF approach can naturally accommodate the distribution of ignition timings in different portions of the charge. Studies have shown that "turbulence can influence ignition timing and emissions for practical HCCI engines," and a PDF method was developed to model exactly these effects [35]. By using a PDF of temperature or composition, one can predict, for example, how a certain variance in equivalence ratio leads to portions of the charge igniting slightly earlier or later, affecting the overall pressure rise and emissions of CO/NO_x [36].

However, implementing PDF transport in engine CFD is computationally expensive (the Monte Carlo particle approach significantly increases the cost). Often, hybrid approaches are used: for example, RANS simulation with an assumed PDF for certain quantities. One common hybrid is the Eulerian stochastic fields method, which retains multiple field variables representing different realizations of the chemistry, effectively a finite approximation to the PDF.

In summary, PDF-based models are an advanced tool allowing a fundamentally correct inclusion of turbulence–chemistry coupling without simplifying the chemistry. They shine in complex regimes like partially premixed or highly turbulent combustion where fluctuations matter. For LTC, they offer a way to incorporate the effects of mixture/thermal stratification on autoignition, which is important for concepts like partially fuel-stratified HCCI (to control the burn rate). The downside is complexity and cost, so they are more common in research than in routine engine design simulations.

4.4.4 G-Equation and Flame Surface Tracking

The G-equation is a modeling approach tailored for premixed flame propagation, representing the flame front as an iso-surface (typically of a level-set function G). It is widely used for SI engine combustion modeling and gas turbine premixed flames. The equation describes the evolution of the flame front in flow, given by:

$$\frac{\partial G}{\partial t} + u \cdot \nabla G = S_L |\nabla G|$$

where $|\nabla G|$is a scalar field whose $\frac{\partial G}{\partial t}$ level-set represents the flame surface, and S_L is the laminar flame speed (which may be modified to an effective turbulent flame speed in turbulent cases). Essentially, the flame is treated as a propagating front with a certain normal velocity (the flame speed) relative to the fluid.

In the context of LTC, one might ask: is the G-equation applicable? Pure LTC modes like HCCI are non-flame propagation combustion—they ignite throughout the chamber rather than a flame front moving from one location. Thus G-equation does not apply to HCCI itself. However, in *transitional or dual-mode* engines (like where you switch between SI and CAI (controlled autoignition) or use spark assist to trigger HCCI), a flame may initiate and then an autoignition completes the combustion. Also, LTC concepts that involve stratified mixtures could have localized flame propagation in richer pockets. An example is Spark-Assisted HCCI or SACI, where a flame kernel from a spark ignites part of charge and the rest autoignites—modeling such a mixed-mode can involve a G-equation for the flame part coupled with an autoignition model for the rest. Another example: in RCCI, if one fuel is highly reactive and the other less so, there might be sequential ignition and possibly some flame propagation in the regions that ignite later (though generally it's mostly autoignition).

The G-equation or flame surface density models (e.g., the level-set or flame wrinkling models) are primarily mentioned here for completeness and to contrast with the other methods. They embody a "front tracking" approach, ideal for flame regimes where flame structure is thin compared to flow scales. For premixed turbulent flames, one often combines G-equation with a correlation for turbulent flame speed or flame wrinkling factor. In the realm of LTC, if we consider lean premixed gas turbine combustors aiming for low NO_x (which is a kind of LTC approach: lean premixed combustion to keep temperatures low), G-equation based models are indeed used in CFD to track the flame and ensure it stays attached etc., while overall flame temperature is low to minimize NO_x. So LTC in a broader sense (not just IC engines) does see G-equation usage.

To summarize this section: We have a suite of combustion modeling techniques: Flamelet/FGM: great for non-premixed or partially premixed LTC, incorporating detailed chemistry via precomputed tables. EDC: allows detailed kinetics in RANS by modeling small-scale reactors in turbulent flow, useful for autoignition-dominated combustion. PDF methods: probabilistic treatment capturing fluctuations, beneficial for stratified or highly turbulent LTC to model scatter in ignition and emissions. G-equation/Flame Surface models: used for flame propagation scenarios, less central in LTC except in mixed combustion modes or gas turbines with lean flames.

Each of these approaches "offers different opportunities and problems" [27]. There is no one-size-fits-all. Often, combustion modelers will choose based on the regime: e.g., for an HCCI engine simulation, one might use EDC with a detailed mechanism (to directly simulate autoignition) or a PDF-based multi-zone model; for a PPCI diesel case, perhaps a flamelet approach or EDC; for a premixed gasoline LTC (if any), maybe G-equation for the flame part. The ongoing development in this field continues to refine these models and even hybridize them (like using flamelet tabulated chemistry within an LES with transported PDFs—a complex multi-fidelity approach). The next section delves into how emissions—the ultimate outcomes of these combustion processes—are modeled and predicted using these simulation frameworks.

4.5 EMISSION MODELING AND PREDICTION

One of the goals of combustion modeling, for automotive engines, is to predict the formation of pollutant emissions such as NO_x, soot (PM), carbon monoxide (CO), and UHC. In the context of LTC, emissions behavior is quite distinct: LTC achieves very low NO_x and soot by keeping combustion temperatures down and fuel–air mixture lean, but it often has higher CO and UHC due to incomplete oxidation [7]. Modeling these emissions requires incorporating additional physics and chemistry into simulations, either through detailed chemistry or via specialized sub-models. Here we discuss how each major emission is modeled and techniques for emissions prediction.

4.5.1 NO_x Modeling

NO_x formation in combustion is highly temperature sensitive. In high-temperature flames, the Zeldovich (thermal NO_x) mechanism dominates: it involves the reaction of atmospheric N_2 with O_2 through a set of reactions producing NO [37]. This mechanism's rate rises exponentially with temperature, especially above ~1800 K. In LTC, since peak temperatures are intended to be low (to avoid NO_x), thermal NO_x is dramatically reduced. Nonetheless, even at lower temperatures, some NO_x can form, and if there are any hotter pockets (e.g., due to residual combustion or nonuniformity), NO_x will come from those zones.

The standard approach in CFD is to use an extended Zeldovich mechanism post-processor or inline model [37]. Extended Zeldovich includes the basic $N_2 + O \rightarrow NO + N$ reactions and considers O, OH radicals' contribution. Many simulations treat NO formation separately from the main hydrocarbon combustion chemistry, under the assumption that NO at ppm levels does not affect the main flame. For example, a simulation might solve all major species with a reduced fuel mechanism but then integrate NO formation along the flow using the local temperature and O/OH concentrations (either in situ or as a post-processing step). This approach is computationally light and reasonably accurate if the local conditions are known. In Fluent, one can enable the NO_x model which typically includes thermal NO_x (Zeldovich), prompt NO_x (from CH radicals in fuel-rich flames, less relevant in LTC as mixtures are lean), and sometimes nitrous oxide intermediate routes—each computed with dedicated formulas.

In LTC where temperatures are low, thermal NO_x is minimal; *prompt NO_x* (Fenimore mechanism) which occurs in rich flames is also usually minimal because LTC avoids rich pockets. Thus, many LTC studies report extremely low NO_x that is near measurement noise levels. Models may predict some NO_x even when experimental values are near-zero, which can be a numerical artifact or trace NO from minor hotspots. If very high accuracy is needed, a detailed chemistry approach could naturally include NO species (some detailed mechanisms do include NO_x chemistry integrated). But including all NO_x reactions makes mechanisms larger; hence a trade-off is often made to compute NO_x separately.

4.5.2 Soot Modeling

Soot (PM) formation is a complex multistep process involving fuel pyrolysis to form precursor species (like acetylene C_2H_2, polycyclic aromatic hydrocarbons (PAHs) like naphthalene), particle inception (nucleation of first solid particles), surface growth, coagulation, and oxidation. Detailed soot models can be extremely complex and coupled with detailed chemistry, making them expensive. In practical CFD, phenomenological soot models are commonly used [38]. One example is the Hiroyasu–NSC (Nagel and Strickland-Constable) model which uses rate equations for soot formation and oxidation calibrated to experiments. Another is the Moss–Brookes model (in Fluent), which uses a two-equation formulation for soot: one for soot mass fraction and one for soot particle number density, with source terms for nucleation, growth, coagulation, oxidation tuned from correlations.

Knop et al. mention using a Phenomenological Soot Kinetics (PSK) model coupled with a CO sub-model [38]. Their PSK likely refers to a multistep model where the formation of soot is tied to a precursor (like C_2H_2 concentration) and oxidation by OH and O_2 is accounted for. The CORK (CO Reduced Kinetics) model they coupled is to ensure CO oxidation and its heat release are properly included [38,39]. This highlights an important point: in fuel-rich combustion, a lot of CO forms and then oxidizes to CO_2 if temperature allows. In LTC, because flame temperatures are low, CO may not fully oxidize, leaving a higher baseline of CO emissions. If one were to simply cut off chemistry at CO and not model its oxidation carefully, the energy release would be slightly off and the predicted CO emissions would be inaccurate.

For diesel LTC (PPCI and RCCI), where mixing is not perfectly homogeneous, some soot can still form in locally rich pockets (like near liquid fuel spray before it fully vaporizes). The aim is to minimize these pockets via high EGR or multiple injections, but modeling can help identify if any regions exceeded the soot threshold. Soot models in LTC simulations need to account for low-temperature oxidation—if the flame never gets hot enough, will soot burn out? Possibly not, which means any soot formed might survive to exhaust. Many LTC studies report orders of magnitude lower soot than conventional combustion, but not necessarily zero. Models such as extended characteristic time-scale models or detailed sectional models (which track soot particle size distribution) have been used in research to predict particle emissions from LTC.

4.5.3 CO and UHC (Unburned Hydrocarbons)

CO and UHC are often the *inverse problem* of NO_x and soot: LTC lowers NO_x and soot by quenching high-temperature reactions, but that same quenching can freeze combustion before completion, leaving CO and UHC. To model CO and UHC, one generally relies on the core chemical mechanism (or a post-processed oxidation calculation). CO is an intermediate in almost all combustion processes; normally, CO is eventually oxidized to CO_2 if temperature stays high enough for long enough. In LTC, temperatures (especially late in expansion stroke) drop such that CO oxidation slows dramatically, and CO gets expelled. A *good chemical kinetic mechanism* is usually sufficient to predict CO—but it must have the CO oxidation pathways (the water–gas shift equilibrium, OH + CO → CO_2, etc.) correct at lower temperatures. If using a skeletal mechanism, one must ensure CO is included (nearly always yes) and that the mechanism was validated for CO oxidation at the relevant conditions [28].

UHC in engines can arise from several sources: (1) overmixing or fuel in lean pockets that never oxidized, (2) crevice cold quenching (fuel getting into piston ring crevices where flame can't reach), (3) wall wetting or oil absorption of fuel, etc. In CFD, capturing UHC exactly is tricky because some sources are subgrid or not modeled (like crevice flows). However, one can infer UHC trends by tracking species like the original fuel or intermediate hydrocarbons that remain. A detailed mechanism can directly predict UHC species if it includes long-chain intermediates; a reduced mechanism might lump "unburned HC" into one surrogate species. Often, post-processing is done: for example, integrate all fuel that didn't burn by end of simulation, which gives an idea of UHC. Some models define a "HC retention" parameter or treat the crevice volume analytically.

In CFD with finite-rate chemistry (like EDC or detailed chemistry), quenching at walls can be captured by the boundary conditions (cold walls absorb radicals and slow reactions near the surface). This leads to a layer of incompletely burned gas near surfaces which contributes to UHC. If grid resolution near walls is fine and the mechanism is detailed, the code may predict a certain amount of unreacted fuel or intermediate remains. Still, in practice, UHC prediction is often semiquantitative.

4.5.4 Emission Prediction Techniques

To get quantitative emissions numbers, simulation results often need further processing: In-situ adaptive tabulation or simplified chemistry might be used for emissions species: For instance, one can use a reduced scheme for NO_x that runs alongside the main mechanism (saving time). Post-processing: Some engine CFD workflows run a detailed chemical kinetic post-processing on computed histories of temperature and mixture. For example, take the pressure-temperature history of a bunch of fluid parcels from a CFD simulation and feed them into a 0D detailed chemistry code to compute final NO_x or UHC that the reduced model might have missed. This approach was suggested in some studies to leverage detailed chemistry without coupling it to CFD everywhere. Emission indices: Emissions are often measured in g/kg-fuel or ppm; simulation provides local concentration fields. By integrating the mass of pollutant in the exhaust region or cylinder at EVO (exhaust valve open), one can estimate emission indices. For multi-cycle simulations or steady state, one might simulate several cycles and sample the exhaust composition statistically.

Knop et al. [9] clearly state the requirement: "The correct prediction of mixture formation and combustion processes are essential for correct pollutant prediction" [28], and that reliable models for NO_x and soot are needed to numerically handle the NO_x/soot trade-off [37]. In their study, they specifically used the extended Zeldovich model for NO_x and a combined PSK + CORK for soot/CO [37], illustrating a typical multifaceted approach.

For LTC engines, a known challenge is that the traditional NO_x–soot trade-off is largely eliminated (both are low), but instead one faces a NO_x–CO trade-off or noise–emissions trade-off. Combustion too early or too fast can reduce UHC/CO but might create more NO_x (if it locally gets hotter) or unacceptable pressure rise (noise). Combustion too late or too cold eliminates NO_x but leaves CO/UHC high and efficiency low. Simulation needs to capture these subtle shifts. Accurate CO prediction is thus as important as NO_x for optimizing LTC.

Finally, soot particle size/number modeling is emerging in importance due to regulations on particle number. Detailed soot models with sections can predict if LTC produces many ultra-fine particles (some LTC modes yield a nucleation mode of very tiny particles even if mass is low). This level of detail often requires coupling CFD with aerosol dynamics models or using advanced techniques like method of moments with interpolative closure for soot. These are active research areas.

In summary, emission modeling in LTC simulations is about augmenting the combustion model with specialized sub-models for NO_x, soot, CO, and HC. Many simulations use a mix of *embedded detailed chemistry* (for CO, UHC inherently, maybe NO if mechanism includes it) and *add-on models* (like Zeldovich NO_x, soot models). The accuracy of emission prediction is a key measure of simulation success in engine development—for LTC, achieving quantitative predictive capability (not just trends) is an ongoing effort [10,11]. Predictive modeling of emissions is difficult because it requires getting the entire combustion process right (mixture formation, ignition, burn rate, and quench), which is why high-fidelity models and validation are crucial. The next section will consider how turbulence and combustion interactions complicate this picture further, especially for LTC which straddles different combustion regimes.

4.6 COMBUSTION–TURBULENCE INTERACTION IN LTC

Turbulence has a profound impact on combustion, and vice versa, forming a two-way interaction. In traditional flames, turbulence can wrinkle and stretch flame fronts, enhancing the burning rate by increasing surface area (in premixed flames) or by enhancing mixing of fuel and oxidizer (in non-premixed flames) [40]. In LTC modes, especially those relying on autoignition (like HCCI), the interaction takes a somewhat different form: turbulence influences the distribution of scalar properties (temperature and concentration) which in turn affect ignition timing and burn rates, rather than flame front surface area per se. Understanding and modeling TCI is crucial for LTC because it often determines the combustion stability and efficiency.

An important concept to frame TCI is the Da—the ratio of flow timescale to chemical timescale. In LTC regimes, we often operate at low temperatures and high dilution, which lengthens chemical timescales (slower kinetics). Thus, Da might be of order 1 or less in some LTC cases (chemical time comparable to or longer than turbulent eddy turnover time). This is quite different from a conventional diesel diffusion flame where chemistry is fast (high Da) or a highly turbulent SI flame where the flame speed still outspeeds small eddies (moderate Da but flamelet assumption holds). When $\mathrm{Da} \approx 1$, turbulence and chemistry strongly interact: turbulent fluctuations can prevent synchronization of ignition (some spots ignite earlier, some later). If $\mathrm{Da} < 1$ (chemistry slow relative to flow), turbulence might homogenize the mixture before it ever ignites (which could actually aid in making HCCI more uniform).

HCCI is often approximated as "zero-dimensional" in that the whole charge ignites nearly simultaneously when modeled in simple terms. However, in reality, slight inhomogeneities and turbulence mean ignition is spatially distributed over a finite duration. Studies have found that "turbulence can influence ignition timing and emissions for practical HCCI engines" [35]. The effect of turbulence in HCCI is primarily through enhancing mixing and heat transfer. For instance, higher turbulence will promote a more uniform temperature field by the time of ignition, which can either delay ignition (by increasing heat loss and mixing hot and cold zones) or advance it (by bringing reactants to contact with residual hot gases) depending on specifics [41]. Turbulence also affects wall heat transfer, which in turn influences charge temperature and ignition.

In partially premixed LTC concepts (PCCI and RCCI), turbulence plays the additional role of mixing multiple fuel streams or creating stratification intentionally. RCCI uses two fuels of different reactivity; turbulence ensures they mix reasonably well but not completely homogeneous. The stratification of reactivity is central to RCCI's control of combustion phasing. Turbulence intensity and length scales will dictate how fine or coarse that stratification is at the time of ignition. If turbulence is too high, it might homogenize the charge too much (reducing the stratification benefits), or it might disrupt the layering of high-cetane and low-cetane fuel. So, there is an optimal turbulence level to maintain sufficient stratification for a controlled burn. Simulation of RCCI often uses LES or fine RANS with multi-zones to capture that effect.

Turbulence–chemistry models for LTC have to address autoignition. Traditional flamelet models or progress-variable models assume either infinitely fast chemistry or flamelet regime; those may break down when chemistry is slow. One approach for TCI in autoignition-dominated combustion is the well-stirred reactor (WSR) concept (also known as perfectly stirred reactor or homogeneous reactor). In highly turbulent flows with slow chemistry, one can imagine the mixture is well-stirred compared to reaction progress—leading to the concept of homogeneous autoignition occurring nearly uniformly (like in HCCI ideal limit). The opposite extreme is ignition in pockets (if turbulence is low and mixture inhomogeneous, different regions ignite at different times).

To capture these, mixed models exist: for example, Partially Stirred Reactor (PaSR) models—similar to EDC—assume that within each computational cell, a fraction of the mass is perfectly mixed and reacts (like a WSR) while the rest is inert. This is analogous to EDC fine-structures and is often used in LES context for combustion with detailed chemistry. In LES, one cannot practically resolve Kolmogorov scales in engines, so a subgrid model like PaSR is used where at each time step a portion of the subgrid volume undergoes reaction based on kinetics (similar in spirit to RANS EDC). PaSR has been applied to autoignition in LES of engines [42]. In fact, one study introduced a PDF approach with PaSR for HCCI, showing improved prediction of ignition timing variability [36,42].

Another phenomenon of TCI in engines is cycle-to-cycle variability: even at the same operating condition, successive cycles might have different ignition timing or burn rates, partly due to turbulent flow randomness. LES can capture this inherently (each cycle simulation being one realization), whereas RANS cannot. Researchers have used techniques like URANS with random perturbations or stochastic mixing models to emulate cyclic variations. The interplay of turbulence and residual gas distribution is a key source of variability in HCCI (since a little pocket of hotter residual can trigger earlier ignition in one cycle vs. another).

In modeling terms, if one aims to capture not just the mean but the statistics of LTC combustion, a statistical approach (PDF or LES) is needed. This is often beyond the scope of routine RANS, but it's an active research area given that LTC's feasibility partly depends on controlling variability.

It's also worthwhile to consider turbulence's effect on emissions: Turbulence can shorten the burnout time of CO by mixing high-CO regions with oxygen-rich regions (provided there's temperature). It can also disperse soot pockets, but if too intense, it may quench flames into soot. So, there are subtle trade-offs. In an LTC diesel flame, for example, a bit more turbulence (via high swirl or adding velocity through multiple injections) might reduce soot by increasing mixing (less local rich), but excessive turbulence could lower temperatures so much that soot oxidation stops. These nuances are often captured by detailed CFD where one can test different swirl ratios or piston geometries that alter turbulence.

Finally, from a modeling perspective, LTC is a regime where neither pure flamelet nor pure WSR assumptions hold globally—there can be pockets of nearly homogeneous ignition and pockets of mixing-controlled combustion. The TCI modeling strategy might thus be hybrid. For example, one could use a tabulated autoignition model for the main combustion but supplement it with a flamelet model for any diffusion flame segments (like around a diesel spray). Or use LES with a simplified kinetics model where subgrid TCI is handled by a model tuned to autoignition.

In summary, TCI in LTC is characterized by the influence of turbulence on ignition phasing and burn rate through mixing and thermal exchange. It is crucial in HCCI and stratified LTC to achieving controlled combustion. Models like EDC/PaSR provide a way to include these interactions by adjusting reaction rates to the local turbulent timescales [43]. Empirical evidence and simulations confirm that turbulence is indeed a *knob* to dial the combustion behavior: *increasing turbulence intensity can retard HCCI combustion slightly and reduce peak pressure, thereby extending the operable load range*, as some experimental studies ([44]; SAE) demonstrated using boosted turbulence levels [41,45]. Capturing these effects in simulations helps researchers design combustion chamber geometries or flow strategies (like intake flow swirl/tumble) to optimize LTC performance.

4.7 APPLICATION EXAMPLES: GASOLINE & DIESEL LTC ENGINES

To illustrate the application of the above modeling techniques, we consider two representative examples of LTC in engines: gasoline LTC (exemplified by HCCI or similar premixed autoignition concepts) and diesel LTC (exemplified by RCCI or highly dilute diesel combustion). These examples highlight how the modeling principles are put into practice in engine CFD simulations, including setup considerations like boundary conditions, cyclic simulations, fuel injection modeling, and the use of multi-zones or sectors.

4.7.1 Example 1: Gasoline HCCI (Homogeneous Charge Compression Ignition)

In a gasoline HCCI engine, a premixed charge of gasoline (or surrogate fuel) and air is compression-heated until it autoignites throughout the cylinder. There is no flame propagation via a spark—ideally the mixture ignites nearly simultaneously. The role of modeling here is to predict the crank angle at which ignition occurs (combustion phasing), the pressure rise rate, and the emissions (especially UHC and CO, since NO_x and soot are very low).

A typical CFD setup for an HCCI cycle might be a sector model of the engine cylinder (exploiting symmetry if it's a central injector or no injector). If the engine is purely HCCI (port fuel injected and homogeneous mixture), there may be no fuel injector to simulate; the fuel is considered uniformly distributed at intake valve closing (IVC). The initial conditions at IVC (start of compression) are crucial: temperature and equivalence ratio fields might have some stratification. In some simulations, researchers deliberately impose a mild temperature stratification (a gradient or random variation of ±10 K, for example) to represent residual gas distribution or thermal boundary layers, because in a perfectly homogeneous simulation HCCI might appear as a mathematical detonation (too abrupt). This stratification helps match experimental observations of a more gradual combustion. Boundary conditions include the wall temperatures (since wall heat transfer affects charge cooling), and often the cylinder volume change is prescribed via moving piston boundary or moving mesh according to the engine geometry.

Multi-zone modeling is a common simplified approach: instead of full 3D CFD, one can divide the charge into zones of different temperatures (or equivalence ratios) and run coupled 0D reactors. This is not CFD, but we mention it as it's widely used for HCCI feasibility studies due to its speed. However, 3D CFD (with models like EDC or detailed chemistry) provides spatial detail that multi-zone cannot.

In a 3D simulation, the combustion model might be as straightforward as detailed chemistry in each cell (if using EDC or well-mixed assumption). For example, researchers have used Converge with a detailed PRF mechanism (like a 40-species mechanism for PRF fuels) to simulate a gasoline HCCI engine, and managed to predict the pressure curve and ignition timing within a few crank angle degrees of experiment [20]. They often validate by comparing cylinder pressure traces and in-cylinder heat release. Another study might use CHEMKIN-Pro coupled with a CFD: for example, first run a CFD fluid mechanics simulation without chemistry to get heat transfer and flow, then post-process with a detailed chemical solver using the CFD-derived temperature history for parcels of fuel/air [46].

An example from literature: Kong et al. [43] performed 3D CFD of an HCCI engine using KIVA code with detailed chemistry and found that slight nonuniformities in π and t lead to spatially distributed ignition, and they had to use a reduced mechanism to make it computationally feasible [21]. The results showed good agreement and underscored the importance of including EGR and wall heat loss in the model.

Key results modeled for HCCI include ignition timing (CA50): The crank angle at which 50% heat is released. CFD should capture how changes in intake temperature, fuel octane, or EGR affect CA50. Pressure rise rate: to assess if it will cause knock or noise. Too fast a rate can be predicted if the model is too homogeneous; turbulence models or stratified initial conditions help spread it out. Efficiency: the indicated mean effective pressure or fuel efficiency can be derived from the pressure trace. Models should predict if an operating point yields stable combustion or misfire. Emissions: extremely low NO_x (often below 10 ppm), which model often confirms. UHC and CO—models might give absolute values or at least trends (e.g., more EGR leads to more CO/UHC).

One case study reported "HCCI has high thermal efficiency and extremely low NO_x and PM emissions… however, there are limitations to be considered" [3]. The limitations are exactly what simulations try to expose: narrow operating range and control difficulties. Using simulations, one can test strategies like combustion phasing control: for example, intake temperature swings, variable valve timing (to retain hot residuals), direct injection of small fuel amounts to stratify mixture (called spark-less stratification). Operating range extension: how far can load be increased before knock or ringing occurs? Simulations can increment fuel and see when pressure rise becomes unacceptable. Effect of fuels: for example, running a gasoline vs. ethanol in HCCI—modeling different kinetics.

4.7.2 Example 2: Diesel RCCI (Reactivity-Controlled Compression Ignition)

RCCI is a dual-fuel LTC strategy where a low-reactivity fuel (e.g., gasoline or ethanol) is premixed (usually port-injected) and a high-reactivity fuel (diesel) is directly injected in small quantities to trigger and control ignition. The result is a partially stratified charge by reactivity. Modeling RCCI in CFD is more complex than HCCI due to the presence of direct injection and two fuels.

In an RCCI simulation (say using ANSYS Fluent or Converge), the domain is typically the full cylinder (or a sector if symmetry allows, though often full 360° is modeled because direct injections can break symmetry). The premixed fuel (port-injected gasoline) can be initialized as a homogeneous background with a given equivalence ratio. The direct-injected fuel (diesel) is modeled with a spray sub-model: typically Lagrangian parcels representing liquid fuel, with breakup, evaporation, and collision models. Models like Kelvin-Helmholtz/Rayleigh-Taylor (KH-RT) breakup are used to simulate diesel spray atomization. Spray modeling is crucial because the distribution of diesel influences the ignition sites. Mesh movement: like any engine simulation, a moving mesh or fixed-mesh with mapping (CONVERGE uses fixed mesh with volume contraction) is needed to simulate the compression and expansion strokes.

Combustion modeling in RCCI often uses detailed kinetics as well, because RCCI's whole idea is controlling ignition by fuel chemistry. A mechanism that can handle both fuels (multi-fuel mechanism) is required, or one may use a combined mechanism (e.g., a merged gasoline + diesel surrogate mechanism). Sometimes, a reduced mechanism is developed specifically for RCCI. For example, there are published reduced mechanisms for PRF (for gasoline) combined with *n*-heptane (diesel surrogate) that are used in RCCI simulations [47]. Patel et al.'s reduced mechanism for HCCI was actually for diesel fuel, but similar efforts exist for dual fuel.

Boundary conditions include not just initial T, P at IVC, but injection parameters (injection pressure, start of injection timing, duration, etc.). In RCCI, multiple injections might be used (e.g., a split diesel injection strategy to optimize combustion phasing). The model must capture how each injection's fuel mixes. This is where turbulence (like swirl in the engine) and spray models intersect.

Key outputs from RCCI simulations: Rate of heat release: RCCI typically has a two-stage heat release (early from first fuel autoignition, second from remaining fuel burning), which the model should capture. Pressure curve and combustion phasing: The goal is often to keep combustion phasing optimal (slightly after top dead centre) and pressure rise moderate. Simulations help tune the ratio of the two fuels and timing. Spatial burn pattern: Unlike HCCI, RCCI might not combust everywhere simultaneously. The diesel spray regions ignite first (due to high cetane), and then that triggers the surrounding premixed fuel. CFD can visualize this, showing for instance high temperature near the spray plumes and then a wave of heat release through the gasoline portion. Emissions: RCCI is known for *very low* NO_x *and soot* (even lower than HCCI in some cases because the stratification allows more gradual heat release). Models should ideally predict extremely low NO_x. Soot is usually negligible because the diesel amount is small and well mixed; any soot model in simulation likely shows almost zero except perhaps tiny at very rich spray core if any exists, but generally RCCI avoids soot formation by design. CO and UHC in RCCI can be higher than conventional diesel, because like HCCI some fuel regions might be too lean or too cool to burn out. The model's chemical kinetics will produce CO/UHC in cooler zones—for example, near walls or far from the diesel spray where ignition might not reach before expansion quenches combustion.

A case from literature: split fuel RCCI where gasoline was premixed and diesel direct-injected was simulated in Converge by Kokjohn et al. They used detailed chemical kinetics (around 40 species mech). Simulation captured how increasing the fraction of diesel advanced combustion (due to more ignition sites) and reduced combustion duration [48]. They also studied piston bowl designs: the piston shape can create different mixing patterns for the two fuels. CFD helped reveal, for example, that a re-entrant piston bowl promoted better mixing of diesel into gasoline, yielding more uniform combustion and lower UHC.

Another example is PCCI diesel (early injection, high EGR to delay mixing-controlled combustion)—which is a single-fuel LTC concept. In Fluent, one might use a flamelet or detailed chemistry model with a spray. One has to capture ignition delay properly, which depends on EGR and temperature. If the CFD predicts the ignition too early or too late compared to experiment, it indicates the mechanism might need tuning or spray model needs adjustment.

Combustion chamber geometry and boundary conditions like crevice flows are also important. In HCCI, a lot of UHC comes from crevices. In RCCI/diesel LTC, wall impingement of fuel can lead to UHC or even slight soot. Advanced models simulate film deposition if spray hits the wall, but that's beyond basic setups.

In summary, these application examples show how the general models are specialized: For HCCI: use of detailed chemistry or reduced mechanism with possibly multi-zone or EDC, no flame propagation model needed, focus on autoignition

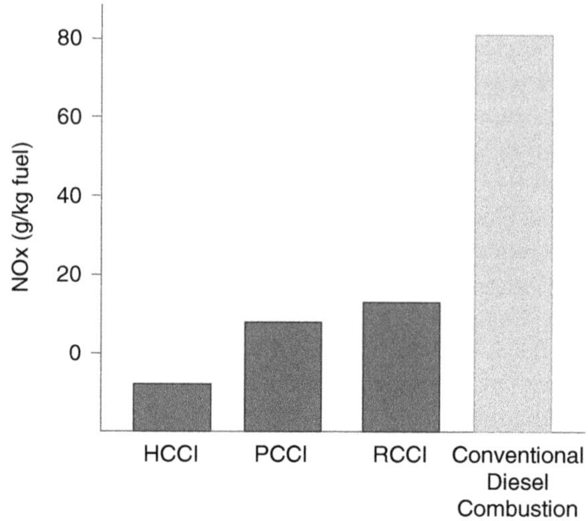

FIGURE 4.2 Homogenous charge compression comparison.

timing. For RCCI/diesel LTC: adding spray models, possibly using multi-fuel chemistry, capturing partially premixed combustion (needs both mixing and kinetics accuracy).

Simulations of these have been pivotal in demonstrating LTC feasibility. For instance, modeling studies have shown that with proper control (like negative valve overlap to retain residuals, or stratified EGR), one can extend HCCI to higher loads without knock [5]. Similarly, RCCI modeling guided experiments on how much diesel vs. gasoline to use for a given compression ratio. Without such modeling, the trial and error in hardware would be expensive and time-consuming (Figure 4.2).

As a final point in this section: model validation is key in these examples. Usually, simulation results (pressure curves, ignition timing, and emissions) are compared with engine experiments or simpler validation data (like ignition delay in a constant-volume bomb). If discrepancies exist, modelers refine the mesh, adjust wall heat transfer (a notoriously uncertain part of CFD), or improve the mechanism. For instance, a model might initially predict HCCI ignition 5 crank degrees too early; sensitivity analysis could find that increasing wall heat transfer (or adding heat loss sub-models) delays ignition to correct timing, meaning the actual engine had more heat loss than initially accounted for. Thus, these case studies are not just blind predictions but part of an iterative calibration of the modeling approach for LTC.

4.8 CHALLENGES AND LIMITATIONS

While significant progress has been made in combustion modeling for LTC, several challenges and limitations persist. These stem from the inherent complexity of the physics/chemistry as well as practical computational constraints and uncertainties in modeling. Key challenges include:

- **Chemical Kinetics Stiffness and Complexity:** LTC involves fuel oxidation chemistry at low-to-intermediate temperatures, which inherently has stiff kinetics (e.g., cool flame chemistry, and Negative Temperature Coefficient, i.e., NTC behavior). Solving large reaction mechanisms in 3D CFD can lead to solver stiffness that slows down simulations or causes convergence difficulties. Even reduced mechanisms can have stiffness due to disparity of timescales. Robust implicit ODE integrators are necessary, but they can be memory- and CPU-intensive. Additionally, detailed fuel mechanisms might not be perfectly tuned for new bio-fuels or fuel blends used in LTC research, necessitating ongoing mechanism development. Mechanism uncertainty (e.g., in rate constants) can translate to simulation uncertainty in ignition timing predictions.
- **Turbulence and Mixing Limitations:** RANS turbulence models (like k–ε) have known limitations in accuracy for transient engine flows, potentially mis-predicting the level of mixing or charge stratification in LTC simulations. LES can overcome some issues but at much higher cost. Also, boundary layer flows and crevice flows are often not fully resolved, leading to underestimation of UHC from crevices, for example. Turbulence–chemistry closure models (EDC, PaSR, etc.) involve empirical constants that may not universally apply to all LTC conditions; tuning them is sometimes necessary, reducing predictive confidence.
- **Mesh Resolution and Numerical Errors:** Adequately resolving important features (spray breakup, ignition kernels, thermal gradients) may require very fine meshes (on the order of microns in sprays, or submillimeter in chamber). However, using such fine meshes in a full engine cycle 3D simulation can be prohibitively expensive. Under-resolved meshes can cause artificial smearing of stratification or may not capture steep temperature gradients, which can alter ignition behavior significantly. AMR is still in development in commercial codes; without it, users rely on static refined regions (e.g., embedding fine mesh in spray path). Even then, numerical diffusion in advection schemes can be an issue: it can cause an initially stratified mixture to artificially homogenize, or a thin reaction zone to overly diffuse harmful for LTC predictions because LTC performance often hinges on stratification.
- **Computational Cost vs. Fidelity Trade-Off:** There is always a tension between how detailed a model is and how long it takes to run. High-fidelity approaches (LES with detailed chemistry, or DNS for fundamental studies) are extremely costly, limiting their use to maybe single-cycle or small domain studies [49]. For design and optimization which require many simulations, engineers often must simplify (e.g., use RANS, coarse grids, skeletal mechanisms). Simplification can erode predictive fidelity. This trade-off means one might not be able to simulate a full transient, like load swings or cold start with full detail. Instead, one simulates a subset and must generalize, which can miss phenomena. As an example, simulating engine knock (an abnormal LTC scenario) might require resolving shock waves and chemistry on very fine scales—beyond current capabilities in a full engine model.

- **Convergence and Stability:** Combustion simulations sometimes suffer from convergence difficulties—the nonlinear coupling of combustion heat release with flow can cause instabilities in the solver. This is aggravated when the combustion is rapid (pressure spikes). In LTC, the issue might appear when trying to simulate an operating point near the high-load boundary where combustion is almost too fast (borderline knock)—numerically, large gradients may cause instability. Solver settings like time-step size, pressure–velocity coupling, etc., need careful adjustment. Furthermore, handling moving meshes adds complexity; interpolation errors when remeshing can perturb solutions. Converging a repeating cycle (for a steady-state periodic solution) may require multiple engine cycles simulated, which multiplies computational effort.
- **Validation and Combustion Model Limitations:** Because LTC often operates in ranges where experimental data are sparse or harder to measure (e.g., it's tricky to measure ignition delay in-cylinder or intermediate species in LTC), validating models is challenging. You might match pressure curves, but how do you ensure the simulation's internal prediction of, say, distribution of combustion is correct? Many sub-models (like soot formation, or wall heat transfer correlations) carry large uncertainties. For instance, most CFD uses empirical heat transfer correlations (e.g., from Woschni's model) to approximate convective heat loss in engines. These might not be accurate for LTC because LTC's heat release pattern differs from the conventional cases those correlations were tuned for. If the heat transfer is off, then the simulated charge temperature is off, leading to wrong ignition timing. Similarly, spray models tuned for high-temperature combustion might not predict spray evaporation correctly under low-temperature (high EGR) conditions.
- **Fuel Variability and Multi-Fuel Issues:** LTC encourages use of alternative fuels (ethanol, methanol, biofuels, hydrogen, etc.). Mechanisms and models must adapt to these fuels. Developing and validating a mechanism for a new fuel is a laborious process. If one uses a surrogate (like n-heptane for diesel, or PRF for gasoline), that is an approximation, the real fuel might have components that behave differently (e.g., diesel fuel has aromatics that produce more soot than pure n-heptane would). Thus, simulation may underpredict soot because the surrogate is "cleaner" than real fuel. This limits absolute accuracy unless mechanisms are updated or certain fudge factors are applied.
- **Dimensionality and Simplification in Models:** Most models have simplifying assumptions that might limit LTC accuracy. For example, the flamelet model assumes thin flames—in LTC autoignition, reaction zones may be more volumetric, so flamelet assumption might break down. PDF models often assume something about mixing at subgrid that might not hold if ignition is rapidly evolving. The G-equation approach assumes a well-defined flame surface, which LTC might not have. Realizing these limitations, modelers sometimes combine models to cover gaps, but that increases complexity and risk of internal inconsistency.

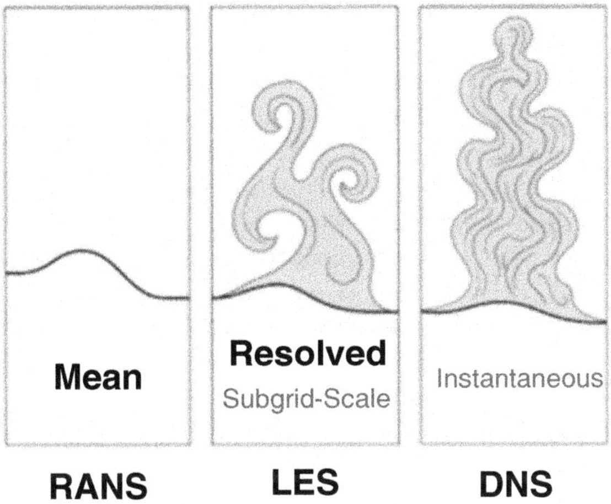

FIGURE 4.3 Turbulence–chemistry interaction models.

In essence, the current limitation is that no single model can capture all aspects of LTC with high accuracy and low cost. Engineers must choose which aspects are most crucial to capture (ignition timing, emissions, efficiency, etc.) and ensure the model is strong in those, possibly at the expense of others. For example, one might accept a less detailed spray model if the focus is on gas-phase kinetics for ignition, or vice versa.

Continued work is addressing these challenges: improving numerical algorithms to handle stiffness better, leveraging parallel computing to allow higher fidelity, developing adaptive meshes, and refining sub-models with new experimental data. As we move forward, some of these limitations will be mitigated, as discussed next in future directions (Figure 4.3).

4.9 FUTURE DIRECTIONS IN COMBUSTION SIMULATION

The field of combustion modeling is rapidly evolving, and several emerging trends promise to enhance our ability to simulate LTC with greater accuracy and efficiency. Here we outline some future directions that are particularly relevant:

4.9.1 Machine Learning and Data-Driven Modeling

ML and artificial intelligence are making inroads into combustion research as tools to handle complexity and big data. ML offers new ways to create surrogate models for combustion chemistry and turbulence closures. For example, researchers are training neural networks to emulate chemical kinetics—taking in local state (T, species) and outputting reaction rates or ignition delay, effectively replacing direct integration of stiff ODEs with a fast approximate model [50,51]. This can dramatically speed up chemistry calculations. There are demonstrations of neural network surrogates for H_2/CO kinetics and for bigger hydrocarbon mechanisms that achieve orders of magnitude speedup while retaining good accuracy over a range of conditions [52]. Similarly, combustion models like flamelet tables can be generated or interpolated using ML, and turbulent sub-models (like SGS models in LES) have been augmented by ML trained on high-fidelity data (DNS results).

For LTC specifically, ML could help in *dynamic mechanism reduction*—on-the-fly adjusting the level of detail in kinetics based on current conditions, something a smart algorithm could learn. Also, controlling LTC is a complex problem; ML-based controllers (like model predictive control using an engine simulation as a digital twin) are being explored [53]. A recent overview noted that "machine learning facilitates data-driven techniques for handling large amounts of combustion data, finding hidden patterns and promoting combustion research" [50]. We can expect ML to assist in optimizing LTC combustion by sifting through large parametric studies or experimental data to find patterns (e.g., correlating fuel properties with LTC behavior).

Furthermore, hybrid physics ML models are on the rise: these embed physical laws (conservation equations) but use ML for unknown closures. An example is using convolutional neural networks to predict subgrid flame wrinkling factor in LES from local flow features [51]. Another is ML-based ignition models that learn from detailed simulation data how ignition occurs under various stratification.

One promising development is using ML for fast emissions prediction, training a model on a set of detailed chemistry runs to predict NO_x outcomes without having to run the full mechanism each time. Overall, ML offers an approach to classical modeling, not replacing first principles but accelerating and guiding them. However, care is needed to ensure ML models obey physical constraints. Hence, the interest in physics-informed ML. Given the complexity of LTC, it's a ripe area for ML to find optimal designs (e.g., piston shapes and injection schedules) by coupling with simulation in optimization loops.

4.9.2 Adaptive Mesh and Hybrid Solvers

AMR and hybrid solver strategies are expected to play a bigger role in combustion CFD. AMR can refine the grid in regions of interest (e.g., where ignition is happening or where steep gradients exist) and coarsen elsewhere, optimizing the use of computational elements. For LTC engines, AMR could, for instance, refine around the piston top where the mixture might be stratified or around the flame kernels in PPCI, and then refined later in the cycle when things smooth out. Some research codes and a few commercial ones have this, but future development will likely make AMR more robust for moving geometries and complex physics. AMR can significantly reduce the cell count while maintaining resolution where needed, which helps simulate more cycles or include more chemistry.

Hybrid solvers refer to combining different numerical methods or models in one simulation. For example, one might use a detailed chemistry solver in one part of the domain and a flamelet model in another. There's also a trend of coupling 1D/0D models with 3D CFD in co-simulation. For instance, one could couple a 3D CFD of the combustion chamber with a 1D gas exchange model of the intake/exhaust to save effort (this is already done in some industry setups). Another hybrid approach is using modal reduction: performing high-fidelity simulation (like LES or DNS) for a short duration or small domain, extracting dominant modes, and then using those modes in a lower-fidelity simulation to represent effects of small scales. Recent work on hybrid reduced order models for reacting flows has shown success in capturing complex dynamics with a fraction of the computational cost [54,55].

Additionally, as combustion simulations push toward exascale computing, new solver paradigms like time-parallelization or matrix-free Newton-Krylov methods for chemistry might become commonplace, effectively acting as hybrid techniques to break the usual time-step-by-time-step bottleneck in an engine cycle simulation.

4.9.3 High-Performance Computing (HPC) and Cloud-Based Simulation

The continual growth in computing power and the onset of exascale supercomputers is opening up “a new horizon in computational modeling” [56] for combustion. High-performance computing enables running simulations that were previously infeasible—such as multi-cycle LES with detailed chemistry, or very fine-grid direct simulations for fundamental understanding. A review noted that with exascale, one can incorporate more complex physics (like multiphase, detailed chemistry, multi-point ignition processes) and still simulate relevant configurations [56]. This means our models can be less drastically simplified, leading to better predictive capability.

For LTC, HPC might allow, for instance, running an ensemble of engine cycle LES to statistically quantify cyclic variability, or running 3D optimization loops (varying injection timing, etc., automatically to find an optimal). Such studies would help engineers not only validate models with high fidelity but also gain insights into uncertainties and stability of LTC operation.

Cloud-based simulation and HPC are also changing how engine simulations are used in design. In the future, rather than running a handful of cases and extrapolating, one could run hundreds of cases (a design of experiments) in parallel on cloud clusters, exploring the LTC operating map thoroughly. HPC also facilitates real-time simulation efforts or digital twins—the idea of running simulations concurrently with an experiment or control system to provide predictive feedback. While true real-time 3D CFD for an engine is not yet practical, faster surrogate models (possibly ML-based as above) could leverage HPC in background to adjust themselves and then run quasi-real-time.

We also see HPC enabling multidisciplinary coupling: for example, combining CFD combustion with computational fluid-structure interaction (for the engine block or injector) or with optimization algorithms. Cloud platforms can host these complex workflows accessible to geographically distributed teams, which might accelerate development of LTC by enabling broader collaboration and data sharing.

Another frontier made possible by HPC is running DNS or fine LES of canonical LTC problems (like autoignition in a turbulent mixing layer) to generate data for model improvement. These first-principles data sets can be used to calibrate models (like PaSR constants or ML networks as mentioned). Already, there have been DNS studies of HCCI-like conditions in simplified domains to see how turbulence changes autoignition, which feed into better understanding of TCI.

4.9.4 Other Emerging Trends

Beyond the three main categories above, a few unsung developments deserve mention: Chemical Kinetics Advances: Automatic mechanism generation and reduction will improve, aided by HPC and ML. We might see "on-demand" chemistry: the CFD code itself deciding which species to track based on local conditions (an advanced form of adaptive chemistry). To account for uncertainties in kinetics, beginning conditions, etc., future simulations are probably going to use advanced uncertainty quantification (UQ). Building polynomial expansions or doing several perturbed simulations—made possible by HPC—could accomplish this. For LTC, where outcomes can be sensitive to some variations, UQ is valuable to provide error bars on predictions. Experimental Computational Synergy: Improved optical diagnostics provide richer validation data (like spatially resolved ignition timing via high-speed chemiluminescence imaging). Coupling this data with simulations (data assimilation) might adjust simulation initial fields or model parameters to better match each experiment. This tightens model reliability. Multi-fidelity modeling frameworks, where different components of the simulation have different fidelity (e.g., a coarse CFD for flow but embedded detailed 0D reactors where needed), could become more user-friendly and automated.

The future of combustion simulation, and LTC modeling in particular, is headed toward integrated, intelligent, and high-fidelity approaches. By leveraging ML to handle complexity, adaptive methods to focus compute effort, and HPC to brute-force what remains, we are moving closer to the long-term goal: a fully predictive "virtual engine" where one can test new LTC concepts purely in silico with confidence in the accuracy. As a 2021 perspective stated, "there is high potential for exploring the combination between ML and combustion research and achieving remarkable results" [44]. The coming years will likely turn many of these possibilities into standard practice.

4.10 CONCLUSION

Modeling and simulation of combustion have become indispensable tools in the development of LTC strategies. In this chapter, we reviewed the fundamental equations and models that underpin reacting flow simulations, and we examined how these are applied and sometimes adapted for the unique context of LTC. The governing equations conservation of mass, momentum, energy, and species provide the framework to capture flow and heat transfer, while turbulence models (RANS, LES, and DES) fill in the gaps for turbulent flows. Chemical kinetics modeling, whether through detailed or reduced mechanisms, is critical for predicting the autoignition-centric combustion of LTC; significant effort goes into mechanism reduction and validation so that complex fuel chemistry can be feasibly included in CFD [20]. We discussed several combustion modeling techniques—flamelet models, the EDC, PDF methods, G-equation—each bringing different strengths to handle the turbulence–chemistry interplay, and often complementary in tackling LTC's mix of premixed and mixing-controlled phenomena [27]. Emission prediction methods were outlined, highlighting how models for NO_x (e.g., Zeldovich mechanism) and soot (phenomenological kinetics) are integrated to meet the quantitative predictive needs of engine design [37].

Throughout, we emphasized the importance of TCI and how it affects LTC combustion. Unlike conventional flames, LTC's distributed combustion and sensitivity to mixture stratification mean that turbulence can alter ignition timing and burn rates considerably [35]. Models such as EDC/PaSR help account for this by embedding finite-rate chemistry within turbulent contexts [43]. Our application examples of HCCI and RCCI illustrated how these modeling concepts come together in practice—from setting up boundary conditions and injection strategies in simulations, to using multi-zone approximations or full 3D detailed chemistry to capture combustion phasing and emissions. These case studies show that when models are properly applied and validated, CFD can replicate key trends in LTC engines (e.g., ultra-low NO_x/soot, the need for stratification to control burn rates) and guide optimization of operating strategies.

We addressed the challenges and limitations such as chemical stiffness, high computational cost, model uncertainties, and the difficulty of achieving predictive accuracy across all metrics simultaneously. These challenges temper our expectations; for instance, while CFD might predict trends well, absolute emission values might require calibration. Recognizing these gaps informs where research should focus.

For further enhancement, this chapter surveyed future directions that will likely shape the next generation of combustion simulations. ML is poised to revolutionize how we handle complex chemistry and multiparameter optimization [50]. Adaptive meshing and advanced solvers will make simulations more efficient and possibly real-time for certain tasks. The increasing power of high-performance computing is breaking barriers, enabling simulations of unprecedented fidelity and breadth [56]. These advancements, combined with better integration of models and experiments, will continuously improve the reliability and utility of combustion modeling.

In summary, the key modeling tools for LTC—detailed/reactive flow equations, tailored kinetics mechanisms, and robust turbulence chemistry models—collectively provide a strong toolkit to investigate and refine LTC concepts. Each tool has its strengths: some capture physics in depth, others provide computational expedience, and together they offer a multifaceted understanding. The strengths of current modeling include the ability to dissect complex in-cylinder processes that are hard to measure (like the evolution of ignition in a stratified charge) and to perform virtual experiments exploring fuels and conditions beyond what is immediately feasible in the lab. The limitations are to interpret simulation results with care and continually validate against real engines.

The research outlook for LTC modeling is exceptionally promising. As emissions regulations are strict and efficiency demands increase, LTC techniques (HCCI, RCCI, and their variants) remain immensely attractive, and advanced simulation will be crucial to surmount their control challenges. With emerging trends such as AI-driven models and exascale computing, we expect simulation to take on an even more major role in engine innovation, not only predicting performance but also actively guiding designs in a closed loop with experimental development. Ultimately, the insights gained from comprehensive combustion modeling will accelerate the realization of cleaner and more efficient combustion engines through LTC and possibly inspire hybrid approaches or new combustion paradigms that we have yet to envision. The ongoing interaction between basic modeling developments and real-world engine applications will propel advancement in this fascinating field of combustion science and engineering.

REFERENCES

1. Agarwal, A. K., Singh, A. P., and Maurya, R. K., "Evolution, challenges and path forward for low temperature combustion engines," *Progress in Energy and Combustion Science*, **61**:1–56, 2017.
2. Yao, M., Zheng, Z., and Liu, H., "Progress and recent trends in homogeneous charge compression ignition (HCCI) engines," *Progress in Energy and Combustion Science*, **35**(5):398–437, 2009.
3. Lu, X., Chen, W., and Huang, Z., "HCCI combustion: Fuel stratification and charge stratification to achieve controllable combustion characteristics," *Energy & Fuels*, **19**(5):1879–1884, 2005.
4. Dec, J. E., "Advanced compression-ignition engines—Understanding the in-cylinder processes," *Proceedings of the Combustion Institute*, **32**(2):2727–2742, 2009.
5. Eng, J. A., "Characterization of pressure waves in HCCI combustion," SAE Technical Paper 2002-01-2858, 2002.
6. Saxena, S., and Bedoya, I. D., "Fundamentals of HCCI combustion and review of hybrid operational strategies," *International Journal of Engine Research*, **15**(7):705–719, 2014.
7. Stanglmaier, R. H., and Roberts, C. E., "Homogeneous charge compression ignition (HCCI): Benefits, compromises, and future engine applications," SAE Technical Paper 1999-01-3682, 1999.
8. Dec, J. E., et al., "Low-temperature gasoline combustion (LTGC) engine research," Sandia National Laboratories, DOE Project Presentation, 2020.
9. Knop, V., et al., "Quantitative pollutant modeling: An essential prerequisite for diesel HCCI and LTC engine design," *Oil & Gas Science and Technology – Revue d'IFP*, **63**(4):495–515, 2008.
10. Colin, O., and Benkenida, A., "The 3-zones extended coherent flame model (ECFM3Z) for complex geometries," *Combustion and Flame*, **137**(1–2):227–244, 2004.
11. Yao, M., Zheng, Z., and Liu, H., "Progress and recent trends in homogeneous charge compression ignition (HCCI) engines," *Progress in Energy and Combustion Science*, **35**(5):398–437, 2009.
12. Warnatz, J., Maas, U., and Dibble, R. W., *Combustion: Physical and Chemical Fundamentals, Modeling and Simulation, Experiments, Pollutant Formation*, 4th ed., Springer, Berlin, 2006.
13. Pope, S. B., *Turbulent Flows*, Cambridge University Press, Cambridge, 2000.
14. Germano, M., Piomelli, U., Moin, P., and Cabot, W. H., "A dynamic subgrid-scale eddy viscosity model," *Physics of Fluids A*, **3**(7):1760–1765, 1991.
15. Wang, B.-L., Lee, C.-W., Reitz, R. D., Miles, P. C., and Han, Z., "Large-eddy simulations for internal combustion engines – A review," *International Journal of Engine Research*, **12**(5):421–451, 2011.
16. Vermorel, O., et al., "Towards the understanding of cyclic variability in a spark ignited engine using multi-cycle LES," *Combustion and Flame*, **156**(8):1525–1541, 2009.
17. Abraham, P., Liu, K., Haworth, D. C., Reuss, D., and Sick, V., "Evaluating large-eddy simulation and high-speed PIV with phase-invariant POD in an IC engine," *Oil & Gas Science and Technology – Revue d'IFP*, **69**(1):41–59, 2014.

18. Law, C. K., Sung, C. J., and Wang, H., "Development of comprehensive detailed and reduced reaction mechanisms for combustion modeling," AIAA Paper 2002-0331, 2002.
19. Lu, T., and Law, C. K., "A directed relation graph method for mechanism reduction," *Proceedings of the Combustion Institute*, **30**(1):1333–1341, 2005.
20. Patel, A., Kong, S.-C., and Reitz, R. D., "Development and validation of a reduced reaction mechanism for HCCI engine simulations," SAE Technical Paper 2004-01-0558, 2004.
21. Yang, B., et al., "Reduced reaction mechanisms for ethanol under ultra-lean conditions in internal combustion engines," *ACS Omega*, **6**(1):206–216, 2021.
22. Ihme, M., and Chung, W. T., "Artificial intelligence as a catalyst for combustion science and engineering," *Proceedings of the Combustion Institute*, **40**(1):105730, 2024.
23. Magnussen, B. F., "On the structure of turbulence and a generalized eddy dissipation concept for chemical reaction in turbulent flow," *19th AIAA Aerospace Sciences Meeting*, St. Louis, MO, AIAA Paper 81-0197, 1981.
24. Luo, Z., Raju, M., and Senecal, P. K., "Dynamic chemical mechanism reduction for internal combustion engine simulations," SAE Technical Paper 2013-01-1110, 2013.
25. Convergent Science Inc., "CONVERGE Chemistry Tools: The simple solution to complex chemistry," *Converge CFD Blog*, 2017.
26. Magnussen, B. F., and Hjertager, B. H., "On the structure of turbulence and a generalized eddy dissipation concept for chemical reaction in turbulent flow," *16th Symposium (International) on Combustion*, St. Louis, MO, pp. 719–729, 1977.
27. Magnussen, B. F., "The eddy dissipation concept – A bridge between science and technology," *Proceedings of the ECCOMAS Combustion Symposium*, Lisbon, 2005.
28. Knop, V., and Jay, S., "Auto-ignition model based on tabulated detailed kinetics (TKI) for HCCI combustion," *Oil & Gas Science and Technology – Revue d'IFP*, **68**(2):279–294, 2013.
29. Mastorakos, E., *Turbulent Combustion: Modeling and Applications*, Cambridge Energy Briefs, Cambridge, 2021.
30. Bösenhofer, M., Wartha, E.-M., Jordan, C., and Harasek, M., "The eddy dissipation concept—Analysis of different fine structure treatments for classical combustion," *Energies*, **11**(7):1902, 2018.
31. Cavaliere, A., and de Joannon, M., "Mild combustion," *Progress in Energy and Combustion Science*, **30**(4):329–366, 2004.
32. Romero-Anthon, A. et al., "New extended eddy dissipation concept model for flameless combustion in furnaces," *Combustion and Flame*, **220**:49–62, 2020.
33. Gran, I. R., and Magnussen, B. F., "A numerical study of a bluff-body stabilized diffusion flame. Part 2: Influence of combustion modeling and finite-rate chemistry," *Combustion Science and Technology*, **119**:191–217, 1996.
34. Najt, P. M., and Foster, D. E., "Compression-ignited homogeneous charge combustion," SAE Technical Paper 830264, 1983; Haworth, D. C., "A PDF method for multidimensional modeling of HCCI engine combustion: effects of turbulence/chemistry interactions on ignition timing and emissions," SAE Technical Paper 2004-01-1984, 2004.
35. Kong, S.-C., and Reitz, R. D., "Use of detailed chemical kinetics to study HCCI engine combustion with consideration of turbulent mixing effects," *Journal of Engineering for Gas Turbines and Power*, **124**(3):702–707, 2002.
36. Moss, J. B., Stewart, C. D., and Young, J. B., "Modeling soot formation and burnout in a high-temperature laminar diffusion flame," *Combustion and Flame*, **101**(4):491–500, 1995.
37. Heywood, J. B., *Internal Combustion Engine Fundamentals*, McGraw-Hill, New York, 1988.

38. Splitter, D. A., Hanson, R. M., Kokjohn, S. L., and Reitz, R. D., "Reactivity controlled compression ignition (RCCI) in heavy-duty engines," SAE Technical Paper 2011-01-0363, 2011.
39. Yu, R., Joelsson, T., Bai, X. S., Vressner, A., and Johansson, B., "Effect of turbulence on HCCI combustion," SAE Technical Paper 2007-01-0183, 2007.
40. D'Amato, M., Viggiano, A., and Magi, V., "On the turbulence–chemistry interaction of an HCCI combustion engine," *Energies*, **13**(22):5876, 2020.
41. Mira, D., et al., "HPC-enabling technologies for high-fidelity combustion simulations," *Proceedings of the Combustion Institute*, **39**(4):5091–5125, 2023.
42. Yu, R., et al., "Effect of turbulence on HCCI combustion" (published in SAE Trans. or related journal, 2007, corresponding to SAE 2007-01-0183). [Same study as ref. 41, alternate source].
43. Kong, S.-C., Patel, A., and Reitz, R. D., "Numerical modeling of diesel combustion and emissions under HCCI-like conditions with high EGR," SAE Technical Paper 2003-01-0752, 2003.
44. Sjöberg, L. (2007). Emotions and Risk Perception. *Risk Management*, *9*(4), 223–237. http://www.jstor.org/stable/4500403
45. Hairuddin, A. A., Yusaf, T., and Wandel, A. P., "A review of hydrogen and natural gas addition in diesel HCCI engines," *Renewable & Sustainable Energy Reviews*, **32**:739–761, 2014.
46. Vasudev, A., Mikulski, M., Balakrishnan, P. R., and Storm, X., "Thermo-kinetic multi-zone modelling of low temperature combustion engines," *Progress in Energy and Combustion Science*, **91**:100998, 2022.
47. Yao, M., et al., "RANS versus scale-resolved approaches for modeling turbulent engine flows," *Metals*, **11**(7):1140, 2021.
48. Zhou, L., et al., "Machine learning for combustion: An energy and AI overview," *Energy and AI*, **7**:100128, 2022.
49. Senecal, P., et al., "Artificial intelligence as a catalyst for combustion science and engineering," *International Journal of Engine Research*, **24**(1):3–10, 2023.
50. Ranasinghe, D., et al., "Accelerating combustion chemistry with deep learning," *Proceedings of the Combustion Institute*, **37**(4):4473–4481, 2019.
51. U.S. DOE Argonne National Lab, "HPC modeling and artificial intelligence to advance hydrogen-fueled gas turbines for power generation," 2021. (Project report/website)
52. Ruth, D., et al., "Time-parallel strategies for reacting flow simulations," *Computers & Fluids*, **173**:73–81, 2018.
53. Abdelgadir, A., et al., "Matrix-free Krylov solvers for implicit integration of stiff reaction kinetics," *Combustion and Flame*, **210**:247–261, 2019.
54. Yao, M., Zheng, Z., and Liu, H., "Progress and recent trends in homogeneous charge compression ignition (HCCI) engines," *Progress in Energy and Combustion Science*, **35**(5):398–437, 2009.
55. Dec, J. E., "Advanced compression-ignition engines—understanding the in-cylinder processes," *Proceedings of the Combustion Institute*, **32**(2):2727–2742, 2009.
56. Reitz, R. D., and Rutland, C. J., "Development and testing of diesel engine CFD models," *Progress in Energy and Combustion Science*, **21**:173–196, 1995.

5 Homogeneous Charge Compression Ignition (HCCI)

Dillip Kumar Mohanta, Kamalakanta Muduli, Surya Narayan Panda, Mahendra Kumar Rath, and Swastik Pradhan

5.1 INTRODUCTION

While electric vehicles are gaining popularity, internal combustion (IC) engines continue to be the most popular choice for powering vehicles. Two commonly used IC engine types are spark ignition (SI) and compression ignition (CI) engines. SI engines rely on a spark plug to ignite the air–fuel mixture, while CI engines use high compression to ignite the fuel. Each type of engine has its own advantages and disadvantages depending on the application and requirements [1] (Table 5.1).

TABLE 5.1
Differences between SI and CI Engines [2–4]

Parameter	SI Engine	CI Engine
Principle	Uses spark to ignite air-fuel mixture	Uses heat of compressed air to ignite fuel
Common fuel	Petrol (gasoline)	Diesel
Operating cycle	Otto cycle (constant volume)	Diesel cycle (constant pressure)
Compression ratio	Low (6:1 to 10:1)	High (14:1 to 22:1)
Method of ignition	Spark plug	Fuel injector (no spark plug)
Air/fuel intake	Mixture of air and fuel	Only air during intake, fuel injected at high pressure
Engine weight	Lighter	Heavier
Engine speed	Higher	Lower
Thermal efficiency	Lower	Higher
Noise & vibration	Less	More
Cost	Lower, easier maintenance	Higher, more complex maintenance
Application	Light vehicles (cars and bikes)	Heavy vehicles (trucks, buses, and ships)
Starting	Easier	Harder, may need preheating in cold
Emissions	Less NO_x, more HC/CO	More NO_x, less HC/CO
Fuel supply	Carburettor or port fuel injection	Direct injection

DOI: 10.1201/9781003633631-5

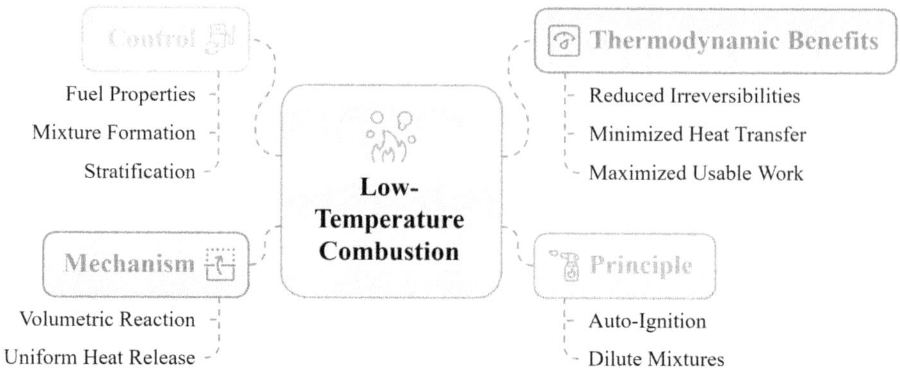

FIGURE 5.1 Principles and benefits of low-temperature combustion.

In conventional CI engines, both NO_x and particulate matter (PM) emissions are high, and the engines are subject to the NO_x–PM trade-off issue. Over the past 20 years, the legal limits of NO_x and PM emissions have significantly decreased globally [5].

Low-temperature combustion (LTC) is an advanced IC engine technology that enables fuel to burn at lower temperatures compared to conventional combustion. This achieves low emissions of nitrogen oxides (NO_x) and soot while maintaining high thermal efficiency. LTC methods alter the in-cylinder combustion process, using autoignition and fuel–air mixture control to produce a more evenly distributed, flameless combustion event [6] (Figure 5.1).

Table 5.2 summarizes the motivations for adopting LTC technologies in IC engines.

Table 5.3 shows a timeline of LTC development from 1970s to 2020s, highlighting major advances in engine research.

The evolution of LTC technology ranges from fundamental chemistry and combustion theory in the late 20th century to advanced engine design and multi-fuel methods in the present day. Significant milestones encompass the demonstration of HCCI, the advancement into premixed charge compression ignition (PCCI) and reactivity controlled compression ignition (RCCI), the integration of control systems,

TABLE 5.2
LTC Research Motivations [7,8]

Motivation	Details
Emissions reduction	Ultra-low NO_x and soot, less need for expensive aftertreatment
Efficiency improvement	Higher thermal efficiency and lower fuel consumption
Fuel adaptability	Works with diesel, gasoline, biofuels, and alternative fuels
Regulatory compliance	Helps meet increasingly strict global emissions standards
Decarbonization	Enables lower lifecycle GHG emissions
Technical innovation	Challenges in control and stability drive new research in materials and engine management

TABLE 5.3
Timeline of Significant Milestones in LTC [9]

Decade	Development	Description
1970s–1980s	Foundational studies in autoignition	Early experiments on premixed charge compression ignition (PCCI) and autoignition chemistry in fuels
1990s	Homogeneous charge compression ignition (HCCI) demonstrations	Systematic research in HCCI: simultaneous reduction of NO_x and soot; first viable HCCI engine tests
Early 2000s	Expansion of LTC concepts (HCCI, PCCI, and RCCI)	Multiple strategies developed for both gasoline and diesel engines. Research on partial & reactivity-control
Mid-2000s	Application to multicylinder engines	LTC demonstrated in four- and multicylinder prototype engines for automotive applications
Late 2000s–2010s	Holistic engine and fuel studies, broader emissions focus	Integration of EGR, advanced fuel injection, and controls. LTC adapted for both spark and compression ignition engines
2010s–2020s	Innovations in engine architecture, control, and emission compliance	Progress includes advanced piston designs, optimized combustion chamber geometry, and multimode engines
2020s	Market-ready technologies, advanced multi-fuel capability, and patents on LTC system innovations	Focus on integrating LTC strategies into production vehicles and power generation. Notable advances in combustion control, low CO/HC emissions, and patented technologies for improved emission and efficiency

and progress toward commercial viability. This advancement underscores LTC's capacity to enhance engine efficiency and emissions while accommodating changing environmental and regulatory requirements [10,11].

Homogeneous charge compression ignition (HCCI) is an advanced combustion technology that blends aspects of both SI and CI engines. In HCCI, a homogeneous mixture of fuel and air is compressed to the point where it autoignites due to elevated temperature and pressure, without the need for a spark plug or direct fuel injection event at the point of combustion. This mode of combustion offers the potential for high thermal efficiency and very low emissions, making it an actively researched topic in engine development [12].

Research into HCCI dates back to the 1970s, initially motivated by the need for cleaner and more efficient engines. Early investigations focused on alternative fuels and nontraditional engine cycles, culminating in experimental demonstrations in gasoline and diesel engines. Over the past few decades, advancements in fuel delivery, engine control, and computational modeling have enabled substantial progress in practical HCCI operation [13] (Table 5.4).

TABLE 5.4
Historical Development of HCCI [14,15]

Period	Events	Significance
1970s	Initial research into HCCI and alternative combustion cycles	Motivated by need for cleaner, more efficient engines
1980s	First laboratory demonstrations for gasoline and diesel fuels	Showed feasibility of autoignition combustion
1990s	Progress in fuel delivery systems and basic electronic controls	Enabled more consistent HCCI operation
Early 2000s	Advancements in computational modeling and sensing	Improved understanding and control strategies
2010s	Demonstration in prototype and limited-production vehicles (e.g., Mazda Skyactiv-X)	Brought HCCI closer to commercial viability
2020s	Integration of advanced diagnostics, hybrid strategies, and AI control	Expanding operational range and reliability

Here is a tabular summary of the key advantages of HCCI over traditional combustion methods based on recent research and authoritative sources.

Table 5.5 summarizes how HCCI offers significant benefits in efficiency and emissions while posing challenges in control, operational range, and robustness. Effective solutions continue to be an active area of research to realize the full potential of HCCI in commercial engines.

TABLE 5.5
Advantages and Disadvantages of HCCI [16–18]

Aspect	Advantages	Disadvantages
Thermal Efficiency	• Achieves diesel-like compression ratios and higher efficiency than conventional spark ignition (SI) engines • Reduced pumping losses (no throttle valve)	• Efficiency declines outside the narrow stable operating range due to ignition timing control challenges
Emissions	• Very low nitrogen oxides (NO_x) due to lower combustion temperatures • Virtually zero particulate matter (PM) emissions • Helps meet stringent emissions regulations	• May produce higher unburned hydrocarbons (HC) and carbon monoxide (CO) emissions compared to SI engines, especially at low loads or unstable combustion
Fuel flexibility	• Can operate with diverse fuels: gasoline, diesel, alcohols, and renewables	• Requires careful fuel and engine management strategies tailored to combustion characteristics, complicating control

(Continued)

TABLE 5.5 (*Continued*)
Advantages and Disadvantages of HCCI [16–18]

Aspect	Advantages	Disadvantages
Combustion characteristics	• Simultaneous autoignition of homogeneous mixture avoids flame fronts, reduces hot spots, and lowers soot formation • Fast heat release improves efficiency	• Rapid pressure rise can cause engine noise, component stress, and potential damage if not precisely controlled
Engine design	• No spark plugs or direct injection timing needed for ignition • Potential reduction in aftertreatment system complexity and cost	• Narrow stable operating window limits engine speed/load range • Cold start and transient operation are difficult to manage • Complex real-time control needed
Control & operation	• Advanced control strategies (e.g., variable valve timing, EGR) can enable HCCI operation within limited ranges	• Precise control of autoignition timing is challenging and sensitive to ambient and operating conditions, affecting stability, and risking knock

5.2 FUNDAMENTAL PRINCIPLES OF HCCI

HCCI combines characteristics of SI and CI engines but follows a unique autoignition-based combustion process. Figure 5.2 is its step-by-step breakdown [19].

Table 5.6 summarizes the critical characteristics of combustion process in HCCI engines. In HCCI engines, the preparation of a uniformly mixed air–fuel charge is a foundational requirement for stable and efficient combustion. Unlike conventional SI engines where direct injection and stratified mixtures are common, HCCI relies on a well-mixed, homogeneous air–fuel mixture throughout the combustion chamber before the start of compression (Table 5.7).

HCCI combustion is governed by the autoignition of a uniformly mixed air–fuel charge at the end of the compression stroke, enabling diesel-like efficiency with much cleaner exhaust, provided that the challenges of controlling the rapid, spontaneous burn are addressed.

HCCI engines possess several distinguishing features that define their operation and benefits [22].

- **Homogeneous Charge:** The air and fuel are mixed thoroughly before entering the combustion chamber, resulting in a uniform (homogeneous) mixture. This even distribution helps achieve more complete and cleaner combustion compared to stratified or direct-injection systems.
- **Autoignition:** Combustion is initiated by compressing the homogeneous air–fuel mixture until it reaches its autoignition temperature, rather than using a spark plug (as in SI engines) or fuel injection (as in CI engines). This method provides a faster and more uniform combustion process, reducing the formation of local hot spots and some pollutants.

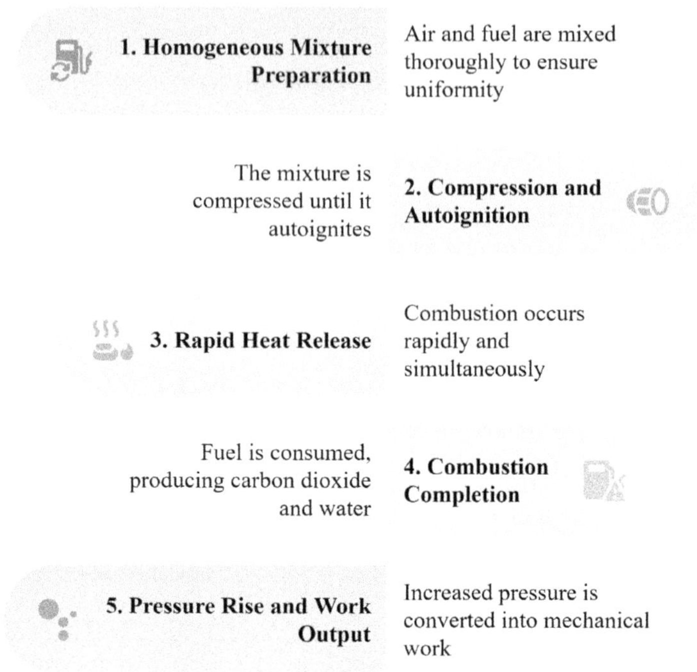

FIGURE 5.2 Primary stages of combustion process in HCCI.

TABLE 5.6
Key Features of HCCI Combustion [20,21]

Feature	Description
Ignition mode	Autoignition (no spark or injector control)
Mixture	Fully premixed air-fuel charge
Flame propagation	None—combustion is spontaneous and distributed
Burn rate	Extremely fast due to volumetric combustion
Combustion temperature	Lower than SI/CI; Reduces NO_x formation
Emissions	Very low NO_x and soot, but HC and CO can be higher

TABLE 5.7
Comparison of Combustion Processes of Different Engine Types

Engine Type	Charge Condition	Ignition Method	Combustion Progression
SI (gasoline)	Homogeneous	Spark ignition	Flame propagation
CI (diesel)	Stratified	Compression ignition	Progressive (fuel spray zones)
HCCI	Homogeneous	Compression autoignition	Simultaneous/multipoint ignition

TABLE 5.8
Advantages and Challenges Associated with HCCI [23]

Aspect	Advantage	Challenge
Efficiency	High theoretical thermal efficiency	Precise combustion control required
NO_x and soot emissions	Very low due to low temperature and lean mixture	Can increase CO and HC emissions
Control	Homogeneous charge leads to uniform combustion	Narrow operating range and sensitive to conditions
Ignition	Simultaneous, multipoint ignition—no spark plug required	Cold starts and variable loads are difficult to manage
Technology complexity	Potential for simpler hardware (no spark or injector for ignition)	Requires advanced control systems and strategies

- **LTC:** HCCI operates at lower peak combustion temperatures compared to traditional engines. Lower combustion temperatures significantly reduce the formation of nitrogen oxides (NO_x), a major pollutant produced at high temperatures in conventional engines.

  HCCI features a homogeneous charge (even air–fuel mix), uses autoignition (no spark or direct injection for ignition), and achieves LTC (reducing emissions and improving efficiency). These features combine to offer high efficiency and low emissions, but they also present unique technical challenges in terms of control and stability.

Table 5.8 summarizes the advantages and challenges associated with HCCI technology. HCCI offers significant benefits in terms of efficiency and emissions, but its practical application is limited by challenges in controlling combustion, ensuring reliability, and expanding its operating range. Ongoing research focuses on overcoming these challenges to make HCCI a viable technology for future engines.

5.3 ENGINE OPERATION AND DESIGN FOR HCCI

Typical engine configurations suitable for HCCI are as follows [24]:

- **Retrofitted SI or CI Engines:** Conventional SI (gasoline) or CI (diesel) engines can often be adapted for HCCI operation with modifications to the fuel system, intake, and control strategies.
- **Dedicated HCCI Engines:** Engines specifically designed for HCCI may include features like variable valve timing (VVT), advanced exhaust gas recirculation (EGR), and direct or port fuel injection (PFI).
- **Dual-Mode Engines:** Many HCCI applications combine HCCI with conventional SI or CI modes, switching between them depending on load and speed to take advantage of HCCI benefits when possible.

TABLE 5.9
Fuel Compatibility Table for HCCI [25]

Fuel Type	Suitability	Special Notes
Gasoline	High	Needs higher intake temperature
Diesel	High	Needs heating and prep
Ethanol	Moderate–Good	Intake heating required
Methanol	Moderate–Good	Intake heating required
Natural gas	Moderate	High compression needed
Hydrogen	Excellent	Special handling needed

HCCI exhibits exceptional fuel versatility and can accommodate a wide spectrum of octane and cetane ratings.

Table 5.9 presents various fuel types evaluated for their suitability in HCCI engines, along with special considerations for each. Alternative fuels like alcohols (e.g., ethanol and methanol), natural gas, and biofuels can be suitable. The key factor is the fuel's autoignition characteristics, which influence combustion timing and stability.

Intake air and mixture preparation strategies include PFI, Direct Injection (Early in Cycle), aerosol and vaporization systems, Intake Air Heating, and EGR. PFI ensures homogeneity by injecting fuel into the intake port, while direct injection allows significant evaporation and mixing for heavy fuels like diesel and biofuels. Heating or assistive systems are used for full evaporation before compression. EGR is used to delay autoignition, increase mixture thermal mass, and extend the operational map.

HCCI engines typically require higher compression ratios than SI engines, typically 12:1–18:1, to achieve desired temperature and pressure for autoignition. Advanced designs use variable compression ratios (VCRs) for performance optimization and operating range expansion. The VCR allows real-time control over autoignition processes, adjusting it based on intake temperature, enabling higher compression for lower intake preheat and vice versa.

Autoignition timing is crucial for a vehicle's performance, and it is influenced by various factors such as temperature, pressure, mixture ratio, and engine geometry. Intake temperature sets the reactivity window for a given charge, which in turn affects ignition timing. Cylinder pressure is determined by engine geometry and load, and fine-tuning of the compression ratio, EGR, or boosting/supercharging can be used for precise control. Mixture composition is a key factor in phasing and emissions. Advanced controls like VVT manage trapped residuals and direct or port injection timing changes for precise phasing. In practice, autoignition timing is sensitive to thermal and pressure history, requiring multiple control levers for robust operation over a wide load and speed range [26].

HCCI engines require careful integration of mixture preparation, advanced fuel/air control, and robust methods of managing temperature and pressure (via intake

TABLE 5.10
Major HCCI Combustion Control Strategies [27]

Strategy	Principle	Control Target	Main Effects
Variable valve timing (VVT)	Alters valve events (timing/lift)	Residuals, compression	Phasing, transient control, EGR, efficiency
EGR (internal/ external)	Recycles exhaust into intake or cylinder	Temperature, reactivity	Delays ignition, reduces NO_x
Fuel stratification/ enrichment	Mixes lean and richer zones, staged injection	Local reactivity/timing	Load range, pressure rise, emission control
Boost & intake temp control	Compresses intake air, heats/cools charge	Density/temperature	Stable combustion, load range extension
Dual-fuel/pilot injection	Introduces second fuel or small pre-injection	Reactivity/phasing	Stable operation, high load, cleaner burn
Advanced feedback control	Sensors and predictive models	All relevant variables	Robust, adaptive cycle-by-cycle optimization

preheating, EGR, VVT, and sometimes VCR) to ensure stable, efficient, and clean combustion. Both engine and fuel design must be considered jointly to exploit HCCI advantages while minimizing its operational and control challenges.

5.4 CONTROL STRATEGIES FOR HCCI COMBUSTION

Controlling HCCI combustion requires a combination of hardware (like variable valve systems and turbochargers), sophisticated fuel and air management, and advanced electronic control systems. These strategies are essential to overcome the inherent challenges of autoignition control, maximize efficiency, and minimize emissions across a wide range of operating conditions.

Strategies in Table 5.10 are often used in combination to offset each other's limitations, providing the means to achieve stable, efficient, and clean HCCI operation under real-world driving conditions.

5.5 FUEL CONSIDERATIONS IN HCCI

Fuel properties especially octane/cetane number and volatility play a crucial role in HCCI operation by influencing ignition timing, combustion stability, and emissions. The use of alternative fuels (alcohols, biofuels, synthetic fuels, and blends) can help optimize HCCI performance, but careful selection and management are necessary to achieve the desired balance of efficiency, stability, and low emissions (Table 5.11).

Higher octane fuels offer flexibility in controlling combustion onset, while low cetane or high-octane fuels prevent early ignition. Homogeneous, well-mixed fuels ensure consistent combustion across all cylinders and cycles, while inappropriate properties may lead to misfire or knocking. The right fuel choice can

TABLE 5.11
Summary of the Fuel Considerations in HCCI Engines [28]

Aspect	Description	Impact on HCCI Combustion
Octane number (ON)	Higher ON fuels resist autoignition, requiring higher intake temperature or compression	Delays ignition timing, improves phasing control, needs intake heating
Cetane number (CN)	Higher CN fuels ignite more easily and earlier under compression	Advances ignition timing, risk of rapid heat release and knock if uncontrolled
Volatility	Ability of fuel to vaporize; high volatility promotes homogeneous mixing	Better mixing and combustion stability; low volatility may require heating
Alcohol fuels (ethanol, methanol, and butanol)	High ON, low CN with oxygen content	Delays ignition, lowers soot and NO_x, may increase CO and HC; requires intake heating
Biofuels (biodiesel, renewable diesels)	Typically, higher CN, lower volatility than diesel	Easier ignition, higher pressure rise rates risk; blending can optimize performance
Hydrogen and synthetic fuels	High reactivity and wide flammability limits (hydrogen)	Fast combustion, requires precise phasing; synthetic fuels tailored for ignition control

reduce NO_x and particulate emissions, but incomplete combustion can increase CO and unburned hydrocarbon emissions if not well matched to engine operating conditions. HCCI operation minimizes NO_x and PM emissions, especially with oxygenated fuels like alcohols or blends. However, low-temperature operation can increase carbon monoxide (CO) and unburned hydrocarbons, especially for fuels with high ON or alcohol blends. Dual-fuel blends, like gasoline with alcohol or diesel with alcohol, allow for a compromise between efficiency, stability, and emission regulations.

5.6 EMISSION CHARACTERISTICS AND AFTERTREATMENT

HCCI engines offer a unique combustion mechanism that blends features of both SI and CI engines. As a result, their emission characteristics and aftertreatment needs differ significantly from conventional engines.

HCCI engines have ultra-low nitrogen oxides and PM emissions compared to SI and conventional diesel engines due to a homogeneous charge and LTC. However, they also produce increased unburned hydrocarbons (HC) and carbon monoxide emissions due to incomplete fuel oxidation at low loads or cold-start conditions. HCCI's high thermal efficiency may result in lower carbon dioxide emissions per unit of power produced, although absolute reductions depend on engine operation and fuel type.

TABLE 5.12
Emission Characteristics in HCCI Engines

Emission Type	Characteristics in HCCI
NO_x (nitrogen oxides)	Very low; due to lower combustion temperatures, NO_x formation is minimized
Particulate matter (PM)	Very low; homogeneous air–fuel mixture prevents soot formation
HC (unburned hydrocarbons)	High; incomplete combustion due to low temperatures in certain zones
CO (carbon monoxide)	High; caused by low combustion temperatures and quenching near walls
CO_2 (carbon dioxide)	Moderate; generally lower than SI/CI engines due to higher efficiency

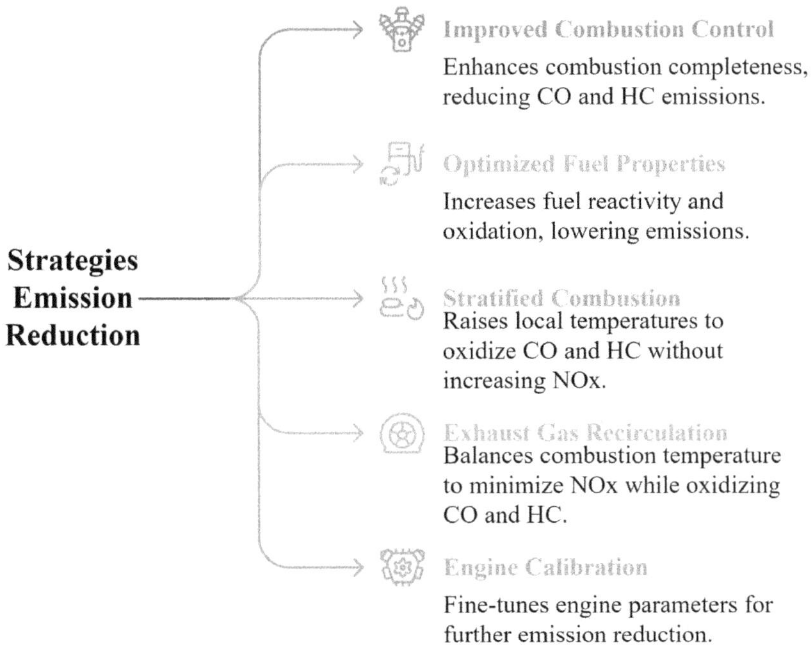

FIGURE 5.3 Strategies for emission reduction.

Table 5.12 outlines the emission behavior of different pollutants in HCCI engines. The table compares five key emission types with explanations for each based on the HCCI combustion process.

Figure 5.3 presents a summary of key methods used to reduce emissions in HCCI engines. These strategies target specific emission challenges (like CO, HC, and NO_x) and aim to enhance combustion efficiency and cleanliness.

Given the unique emission profile, specialized aftertreatment strategies are required. Three-Way Catalysts are suitable for stoichiometric or slightly rich operation, capable of simultaneous NO_x, HC, and CO removal [29]. They are limited by oxygen content and exhaust temperature for lean-burn or low-temperature HCCI cycles. Diesel oxidation catalysts are often required to oxidize HC and CO under

TABLE 5.13
Emission Profiles and Aftertreatment for HCCI

Emission	HCCI Profile	Reduction Strategy	Compatible Aftertreatment
NO_x	Ultra-low	In-cylinder (EGR, mixture, temp)	Usually not required (or minimal)
Particulate	Ultra-low	Homogeneous mixing, O_2-rich, temp	Rarely needed
HC	Elevated	Improved mixing, stratification, cat.	Oxidation catalyst, in-cylinder cat.
CO	Elevated	Mixture control, catalyst heating	Oxidation catalyst
CO_2	Reduced per kWh	High efficiency operation	Not aftertreatment; design focus

HCCI's lean and low-temperature exhaust. Their effectiveness is highly temperature dependent, and supplemental heating may be employed to ensure light-off at low engine loads or during cold starts. Selective catalytic reduction and lean-NO_x traps are less critical due to already low NO_x, but may be included for regulatory compliance or under mixed load/transitional conditions. Particulate filters are rare for pure HCCI due to minimal soot, but may be present in multimode engines or to ensure regulatory margins under all operational conditions. Table 5.13 summarizes the emission behavior of NO_x and particulates in HCCI. It outlines the emission levels, reduction strategies, and the need for aftertreatment.

5.7 RECENT ADVANCES AND RESEARCH TRENDS

The recent trends in HCCI research are pushing the boundaries of combustion control, system integration, and operational adaptability. The focus is now on overcoming HCCI's traditional limitations like ignition control and load range through simulation, hybrid strategies, and sensor-driven real-time management, making it a promising path toward cleaner and more efficient engines [30].

Advanced computational models are being developed to accurately predict ignition timing, combustion phasing, and emissions of HCCI combustion. Computational Fluid Dynamics tools simulate mixture formation, temperature distribution, and in-cylinder turbulence, crucial for understanding and optimizing HCCI processes. Reduced-order models are being created for real-time application in engine control units, enabling better combustion event prediction and management.

Novel combustion concepts integrating HCCI include Partially Premixed Combustion, a hybrid approach that combines features of HCCI and conventional combustion, offering more control over ignition and reduced emissions. RCCI uses two fuels with different reactivities to control combustion timing and progression. Research is ongoing to explore dual-fuel HCCI, stratified HCCI, and mode-wwswitching strategies.

Advancements in sensors and real-time control algorithms have revolutionized the automotive industry. Advanced sensors, such as real-time in-cylinder pressure and fast temperature sensors, offer a cost-effective and efficient way to monitor combustion phasing. Model Predictive Control and other advanced algorithms enable real-time management of combustion timing, load, and emissions in HCCI engines, accounting for actuator and process constraints. Machine learning and data-driven calibrations also facilitate onboard emission diagnostics and adaptive control, improving robustness to environmental conditions, fuel properties, and engine wear.

Hybrid and multimode engine systems are being developed that incorporate HCCI. Hybrid powertrains integrate HCCI with electric propulsion systems, allowing engines to operate efficiently in their most efficient range. Multimode engines seamlessly switch between HCCI and conventional SI or CI modes, maximizing efficiency and minimizing emissions. Prototype vehicles and test engines have demonstrated the practical benefits of HCCI, paving the way for commercialization.

Table 5.14 summarizes the recent academic and industry research, reflecting the state-of-the-art strategies to realize practical and efficient HCCI combustion through advanced simulation tools, innovative combustion modes, real-time control, and flexible multimode powertrains.

TABLE 5.14
Recent Advances and Research Trends in HCCI Combustion

Area	Description	Key Notes and Impact
Modeling and simulation	Multi-scale, multi-zone, and reduced-order kinetic models simulate HCCI ignition and combustion chemistry with improved accuracy and computational efficiency	Enables prediction of ignition timing (SOC), heat release, emissions; supports control-oriented design
	Coupled 3D CFD and chemical kinetics models provide detailed combustion and emissions insight but are computationally intensive	Useful for detailed design and validation; limited real-time use
	Tabulated chemistry and model predictive control (MPC) frameworks link simulations to real-time control applications	Allow fast, adaptive combustion phase control and robustness in practical engines
Novel combustion concepts	Premixed Charge Compression Ignition (PCCI) uses early injection to better control lean premixed combustion	Reduces emissions and pressure rise, extends HCCI operability
	Reactivity Controlled Compression Ignition (RCCI) uses two fuels with different reactivities for fine control on ignition timing and emissions	Enables smoother combustion and higher load range than pure HCCI or PCCI

(Continued)

TABLE 5.14 (*Continued*)
Recent Advances and Research Trends in HCCI Combustion

Area	Description	Key Notes and Impact
	Hybrid strategies like Pre-Chamber HCCI (PC-HCCI), Spark-Assisted Compression Ignition (SACI) help expand stable operating regions	Improves ignition control and load extension
Sensors and real-time control	Fast-response in-cylinder pressure and temperature sensors facilitate cycle-to-cycle combustion monitoring	Provide feedback for precise ignition timing control and transient response
	Advanced algorithms (MPC, adaptive control, and machine learning) optimize combustion timing and emissions control in dynamic conditions	Enhance robustness, reduce emissions, and improve efficiency under real-world driving conditions
	Novel sensing techniques, including ion current sensors and low-cost pressure transducers, are being developed	Increase feasibility of deployment in commercial engines
Hybrid and multimode systems	Integration of HCCI in multimode engines allows switching among SI, CI, HCCI, RCCI, and PCCI based on load and speed for optimal performance	Enables near-optimal efficiency and emissions across operating conditions
	HCCI engines incorporated in hybrid vehicles leverage steady-state operation for high efficiency and low emissions	Useful in series hybrids and stationary power generation for extended range and efficiency benefits

5.8 APPLICATIONS AND FUTURE PROSPECTS

HCCI engines offer a transformative potential for cleaner combustion and higher fuel efficiency. While commercial adoption is currently limited by technical challenges, ongoing advancements in combustion control, fuel technology, and hybrid integration are rapidly enhancing its feasibility. HCCI stands as a promising bridge toward a more sustainable, low-emission transportation and power generation future. Table 5.15 provides a comparative overview of applications of HCCI technology across different sectors. The table categorizes each sector by suitability, advantages, and the main challenges associated with HCCI implementation.

Future research in HCCI aims to improve ignition timing and heat release rate, enable fuel flexibility and reactivity control, overcome challenges during engine startup and low-speed conditions, enhance real-time monitoring of in-cylinder processes, and integrate with SI/CI or hybrid powertrains. These advancements will expand the operation range and improve stability under transient conditions. Additionally, HCCI will explore dual-fuel strategies, tailored ignition enhancers,

TABLE 5.15
Current Industrial Applications of HCCI [31]

Sector	Suitability	Advantages	Main Challenges
Automotive	High (hybrid modes)	Fuel efficiency, NO_x/PM reduction, and meets regulations	Transient control, cost, cold operation
Stationary power	Very high	Efficiency, low emissions, fuel flexibility	Initial investment, fuel quality
Aviation	Moderate	Potential for low-emission auxiliary power units	Weight, stability, and certification
Commercial/Heavy	High (steady-state)	Fuel economy, emissions compliance	Advanced control, fuel adaptability

and surrogate fuels to improve fuel adaptability and sustainability. The integration of sensors and diagnostics will enhance closed-loop combustion control and reduce emissions further. The ultimate goal is to achieve the best of all combustion modes for performance and emissions [32,33].

HCCI's low NO_x and particulate emissions make it suitable for meeting strict environmental regulations. Its high thermal efficiency contributes significantly to corporate average fuel economy standards. HCCI's compatibility with renewable and alternative fuels supports the transition toward low-carbon transportation and power solutions, promoting sustainable mobility.

5.9 CONCLUSIONS

HCCI is a hybrid engine that combines the benefits of SI and CI engines, using a homogeneous air-fuel mixture and autoignition without spark plugs. It results in rapid, uniform burning and lower peak combustion temperatures, reducing nitrogen oxides and PM emissions. However, it typically produces increased carbon monoxide and unburned hydrocarbon emissions.

HCCI engines achieve thermal efficiency comparable to or better than conventional diesel engines due to lean combustion and high compression ratios. They significantly reduce NO_x and soot without complex aftertreatment systems, but challenges remain in managing CO and HC emissions.

HCCI is compatible with various fuels like gasoline, diesel, alcohols, biofuels, hydrogen, and synthetics. Fuel properties like octane number, cetane number, and volatility affect ignition timing, combustion stability, and emissions. HCCI engine designs use VVT, intake air heating, EGR, and VCR to control combustion phasing and extend operating range.

The high sensitivity of autoignition in HCCI engines to engine conditions, mixture composition, temperature, and pressure necessitates precise control of combustion timing. However, these engines have limited stability, especially under transient conditions, low loads, and cold starts, necessitating advanced control strategies.

HCCI produces low NO_x and PM emissions, reducing reliance on complex aftertreatment systems. Oxidation catalysts control CO and HC emissions during cold-start and low-load operation.

Advanced modeling and simulation techniques enable real-time control of HCCI combustion, with new strategies like PCCI, RCCI, and spark-assisted HCCI improving operable load range and stability. Advanced sensors and predictive control algorithms enhance combustion timing and emissions performance. Multimode and hybrid engine architectures maximize efficiency.

HCCI technology is being commercialized in automotive engines, stationary power generation, hybrid powertrains, aviation auxiliary power units, and heavy-duty commercial engines, but faces challenges in cost, complexity, and transient control.

HCCI is a promising solution for cleaner IC engines, reducing NO_x and soot emissions while improving fuel efficiency. It operates on diverse fuels, supports energy security, reduces environmental footprint, and aligns with global emissions regulations.

Progress in combustion diagnostics, fuel formulation, and engine control is crucial for overcoming HCCI challenges like narrow operating range and transient response. Integration with electrified powertrains can accelerate commercial viability. Multimode combustion strategies enhance drivability and emissions compliance. Wider adoption depends on reducing hardware complexity and improving control algorithms. Research in fuels, sensors, and aftertreatment will broaden HCCI's applicability to various sectors, supporting sustainable evolution of IC engines in a carbon-constrained future.

Abbreviation	**Full Form**
CFD	Computational Fluid Dynamics
CI	Compression Ignition
CO	Carbon Monoxide
CO_2	Carbon Dioxide
DOC	Diesel Oxidation Catalyst
ECU	Engine Control Unit
EGR	Exhaust Gas Recirculation
GTL	Gas-to-Liquid (synthetic fuel)
HC	Hydrocarbons (Unburned Hydrocarbons)
HCCI	Homogeneous Charge Compression Ignition
LNT	Lean-NO_x Trap
LTC	Low Temperature Combustion
MPC	Model Predictive Control
NO_x	Nitrogen Oxides
PCCI	Premixed Charge Compression Ignition
PC-HCCI	Pre-Chamber Assisted Homogeneous Charge Compression Ignition
PFI	Port Fuel Injection

PM	Particulate Matter
PPCI	Partially Premixed Compression Ignition
RCCI	Reactivity Controlled Compression Ignition
SACI	Spark-Assisted Compression Ignition
SCR	Selective Catalytic Reduction
SI	Spark Ignition
SI (in context of catalyst)	Three-Way Catalyst (TWC)
TDC	Top Dead Center
UAV	Unmanned Aerial Vehicle
VCR	Variable Compression Ratio
VVT	Variable Valve Timing

REFERENCES

1. Maurya, R. K. (2017). *Characteristics and Control of Low Temperature Combustion Engines. Mechanical Engineering Series.* Springer, Cham. https://doi.org/10.1007/978-3-319-68508-3
2. Heywood, J. B. (2018). *Internal Combustion Engine Fundamentals* (2nd ed.). McGraw-Hill Education, New York.
3. Pulkrabek, W. W. (2014). *Engineering Fundamentals of the Internal Combustion Engine* (2nd ed.). CRC Press, Boca Raton, FL.
4. Pundir, B. P. (2019). *IC Engines: Combustion and Emission* (3rd ed.). CRC Press, Boca Raton, FL.
5. Zheng, M., Reader, G. T., & Hawley, J. G. (2004). Diesel engine exhaust gas recirculation—A review on advanced and novel concepts. *Energy Conversion and Management*, 45(6), 883–900.
6. Singha, S., Rao, D. G., & Dixit, M. (2024). Low temperature combustion engines for future transport—A review. In: Pandey, S. M., Maurya, A., Hirwani, C. K., & Shukla, O. J. (Eds.) *Challenges and Opportunities in Industrial and Mechanical Engineering: A Progressive Research Outlook* (pp. 53–60). CRC Press, London.
7. Zhao, H. (Ed.). (2007). *HCCI and CAI Engines for the Automotive Industry.* CRC Press, Boca Raton, FL.
8. Agarwal, A. K., Singh, A. P., & Maurya, R. K. (2017). Evolution, challenges and path forward for low temperature combustion engines. *Progress in Energy and Combustion Science*, *61*, 1–56.
9. Saxena, S., & Bedoya, I. D. (2013). Fundamental phenomena affecting low temperature combustion and HCCI engines, high load limits, and strategies for extending these limits. *Progress in Energy and Combustion Science*, 39(5), 457–488. https://doi.org/10.1016/j.pecs.2013.05.002
10. Manente, V., Tunestal, P., Johansson, B., & Cannella, W. J. (2010). Effects of different EGR types and fuel properties on LTC. *SAE Technical Paper 2010-01-1471.* https://doi.org/10.4271/2010-01-1471
11. Agarwal, A. K., Singh, A. P., & Maurya, R. K. (2017). Evolution, challenges and path forward for low temperature combustion engines. *Progress in Energy and Combustion Science*, 61, 1–56. https://doi.org/10.1016/j.pecs.2017.02.001
12. Pundir, B. P. (2019). *IC Engines: Combustion and Emission* (3rd ed.). CRC Press, Boca Raton, FL.

13. Zhao, H. (Ed.). (2007). *HCCI and CAI Engines for the Automotive Industry*. Woodhead Publishing, New York.
14. Saxena, S., & Bedoya, I. D. (2013). Fundamental phenomena affecting low temperature combustion and HCCI engines. *Progress in Energy and Combustion Science*, 39(5), 457–488. https://doi.org/10.1016/j.pecs.2013.06.001
15. Zhao, H., Ladommatos, N., Ma, T., & Balachandran, R. (2002). Engine performance and combustion characteristics of a gasoline HCCI engine with a two-stage ignition fuel. *Proceedings of the Institution of Mechanical Engineers, Part D: Journal of Automobile Engineering*, 216(7), 703–713.
16. SAE International. (n.d.). *HCCI Engine Research*. https://www.sae.org
17. Bendu, H., & Murugan, S. (2014). Homogeneous charge compression ignition (HCCI) combustion: An overview and experimental investigation using biofuels in diesel engines. *Renewable and Sustainable Energy Reviews*, 43, 49–67. https://doi.org/10.1016/j.rser.2014.11.025
18. Zhao, H., & Ladommatos, N. (2001). A review of HCCI combustion research at the same oil company. *SAE Technical Paper Series*, 2001-01-2076.
19. Kattela, S., Surapaneni, S. R., & Vysyaraju, R. K. R. (2022). Enhancement of mixture homogeneity for DI-CI engine to achieve homogeneous charge compression ignition (HCCI) combustion characteristics: A numerical approach. *Energy Sources, Part A: Recovery, Utilization, and Environmental Effects*, 44(1), 1539–1551.
20. Dec, J. E. (2009). Advanced compression-ignition concepts: Homogeneous charge compression ignition (HCCI) and others. *Proceedings of the Combustion Institute*, 32(1), 2727–2742.
21. Pickett, L. M., & Siebers, D. L. (2009). Autoignition of surrogate diesel fuels: Effects of fuel composition on low-temperature combustion in HCCI engines. *Combustion and Flame*, 156(5), 1153–1166. https://doi.org/10.1016/j.combustflame.2009.01.011
22. Reitz, R. D. (2013). Control of combustion phasing in HCCI engines. *SAE International Journal of Engines*, 6(2), 545–557. https://doi.org/10.4271/2013-01-1465
23. Yin, P., & Sun, H. (2021). Homogeneous charge compression ignition (HCCI) combustion: Combustion process and control strategies. *International Journal of Frontiers in Sociology*, 3(11), 107. https://doi.org/10.25236/ijfs.2021.031107
24. Fei, Y. (2025). HCCI engines: Benefits, challenges and future advancements. *E3S Web of Conferences*, 606, 01006. https://doi.org/10.1051/e3sconf/202560601006
25. Olsson, J., Tunestål, P., Haraldsson, G., & Johansson, B. (2001). A turbo charged dual fuel HCCI engine. *SAE Technical Papers on CD-ROM/SAE Technical Paper Series*. https://doi.org/10.4271/2001-01-1896
26. Anh, T. L., Duy, V. N., Thi, H. K., & Xa, H. N. (2018). Experimental investigation on establishing the HCCI process fueled by N-Heptane in a direct injection diesel engine at different compression ratios. *Sustainability*, 10(11), 3878. https://doi.org/10.3390/su10113878
27. Jun, D., Ishii, K., & Iida, N. (2003). Autoignition and combustion of natural gas in a 4 stroke HCCI engine. *JSME International Journal Series B*, 46(1), 60–67. https://doi.org/10.1299/jsmeb.46.60
28. Flowers, D., Aceves, S., Westbrook, C. K., Smith, J. R., & Dibble, R. (2000). Detailed chemical kinetic simulation of natural gas HCCI combustion: Gas composition effects and investigation of control strategies. *Journal of Engineering for Gas Turbines and Power*, 123(2), 433–439. https://doi.org/10.1115/1.1364521
29. Liu, H., Yao, M., Zhang, B., & Zheng, Z. (2009). Influence of fuel and operating conditions on combustion characteristics of a homogeneous charge compression ignition engine. *Energy & Fuels*, 23(3), 1422–1430. https://doi.org/10.1021/ef800950c

30. Da Costa, S., Bukkarapu, K. R., Fernandes, R., Krishnasamy, A., & Morajkar, P. P. (2024). Advancements in in-cylinder and after-treatment strategies for mitigating pollutants from diesel engines: A critical review. *Energy & Fuels*, 38(11), 9218–9261. https://doi.org/10.1021/acs.energyfuels.4c00705
31. Fei, Y. (2025b). HCCI engines: Benefits, challenges and future advancements. *E3S Web of Conferences*, 606, 01006. https://doi.org/10.1051/e3sconf/202560601006
32. Mogra, A., & Gupta, K. K. (2020). A review on new technology in internal combustion engine- HCCI engine. *IOP Conference Series Materials Science and Engineering*, 810(1), 012054. https://doi.org/10.1088/1757-899x/810/1/012054
33. Najafabadi, M. I., & Aziz, N. A. (2013). Homogeneous charge compression ignition combustion: Challenges and proposed solutions. *Journal of Combustion*, 2013, 1–14. https://doi.org/10.1155/2013/783789

6 Emissions and Power in Stratified Charge Compression Ignition

Jaya Gupta, Shubham Mishra, and Geeta Arora

6.1 INTRODUCTION

The contemporary challenges in the domain of internal combustion engines are primarily characterized by the rapid depletion of crude oil reserves and the environmental impact of engine emissions. The escalating demand for power sources with more efficiency and least environmental impact stems from concerns about potential petroleum shortages, the greenhouse effects as well as carbon dioxide, and the pollution caused by NO_x. By utilizing alternative fuels, we can mitigate fuel scarcity, and by modifying engine operating cycles, we can achieve emission control. The need for green and environmentally friendly technologies is increasing. This drives continuous research into methods that enhance air quality and decrease fuel usage. The extent and characteristics of fuel oxidation, and consequently emissions, in compression ignition engines are determined by crucial combustion characteristics, including compression ratio, spray pattern, combustion chamber design, and swirl. The combustion phenomenon within engines exhibits increased effectiveness when the air–fuel mixture is diluted to its upper limit. Over the past decade, the homogeneous charge compression ignition (HCCI) engine has garnered global interest due to its good efficacy and low NO_x and soot emissions, combining the advantages of spark ignition (SI) and direct injection compression ignition (DicI) engines. Limitations such as a narrow operating window, increased CO and unburned hydrocarbon emissions, and challenges in switching to traditional combustion prevent the successful commercialization of HCCI engines. To leverage the benefits of HCCI combustion, compound HCCI methods, which utilize mode transitions or mixture stratification, have been developed. Specifically, compound HCCI with mode transitions involves operating in HCCI mode for low to medium loads and switching to SI or DicI mode for cold starts and then idle after high loads. While the transition from HCCI to SI can occur rapidly within 10 cycles, the reverse transition requires optimization. Alternatively, SCCI, a mixture stratification-based compound HCCI, allows real-time control of in-cylinder temperature, mixture concentration, and fuel composition through adjustments to injection timing, spray mode, combined port and direct injection, fuel properties, and Exhaust Gas Recirculation (EGR). Low-temperature combustion technology and SCCI represent viable and preferred combustion system options. The technology offers significant environmental benefits, minimizing NO_x, emission of soot particles, and particulate matter (PM).

DOI: 10.1201/9781003633631-6

Furthermore, it improves heat efficiency over conventional diesel engines. Despite its benefits, this technology faces challenges, particularly a fall in combustion efficiency at less engine loads. The study presents SCCI as a promising development avenue, expanding its applications through diverse combustion modes via direct fuel injection and extending operational capacity to full load conditions. The integration of multistage induction and EGR with SCCI is highlighted as a viable strategy for reducing emissions and enhancing engine power output and efficiency.

In comparison to existing literature on HCCI, this chapter positions SCCI as a compound HCCI method that utilizes mixture stratification to overcome limitations of HCCI, such as a narrow operating window and challenges in switching to traditional combustion. The novelty of SCCI lies in its ability to achieve real-time control of in-cylinder temperature, mixture concentration, and fuel composition through adjustments to injection timing, spray mode, combined port and direct injection, fuel properties, and EGR. High thermal efficiency and ultra-low emissions are maintained throughout the engine's operating range, as HCCI and traditional combustion modes are indistinguishable. Furthermore, this chapter discusses factors affecting SCCI performance, including optimization of injection timing, water injection technology, analysis of pressure rise rate, and the impact of nitrogen oxides. It suggests strategies to minimize difficulties faced by SCCI, such as designing physical and chemical properties based on operating parameters, applying various methods for fuel stratification, selecting test fuels with contrasting chemical properties, and controlling combustion rate and peak temperature with EGR.

6.2 HOMOGENEOUS CHARGE COMPRESSION IGNITION

Conventional compression ignition engines produce high particulate and NO_x emissions. To address this, researchers are increasingly interested in HCCI, a hybrid combustion process utilizing advanced low-temperature combustion (LTC) technology. HCCI is a versatile, clean, and efficient combustion technology, suitable for any size class of transportation engine and stationary applications, achieved through lean/dilute operation. The late 1990s saw the introduction of the earliest HCCI combustion systems, which relied on early fuel injection into the combustion chamber for homogeneous mixture preparation. HCCI combustion is considered a highly promising concept for future internal combustion engines. The HCCI process is characterized as a hybrid combustion mechanism, bridging Otto and Diesel engine technologies. The efficiency of HCCI is well established. Early 20th-century hot-bulb engines, functioning with an HCCI-like combustion, achieved brake efficiencies greater than gasoline engines and equivalent to diesel engines. HCCI engines deliver excellent engine performance, extremely low NO_x emissions, and reduced PM. The volumetric autoignition of a compressed lean air–fuel mixture is the source of these profits. HCCI engines also have several drawbacks, including low specific output, a narrow operating range, long start-up times, lack of ignition control, and high CO and UHC emissions. Catalytic converters can be used to treat CO and UHC emissions [1–3].

6.2.1 The Foundational Principles of HCCI Combustion

The complexity of HCCI diesel combustion, involving numerous species and reactions, manifests as a two-stage heat release. HCCI diesel combustion exhibits two distinct heat release stages. The first, driven by low-temperature reactions, releases a small part of the total energy. The second stage, caused by high-temperature reactions, releases the majority of the energy (approximately 90%). Low-temperature reactions involve the breakdown in the beginning of hydrocarbon fuel molecules, producing hydrocarbon radicals. These radicals then react with oxygen to form alkyl peroxy radicals. Through isomerization, alkyl peroxy radicals become hydroperoxy alkyl radicals. A subsequent reaction with oxygen yields oxo-hydroperoxy radicals, which then decompose into keto-hydroperoxides and OH radicals. The period between the low-temperature and high-temperature reaction stages is known as the negative temperature coefficient regime. Around 700 K, alkyl peroxy radicals break down into their original components, forming olefins and hydroperoxyl radicals through moderate reactions. The heat generated during this phase increases the mixture's temperature, triggering the high-temperature reaction at 900 K. In this final stage, the mixture temperature quickly rises as oxygen and fuel are completely consumed, yielding carbon dioxide, water, hydrogen, and carbon monoxide.

6.2.2 Processes Involved in NO_x and Soot Generation

To grasp HCCI combustion, it's essential to understand NO_x and soot formation. NO_x, a collective term for nitric oxide and nitrogen dioxide, forms from complex reactions involving nitrogen in the combustion air and fuel. Since NO_x is thermally produced, its formation is highly dependent on controlling the combustion temperature, typically occurring at low equivalence ratios and high flame temperatures. To minimize NO_x, flame temperatures should be kept below 2200 K. While NO_x forms at high temperatures and low equivalence ratios, soot forms in fuel-rich regions with moderate temperatures. Reducing soot emissions, a balance of formation and oxidation, requires improved mixing or increased oxidation. In traditional diesel engines, fuel combustion typically results in both NO_x and soot emissions. Efforts to reduce one often increase the other, creating a trade-off. Since NO_x and soot emissions are strongly influenced by equivalence ratio as well as by temperature, simultaneous reduction can be achieved by carefully controlling these parameters. HCCI combustion, adhering to this principle, seeks to achieve simultaneous reductions in flame temperature and improvements in charge homogeneity through enhanced air–fuel mixing. Conceptually, a perfectly homogeneous in-cylinder mixture leads to simultaneous autoignition throughout the cylinder due to pressure and temperature increases during compression. HCCI maintains low local temperatures, eliminating high-temperature flame fronts and preventing NO_x formation. The homogeneous lean mixture also avoids soot formation by lowering local equivalence ratios [4–6].

6.2.3 Operational Characteristics of HCCI Engines

HCCI engine combustion, which follows the four-stroke Otto cycle, critically relies on precise fuel delivery. Fuel is directly injected into the cylinder's combustion chamber by head-mounted injectors during the intake stroke, operating independently of

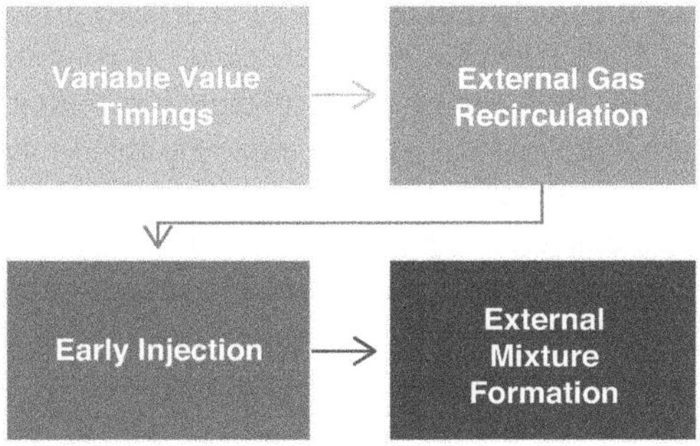

FIGURE 6.1 A flow diagram to achieve HCCI combustion.

the air intake plenum. A fully mixed air–fuel charge is present within the cylinder's combustion chamber at the conclusion of the intake stroke. During the compression stroke, the upward movement of the piston leads to a critical heat buildup in the combustion chamber. This heat is precisely what triggers the air–fuel mixture's autoignition at the end of compression, initiating the power stroke and driving the piston down, with no spark required. HCCI combustion distinguishes itself from traditional SI and diesel engines by producing a lean, low-temperature, and flameless energy release across the chamber in which combustion takes place. Simultaneous combustion of the entire air–fuel mixture in HCCI engines produces comparable power while consuming low amount of fuel and generating significantly fewer emissions. Immediately following the power stroke, the piston starts the exhaust stroke. However, exhaust valves close early, trapping residual combustion heat. To sustain this thermal energy and regulate combustion temperatures and emissions, a small fuel pre-charge is injected into the combustion chamber before the upcoming intake stroke (Figure 6.1).

6.2.4 Advantages of HCCI

1. HCCI engines, operating with lean fuel mixtures and diesel-like compression ratios, achieve higher efficiencies compared to conventional SI gasoline engines.
2. The absence of a flame front, flame spread, or localized high temperatures and rich mixtures results in significantly reduced soot and NO_x generation.
3. It offers cleaner combustion and reduced emissions, particularly achieving almost negligible NO_x levels [7–9].

6.2.5 Challenges Faced in Proposing HCCI

Despite offering superior thermal efficiency and reduced NO_x emissions compared to conventional engines, HCCI combustion faces significant challenges, notably controlling combustion timing, achieving sufficient power output, minimizing unburned hydrocarbon, and ensuring homogeneous mixture formation, carbon monoxide emissions, and enabling robust cold-start performance. Compared to diesel engines,

HCCI engines produce relatively higher levels of hydrocarbon and carbon monoxide emissions. To address high-load emissions, to create partial-charge stratification a direct in-cylinder fuel injection can be used, offering a potential mitigation strategy. Exhaust emission control devices are usually indispensable for controlling hydrocarbon and carbon monoxide emissions in HCCI engines, unless fuel optimization is implemented. Catalytic converters, which effectively reduce hydrocarbon and carbon monoxide emissions, are a well-established and long-standing technology in the automotive industry. Catalysts may suffer from increased light-off times and reduced average effectiveness in HCCI engines due to the engines' cooler exhaust temperatures. The necessity to overcome prolonged catalyst light-off and reduced efficiency in HCCI engines, due to their lower exhaust temperatures, mandates the development of improved oxidation catalysts tailored for low-temperature exhaust. The relative ease of hydrocarbon and carbon monoxide emission control in HCCI engines, due to simpler and more robust oxidation catalysts, contrasts sharply with the difficulty of simultaneous NO_x reduction and PM oxidation in CIDI engines. During a cold start, HCCI engines experience a decrease in compressed-gas temperature due to the absence of intake manifold preheating and rapid heat loss to the cold combustion chamber walls. To prevent ignition failure due to low compressed-charge temperatures, HCCI engines need compensation, especially for cold starting. Various strategies have been proposed to address this, such as increasing the compression ratio through variable compression ratio or variable valve timings or using glow plugs, alternative fuels or additives.

6.2.6 Limitation of HCCI

Despite its advantages, HCCI technology has drawbacks, including high unburned hydrocarbon and carbon monoxide emissions, as well as knocking under specific operating conditions [10]. Diesel engines generate high NO_x and PM emissions, posing health and environmental risks that necessitate control strategies. HCCI combustion engines offer a potential solution [11]. HCCI's lean mixture operation results in lower peak temperatures compared to SI and CI engines, effectively reducing NO_x emissions. However, this also creates a trade-off, as the cooler temperatures can lead to incomplete fuel combustion, especially near the combustion chamber walls. The incomplete combustion results in high carbon monoxide as well as hydrocarbon emissions. However, these emissions can be addressed using an oxidizing catalyst, as the exhaust remains oxygen rich. The transition from the power stroke to the exhaust stroke involves the piston reversing direction. However, the exhaust valves close prematurely, trapping residual combustion heat before all exhaust gases are expelled. To optimize thermal efficiency, residual heat is retained, and a small fuel pre-charge is introduced into the combustion chamber to regulate temperatures and emissions before the subsequent intake stroke.

Stratified charge compression ignition is a compound HCCI combustion method that uses mixture stratification to create varying in-cylinder temperature and mixture concentration levels throughout combustion. Real-time control of temperature and fuel composition in engines is achieved through various strategies. These include adjusting in-cylinder injection timing and spray mode, combining port and direct fuel injection, optimizing fuel component design and management, and manipulating EGR and its distribution. The comprehensive control enables ultra-low emissions and high thermal efficiency across the entire engine operating range. The advanced combustion mode blurs the lines between HCCI and traditional combustion, effectively combining the benefits of both. It offers direct control over combustion phasing and burn rate [12].

Further research by Sankaran et al. [13] and Bisetti et al. [14] analytically demonstrated that mixture stratification as well as temperature are effective in controlling the beginning of combustion and mitigating sharp pressure increases at high loads in HCCI engines. Real HCCI engines inherently possess thermal or mixture inhomogeneities [15]. Furthermore, these stratification strategies are also applicable to diesel and alternative fuel HCCI combustion (Figure 6.2).

6.2.7 Advantages of SCCI over HCCI

The SCCI combustion system expands the operational range of HCCI by employing stratified mixture compression ignition. Unlike HCCI, which relies on a uniform mixture and high-cetane fuel for compression ignition, SCCI utilizes a heterogeneous mixture formed with lower-cetane fuels, such as gasoline. Essentially, the key distinction lies in the mixture's homogeneity: HCCI is homogeneous, while SCCI is heterogeneous. Achieving successful SCCI combustion is significantly influenced by intake temperature and equivalence ratio.

6.3 SCCI

Stratification in the cylinder is generally understood to involve three coupled aspects: temperature field, mixture concentration, and fuel composition. The formation of a homogeneous mixture is a critical prerequisite for LTC in compression ignition engines. Injecting the complete fuel quantity into each cylinder is a method for achieving this. The growing demand for LTC stems from its characteristics: a longer ignition delay, increased exhaust gas recirculation, and reduced oxygen content. The formation of a uniform mixture within the cylinder is contingent upon the generation of substantial turbulence. Compared to traditional engine designs, HCCI and LTC technologies provide unique benefits. To integrate these advantages into conventional combustion systems, fuel was introduced in a stratified manner, resulting in a technology known as SCCI.

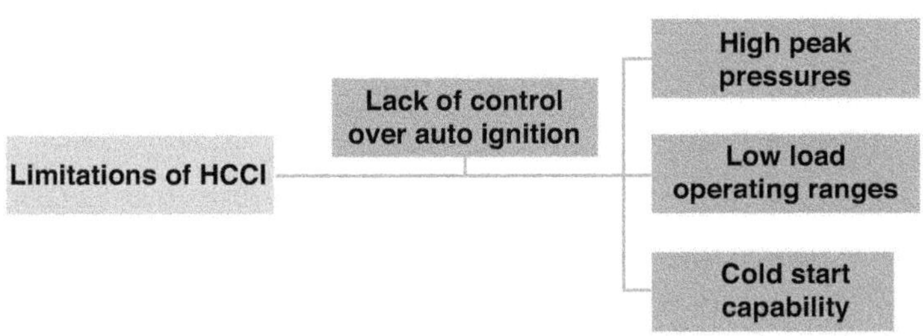

FIGURE 6.2 Limitations of the homogeneous charge compression ignition process.

By employing stratified fuel mixtures, SCCI technology, or compound HCCI, combines the benefits of both HCCI and LTC. Using this technique, combustion phasing and rate can be directly and easily controlled. Despite various mixture preparation methods, SCCI engines exhibit an unavoidable degree of inhomogeneity. A uniform fuel–air mixture is not achievable. Stratification, however, leads to three distinct aspects within the cylinder: variations in fuel composition, mixture concentration, and temperature. By controlling the homogeneous charge, fuel composition stratification directly influences autoignition characteristics. Increased fuel stratification leads to enhanced combustion efficiency, according to published studies. While stratification can improve combustion efficiency, excessive stratification creates a trade-off, leading to higher NO_x emissions and reduced power output. An experimental study using a single-cylinder engine with two gasoline direct injectors investigated the effects of fuel stratification. This research revealed that fuel stratification significantly increased indicated mean effective pressure and emissions, particularly under lean limit conditions. It also showed that fuel stratification reduces the rate of pressure rise and is sensitive to the local equivalence ratio, suggesting its potential for controlling SCCI combustion [16–22]. Furthermore, increased mixture concentration stratification advances autoignition. Increasing late-cycle fuel injection advances autoignition. Mixture stratification is essential for predicting CO, unburned hydrocarbon, and soot emissions. Furthermore, NO_x emissions are affected by mixture concentration stratification.

6.3.1 Steps to be Taken to Minimize the Difficulties the SCCI Faces

1. The properties (physical as well as chemical) should be tailored to the operating parameters and selected test fuels.
2. To create the target fuel stratified concentration, several techniques can be used: port fuel injection with direct injection, diverse injection modes, in-cylinder injection with multi-hole injectors.
3. When selecting test fuels, ensure they have contrasting chemical properties and can be used with fuel injection.
4. To control the combustion rate and peak combustion temperature, exhaust gas recirculation in a very controlled amount should be used (Figure 6.3).

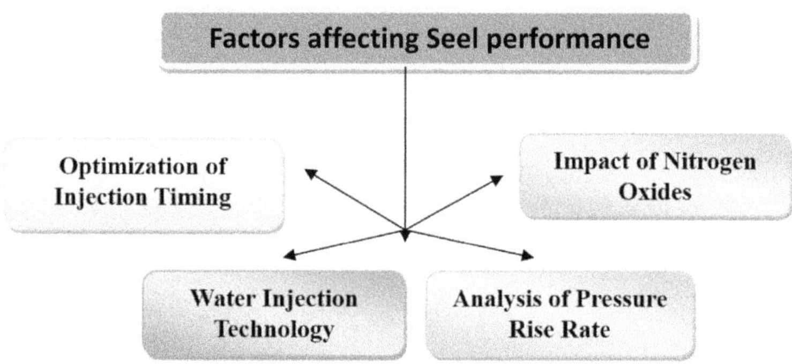

FIGURE 6.3 Factors on which performance of SCCI depends.

6.3.2 Analysis of Engine Parameter Effects on SCCI

Port fuel injection is the most effective in-cylinder method in order to achieve proper SCCI combustion, and engine parameters significantly influence overall engine performance. The chemistry of fuel mixture concentration, influenced by early or port fuel injection, is key to controlling SCCI combustion phasing. There is a direct proportionality between the equivalence ratio and the quantity of heat released at low temperature. Therefore, this phenomenon occurs. Adjusting the timing of the injection within the cycle allows for control of the stratification mixture. Combustion phasing can be managed using this technique. Furthermore, various fuel injection timings can be applied to adjust the combustion duration of stratified mixtures. As a result of partial fuel stratification, combustion duration lengthened, and the rate of pressure rise decreased. Stratified charge compression ignition engines have lower indicated thermal efficiency than conventional combustion engines due to the need for optimized combustion parameters, including heat release rate, phasing control, and combustion chamber design. Consequently, the primary challenges are achieving low NO_x emissions and maximizing efficiency in stratified charge combustion (Scc) engines. To expand the operating range of SCCI engines from low to high loads, partial fuel stratification is necessary. This partial fuel stratification enables staged combustion by reducing pressure rise rates.

6.3.3 Optimization of Injection Timing

To effectively manage combustion rate and phasing, achieve a homogeneous mixture, and extend the operating load of SCCI combustion, researchers suggest utilizing two-stage gasoline direct injection. Control of diesel fuel stratification is achievable through multi-pulse fuel injection technology by manipulating the number of injections, injection pulse width, and dwell time between adjacent pulses. Improved fuel mixing is achieved through two injections per cycle. By injecting at 80° before TDC and 10° after TDC, smoke is significantly reduced. The technology not only brings SCCI engine thermal efficiency to conventional diesel engine levels but also lowers NO_x, soot, and PM emissions. Partial fuel injection creates a homogeneous stratified mixture through a two-stage process: some fuel is injected during the intake stroke, and the rest during the compression stroke. While this technique, when applied to stratified charge compression ignition, can lower pressure rise rates and boost load capacity, research indicates that increasing the amount of stratified fuel leads to higher NO_x emissions. In closed valve injection, a large portion of the fuel impacts the port walls before entering the cylinder. Therefore, injection details are less critical. Conversely, during open valve injection, hydrocarbon emissions rise as fuel droplets are introduced into the cylinder. Furthermore, experiments on stratified mixture concentration revealed that achieving proper stratification is crucial for enhancing the load capacity of SCCI engines [22–25]. A crucial aspect of stratified combustion is gasoline direct injection, where combustion stability hinges on the symmetry of fuel injection with respect to the inlet port and stratified mixture. To optimize this, researchers utilized a system with two injectors. One injector was responsible for establishing a homogeneous charge, while the other directly injected fuel into the combustion chamber to control the stratification. This innovative setup resulted in substantial increases in effective pressure and a decrease in emissions when operating at the lean limit. In a separate approach to create fuel stratification, they implemented a dual injection strategy, their research also demonstrated that adjusting fuel ratios allows for ignition timing control in dual-fuel stratified combustion [26–28].

The practical success of SCCI ultimately depends on balancing these implementation challenges against the performance and emissions benefits, with cost-effectiveness being a critical factor in determining commercial viability.

6.3.3.1 Fuel Injection Timing

Injection timing emerges as the most critical variable in SCCI operation. Early injection timing (during compression stroke) promotes better fuel–air mixing and stratification, while late injection timing affects the combustion phasing with heat release rate. The timing directly influences:

- Charge stratification quality
- Ignition delay period
- Combustion duration and phasing
- Knock tendency and combustion stability

Research shows that retarded injection timings combined with moderate EGR rates provide better efficiency, indicating the complex interaction between injection timing and other parameters.

6.3.3.2 Swirl Ratio

Swirl ratio significantly impacts mixing characteristics and combustion quality. Increasing swirl ratio from 1.0 to 4.0 achieved significant reduction in soot emissions, especially for advanced injection timing. The swirl ratio affects:

- Fuel–air mixing quality
- Stratification maintenance
- Turbulence intensity during combustion
- Heat transfer characteristics

However, higher swirl ratio does not always imply increased mixing during late combustion, demonstrating the complex relationship between swirl and combustion phases.

6.3.3.3 EGR Rate

EGR rate is crucial for controlling combustion temperature and phasing. EGR reduces NO_x through lowering oxygen concentration and heat absorption. In SCCI applications, EGR affects:

- Combustion temperature reduction
- Ignition delay extension

- Knock suppression
- Emission formation control

The analysis reveals strong coupling between key variables:

- **Injection Timing × EGR Rate:** Early injection with high EGR extends ignition delay, while late injection with low EGR may cause incomplete combustion
- **Swirl Ratio × Injection Pressure:** Higher swirl ratios require adjusted injection pressures for optimal mixing
- **EGR Rate × Compression Ratio:** EGR effects are amplified at higher compression ratios

6.3.4 Water Injection Technology

Water injection has a long-standing history in internal combustion engines. It has been employed in SI combustion to reduce knock, lower temperatures, and minimize NO_x emissions. Water injection has been previously studied in combustion techniques such as HCCI and Premixed Charge Compression Ignition, and more recently, in SCCI engines. Experimental analysis was conducted to gain a deeper understanding of SCCI engine operation with and without water injection. Water injection significantly increased the combustion rate in a 2.0 L four-cylinder gasoline SI engine, specifically between 20° and 70° before top dead center (TDC), compared to operation without water injection. An initial direct water injection timing of 50° before TDC, coupled with a 153° crank angle increase, caused excessive combustion retardation, leading to misfiring. Using this injection system resulted in a 44% rise in CO emissions and a 103% rise in unburned hydrocarbon emissions when 9.0 mg/cycle of water was injected [29,30].

6.3.5 Analysis of Pressure Rise Rate

The rate of pressure rise is a crucial parameter to consider when investigating SCCI combustion. It defines the upper limit of SCCI combustion. Increasing combustion chamber pressure reduces NO_x and smoke formation due to the resulting dilute mixture. However, split injection decreases mean effective pressure without affecting NO_x or smoke formation.

6.3.6 Impact of Nitrogen Oxides

Injection timing and fuel quantity directly affect NO_x emissions. While conventional engines produce 30–35 ppm of NO_x, split injection reduces this to 22 ppm, a significant decrease. While overall pressure increases, it influences fuel quantity. As fuel quantity rises, mean effective pressure also increases. Hydrocarbon emissions change based on when fuel is injected and how much fuel is injected per stroke. Multiple fuel injection systems with homogeneous charge operation minimize NO_x concentration. This is because early injections enhance mixing intensity before the second injection stage. As a result, NO_x formation is reduced, and temperature is kept within acceptable levels. Furthermore, a Port Fuel Injection (PFI) system allows for easy control of Hc emissions in SCCI engines [31,32].

6.4 PRACTICAL IMPLICATIONS FOR SCCI

Here are the key practical implications for SCCI:

a. **Manufacturability Challenges**
   SCCI engines face significant manufacturing complexity due to their sophisticated control requirements. The technology demands precise fuel injection systems capable of creating stratified charge distributions, advanced engine control units with real-time combustion monitoring, and high-precision sensors for temperature, pressure, and air–fuel ratio monitoring. These components require tighter manufacturing tolerances than conventional engines, increasing production complexity and quality control requirements.
b. **Cost Implications**
   The implementation costs for SCCI are substantial. The technology requires advanced direct injection systems, sophisticated control algorithms, and additional sensors beyond those found in conventional engines. Mazda's SKYACTIV-X implementation demonstrates the commercial viability but the added complexity translates to higher development and manufacturing costs. The control systems alone represent a significant cost increase, as they must manage the delicate balance between SI and compression ignition modes.
c. **Compatibility with Existing Platforms**
   SCCI integration into existing engine platforms presents notable challenges. Traditional engine architectures require substantial modifications to accommodate the precise fuel injection timing, advanced combustion chamber designs, and sophisticated control systems necessary for SCCI operation. The technology typically requires higher compression ratios and modified piston designs compared to conventional gasoline engines, making retrofit applications difficult and expensive.
d. **Control System Complexity**
   The most significant practical hurdle is the control system sophistication required. SCCI engines must seamlessly transition between different combustion modes based on operating conditions, requiring real-time combustion feedback and predictive control algorithms. This complexity extends to calibration requirements, as the system must be precisely tuned for various fuel qualities, ambient conditions, and engine loads.
e. **Market Adoption Barriers**
   Despite the potential for improved fuel efficiency and reduced emissions, SCCI faces adoption barriers including higher initial costs, increased maintenance complexity, and the need for specialized service training. The technology's sensitivity to fuel quality and operating conditions also presents challenges for widespread commercial deployment.

6.5 DIFFICULTIES IN ACHIEVING SCCI COMBUSTION

Due to lean mixtures or high exhaust gas recirculation, SCCI engines experience a reduction in mean effective temperature, causing a simultaneous rise in unburned hydrocarbons (UHC) and CO emissions relative to conventional engines. A challenge with split injection, where part of the fuel is inducted into the suction line, is that diesel fuel doesn't readily vaporize under normal conditions. While a turbocharger can increase fuel pressure and temperature, it's insufficient for complete vaporization. To address this, a heat exchanger, EGR, or a combination of both can be installed, depending on operating conditions, to raise incoming air temperature for rapid fuel vaporization. Alternatively, EGR can be used to heat the fuel itself to its boiling point.

6.6 COMPARISON OF SCCI AND HCCI

To potentially improve the thermal efficiency of the working cycle, stratified fuel injection can be used in place of traditional homogeneous fuel injection. Scc offers several advantages: minimized pumping losses from unthrottled part-load operation, reduced cylinder wall heat loss, and increased compression ratio enabled by lower end-gas temperatures. The lower end-gas temperatures associated with Scc lessen the risk of autoignition, even though these engines generally operate where autoignition is already unlikely.

Fuel stratification introduces two primary drawbacks: elevated NO_x emissions relative to homogeneous lean burn engines, and the complexity of precisely timing fuel spray delivery to the spark plug. Because Scc results in cooler exhaust temperatures compared to homogeneous lean burn, careful consideration of aftertreatment system performance at these lower temperatures is essential. Wirth et al.'s study confirmed that fuel stratification is an effective combustion strategy, leading to lower fuel consumption and nitrogen oxide emissions [33–35].

6.7 ENHANCED HEALTH AND ENVIRONMENTAL MOTIVATIONS FOR EMISSIONS REDUCTION

The global automotive industry operates under increasingly stringent emission regulations that directly address public health and environmental concerns. These regulatory frameworks provide the legal foundation for the technological advances discussed in SCCI and HCCI combustion systems.

6.7.1 Key International Emission Standards

6.7.1.1 European Union—Euro Standards

- **Euro 6 (2014-Present):** Represents the current European standard with NO_x limits of 80 mg/km for diesel passenger cars and 60 mg/km for gasoline vehicles
- **Euro 6d-TEMP/Euro 6d:** Introduced Real Driving Emissions (RDE) testing to ensure emissions control under actual driving conditions

- **Future Euro 7:** Proposed to further reduce NO_x limits to 35 mg/km for all vehicle types by 2025–2027

6.7.1.2 India—Bharat Stage Standards

- Aligned with Euro 6 standards, representing a significant leap from BS-IV
- Reduced NO_x limits from 250 mg/km (BS-IV) to 80 mg/km (BS-VI) for diesel vehicles
- Introduced On-Board Diagnostics and RDE testing

6.7.1.3 United States—EPA Standards

- NO_x limits of 30 mg/mile for light-duty vehicles
- Progressive reduction in NO_x emissions from heavy-duty engines

6.7.2 Health Impact Justification

- Respiratory irritation and asthma exacerbation
- Cardiovascular stress and increased heart disease risk
- Premature mortality in vulnerable populations (children, elderly, individuals with pre-existing conditions)
- Increased lung cancer risk
- Chronic obstructive pulmonary disease
- Developmental issues in children exposed to high pollution levels

6.7.2.1 Environmental Motivations–Climate Change Mitigation

- Transportation sector contributes approximately 24% of global CO_2 emissions
- NO_x emissions contribute to ground-level ozone formation and acid rain
- PM affects global warming through complex atmospheric interactions
- Cities like Delhi, Beijing, and Los Angeles face severe air quality challenges primarily due to vehicular emissions
- The concept of "diesel ban" in city centers across Europe reflects the urgency of emission reduction

6.8 CONCLUSION

This chapter presents a systematic approach to understanding SCCI as a distinct advancement over HCCI. The novelty lies in positioning SCCI not merely as a variation of HCCI, but as a "compound HCCI method" that fundamentally addresses HCCI's operational limitations through controlled mixture stratification. SCCI offers a significant advantage in improving the thermal efficiency of internal combustion engines. SCCI technology can be viewed as an extension of HCCI engines. The SCCI combustion concept optimizes performance and mixture preparation by adjusting operating and design parameters such as port fuel injection, EGR, fuel management, and component design. SCCI combines the advantages of HCCI combustion, allowing for direct and flexible control of combustion phasing and rate. SCCI combustion can be controlled by utilizing fuels with appropriate physical as well as chemical properties and by managing design parameters such as split injection, port fuel injection, and EGR. A common feature of these advanced combustion techniques is the preparation of either a homogeneous or stratified mixture before ignition [36–38].

REFERENCES

1. Aceves, S.M., Smith, J.R., Westbrook, C.K. and Pitz, W.J., 1999. Compression ratio effect on methane HCCI combustion. *Journal of Engineering for Gas Turbines and Power*, 121(3), pp. 569–574.
2. Ali, M.S., Islam, S.N., Sapuan, S.M. and Megat, H.M., 2001. Effects of compression ratio on turbulence kinetic energy and dissipation in the suction stroke of a four-stroke internal combustion engine. In: *4th International Conference on Mechanical Engineering*, Dhaka, Bangladesh, 26–28 Dec 2001. pp. III-165–172.
3. Antunes, J.G., Mikalsen, R. and Roskilly, A.P., 2008. An investigation of hydrogen-fuelled HCCI engine performance and operation. *International Journal of Hydrogen Energy*, 33(20), pp. 5823–5828.
4. Sharma, T.K., Rao, G.A.P. and Murthy, K.M., 2015. Effective reduction of NO_x emissions of a HCCI engine by enhanced rate of heat transfer under varying conditions of operation. *Energy*, 93(2), pp. 2102–2115.
5. Pourfallah, M., Armin, M. and Ranjbar, A.A., 2019. A numerical study on the effect of thermal and charge stratification on the HCCI natural gas engine. *International Journal of Ambient Energy*, 42(12), pp. 1386–1395.
6. Telli, G., Altafini, C., Costa, C. and Rosa, J., 2021. A comprehensive review of homogeneous charge compression ignition (HCCI) engines: Advantages, challenges and evolution. *SAE Technical Paper*, 2020-36-0042.
7. Killingsworth, N.J., Aceves, S.M., Flowers, D.L. and Krstic, M., 2006. A simple HCCI engine model for control. In: *IEEE International Symposium on Intelligent Control*, Munich, Germany, pp. 2424–2429.
8. Mack, J.H., Aceves, S.M. and Dibble, R.W., 2009. Demonstrating direct use of wet ethanol in a homogeneous charge compression ignition (HCCI) engine. *Energy*, 34(6), pp. 782–787.
9. Warnatz, J., Maas, U. and Dibble, R.W., 2006. *Combustion: Physical and Chemical Fundamentals, Modeling and Simulation, Experiments, Pollutant Formation*. Berlin: Springer.
10. Kong, S.C. and Reitz, R.D., 2003. Numerical study of premixed HCCI engine combustion and its sensitivity to computational mesh and model uncertainties. *Combustion Theory and Modelling*, 7(3), pp. 417–433.
11. Gan, S., Ng, H.K. and Pang, K.M., 2011. Homogeneous charge compression ignition (HCCI) combustion: Implementation and effects on pollutants in direct injection diesel engines. *Applied Energy*, 88(3), pp. 559–567.
12. Ma, J.J., Lu, X.C., Ji, L.B. and Huang, Z., 2008. Evaluation of SCCI potentials in comparison to HCCI and conventional DICI combustion using *n*-heptane. *Energy & Fuels*, 22(2), pp. 954–960.
13. Sankaran, R., Im, H.G. and Hewson, J.C., 2007. Analytical model for auto-ignition in a thermally stratified HCCI engine. *Combustion Science and Technology*, 179(10–11), pp. 1963–1989.
14. Bisetti, F., Chen, J.Y., Hawkes, E.R. and Chen, J.H., 2008. Probability density function treatment of turbulence/chemistry interactions during the ignition of a temperature stratified mixture for application to HCCI engine modeling. *Combustion and Flame*, 155(4), pp. 571–584.

15. Dec, J.E., 2009. Advanced compression-ignition engines-understanding the in-cylinder processes. *Proceedings of the Combustion Institute*, 32(2), pp. 2727–2742.
16. Sjöberg, M. and Dec, J.E., 2006. *Smoothing HCCI heat-release rates using partial fuel stratification with two-stage ignition fuels* (No. 2006-01-0629). SAE Technical Paper.
17. Jung, G.S., Sung, Y.H., Choi, B.C. and Lim, M.T., 2009. Effects of mixture stratification on HCCI combustion of DME in a rapid compression and expansion machine. *International journal of automotive technology*, 10(1), pp. 1–7.
18. Mukhopadhyay, S. and Abraham, J., 2011. Influence of compositional stratification on autoignition in *n*-heptane/air mixtures. *Combustion and Flame*, 158(6), pp. 1064–1075.
19. Dec, J.E., Yang, Y. and Dronniou, N., 2011. Boosted HCCI-controlling pressure-rise rates for performance improvements using partial fuel stratification with conventional gasoline. *SAE International Journal of Engines*, 4(1), pp. 1169–1189.
20. Hunicz, J., Tmar, A. and Krzaczek, P., 2017. Effects of mixture stratification on combustion and emissions of boosted controlled auto-ignition engines. *Energies*, 10(12), pp. 2172.
21. Li, L., Xie, H., Chen, T., Yu, W. and Zhao, H., 2012. *Experimental study on spark assisted compression ignition (SACI) combustion with positive valve overlap in a HCCI gasoline engine* (No. 2012-01-1126). SAE Technical Paper.
22. Lawler, B., Splitter, D., Szybist, J. and Kaul, B., 2017. Thermally stratified compression ignition: A new advanced low temperature combustion mode with load flexibility. *Applied Energy*, 189, pp. 122–132.
23. Kolodziej, C.P., Ciatti, S., Vuilleumier, D., Adhikary, B.D. and Reitz, R.D., 2014. *Extension of the lower load limit of gasoline compression ignition with 87 AKI gasoline by injection timing and pressure* (No. 2014-01-1302), SAE Technical Paper.
24. Dempsey, A.B., Curran, S.J. and Wagner, R.M., 2016. A perspective on the range of gasoline compression ignition combustion strategies for high engine efficiency and low NO_x and soot emissions: Effects of in-cylinder fuel stratification. *International Journal of Engine Research*, 17(8), pp. 897–917.
25. Kalaskar, V.B., Splitter, D.A. and Szybist, J.P., 2014. Gasoline-like fuel effects on high-load, boosted HCCI combustion employing negative valve overlap strategy. *SAE International Journal of Fuels and Lubricants*, 7(1), pp. 82–93.
26. Scaringe, R., Wildman, C.B. and Cheng, W., 2010. On the high load limit of boosted gasoline HCCI engine operating in NVO mode. *SAE International Journal of Engines*, 3(1), pp. 35–45.
27. Storey, J.M., Curran, S.J., Lewis, S.A., Barone, T.L., Dempsey, A.B., Moses- DeBusk, M. and Northrop, W.F., 2017. Evolution and current understanding of physicochemical characterization of particulate matter from reactivity-controlled compression ignition combustion on a multicylinder light-duty engine. *International Journal of Engine Research*, 18(5–6), pp. 505–519.
28. Lawler, B., Splitter, D., Szybist, J. and Kaul, B., 2017. Thermally stratified compression ignition: A new advanced low temperature combustion mode with load flexibility. *Applied Energy*, 189, pp. 122–132.
29. Dronniou, N. and Dec, J., 2012. Investigating the development of thermal stratification from the near-wall regions to the bulk-gas in an HCCI engine with planar imaging thermometry. *SAE International Journal of Engines*, 5(3), pp. 1046–1074.
30. Reitz, R.D. and Duraisamy, G., 2015. Review of high efficiency and clean reactivity-controlled compression ignition (RCCI) combustion in internal combustion engines. *Progress in Energy and Combustion Science*, 46, 12–71.
31. Manofsky, L., Vavra, J., Assanis, D.N. and Babajimopoulos, A., 2011. *Bridging the gap between HCCI and SI: Spark-assisted compression ignition* (No. 2011-01-1179). SAE Technical Paper.

32. Zhao, F., Lai, M.C. and Harrington, D.L., 1999. Automotive spark-ignited direct-injection gasoline engines. *Progress in Energy and Combustion Science*, 25(5), pp. 437–562.
33. Schmieg, S.J., 2010. Aspects of HC-SCR catalyst durability for lean-burn engine exhaust aftertreatment. *SAE* International Journal of Fuels and Lubricants, 3(2), pp. 691–709.
34. Wirth, M., Piock, W.F., Fraidl, G.K., Schoeggl, P. and Winklhofer, E., 1998. *Gasoline DI engines: The complete system approach by interaction of advanced development tools*. SAE Technical Paper 980492.
35. Gupta, K.K. and Mogra, A., 2020. Effect of stratified charge compression ignition on emission and power output- A review. *IOP Conference Series: Materials Science and Engineering*, 810, pp. 012026.
36. Cakir, M. 2025. Effect of stratified charge combustion chamber design on natural gas engine performance. *Energies*, 18(9), pp. 2187.
37. Morales Rojas, A.D., Quintana, S.H. and Caro I.D.B., 2024. Experimental study of a homogeneous charge compression ignition engine using hydrogen at high-altitude conditions. *Sustainability*, 16(5), pp. 2026.

7 Dual-Fuel Combustion Strategies

Satsimran Singh, Amit Kumar,
Sarbjot Singh Sandhu, Sumit Kanchan,
and Anara Yegzekova

7.1 INTRODUCTION

The internal combustion (IC) engine continues to dominate many sectors, including transportation, agriculture, and power generation, due to its robustness, reliability, and relatively high energy density of liquid fuels. However, increasing environmental concerns, tightening emission norms (such as Euro 7, EPA Tier 4, IMO Tier III, and BS-VI), and the global shift toward decarburization have driven a renewed focus on advanced combustion approaches. To illustrate the increasing stringency of these global environmental mandates, Figure 7.1 visually compares the stringent NO_x emission limits between earlier Euro 3 and the more recent Euro 6 standards across different driving conditions and fuel types. Out of these, low-temperature

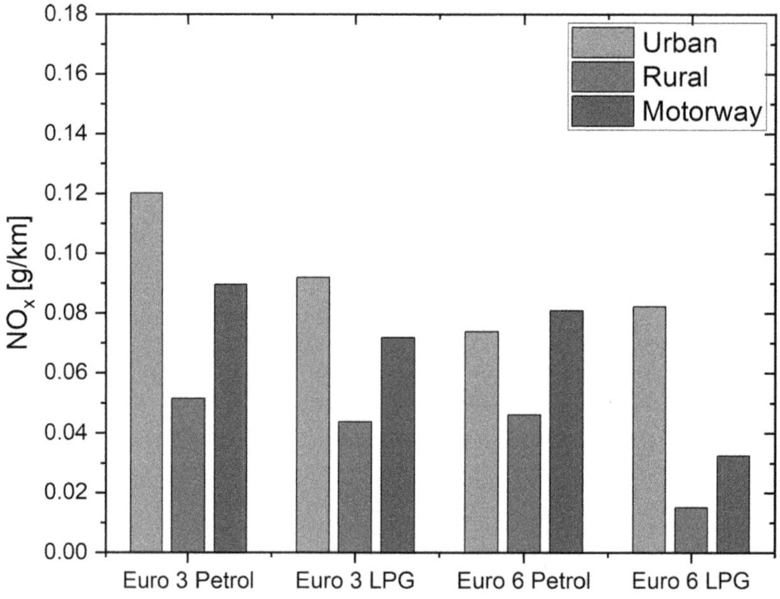

FIGURE 7.1 Comparison of the emission limits between Euro 3 and Euro 6 [2].

DOI: 10.1201/9781003633631-7

combustion (LTC) represents a notable advancement in combustion strategies for IC engines. Its core advantage is the ability to decouple the traditional trade-off between nitrogen oxides (NO_x) and particulate matter (PM) emissions [1]. By carefully controlling factors such as in-cylinder temperature and mixture preparation, LTC promotes combustion regimes that inherently limit the formation of both pollutants, all while preserving the engine's thermal efficiency. Building upon the advancements in combustion strategies like LTC, which aims to optimize emissions and efficiency through sophisticated control of the combustion process, another promising avenue in IC engine technology is the exploration of dual-fuel operation.

Dual-fuel combustion, a strategy where two fuels are introduced into the engine—typically one with high autoignition resistance and another with high reactivity—offers a flexible platform to achieve LTC characteristics. This approach enables control over combustion phasing, ignition delay, and temperature distribution, which are critical for achieving clean and efficient combustion [1]. A typical dual-fuel system might employ a gaseous fuel such as methane or hydrogen as the main energy source and use a smaller quantity of a liquid fuel like diesel to initiate ignition. Depending on engine type and operational strategy, combinations may also include gasoline, ethanol, dimethyl ether (DME), or biodiesel. This inherent flexibility offers significant advantages by enabling the use of readily available or potentially cheaper fuels, thereby reducing reliance on single fuel sources and enhancing energy security. In recent years, Reactivity Controlled Compression Ignition (RCCI) has appeared as a more refined form of dual-fuel LTC, offering high efficiency with ultra-low emissions. This method relies on precise control of fuel reactivity gradients within the cylinder to tailor the combustion process. This chapter explores the fundamental principles, implementation strategies, and practical challenges associated with dual-fuel combustion in IC engines, with a specific focus on how it aligns with the goals of LTC. It covers fuel selection criteria, control strategies, emissions performance, and future research directions, aiming to provide a comprehensive understanding of how dual-fuel systems contribute to next-generation clean combustion technologies.

7.2 FUNDAMENTALS OF DUAL-FUEL COMBUSTION

The fundamental principle behind dual-fuel combustion in IC engines lies in the simultaneous utilization of two distinct fuels within the combustion chamber to optimize engine performance, efficiency, and emissions. This approach leverages the complementary properties of different fuels, enabling the tailoring of the combustion process in ways not achievable with a single fuel source. The increasing demand for cleaner and more sustainable energy sources, coupled with stringent emission regulations, has driven significant research and development in alternative combustion strategies. Dual-fuel combustion emerges as a promising resolution, presenting a pathway to decrease dependence on conventional fossil fuels while potentially lowering greenhouse gas emissions and improving air quality.

The key is the controlled introduction and interaction of a primary fuel, which typically provides the majority of the energy, and a secondary fuel, which plays a crucial role in initiating or controlling the combustion of the primary fuel.

Common fuel combinations in dual-fuel engines include natural gas or biogas as the primary fuel with diesel as the pilot, or even combinations of gasoline and diesel in more advanced strategies like RCCI. The specific choice of fuel pair is dictated by factors such as availability, cost, desired emissions characteristics, and engine application. The specific manner in which these two fuels are introduced, mixed, and combusted defines the different strategies within the broader realm of dual-fuel technology. This strategic combination of desirable fuel properties aims to optimize the combustion process. The successful implementation of dual-fuel combustion requires careful optimization of various parameters, including the ratio of the two fuels, their injection timing and strategy, and the overall combustion chamber design. Achieving stable and efficient combustion across a wide range of engine operating conditions presents a unique set of engineering challenges. The aim is to harness the benefits of both fuels to achieve improved thermal efficiency, lower emissions, and better control over combustion characteristics [3]. Building upon the fundamental principles of utilizing two fuels to tailor combustion, the specific characteristics and interactions of these fuels within the dual-fuel system give rise to several key mechanisms that govern the combustion process and enable different operational modes. These include:

- **Stratified Reactivity:** The difference in ignition properties between the two fuels allows for more precise combustion phasing [4].
- **Ignition Delay Control:** The pilot fuel controls the ignition timing, which is critical in managing combustion temperature and pressure rise rate.
- **Combustion Modes:** Depending on timing, fuel quantity, and air–fuel mixing, dual-fuel systems can operate in conventional CI mode, partially premixed combustion, or advanced LTC regimes [5].

The successful implementation of dual-fuel combustion requires careful optimization of various parameters, including the ratio of the two fuels, their injection timing and strategy, and the overall combustion chamber design. Achieving stable and efficient combustion across a wide range of engine operating conditions presents a unique set of engineering challenges. In practical applications, the dual-fuel concept, particularly in the context of achieving LTC, is often integrated into a broader engine management strategy known as dual-mode operation (DMO), which combines it with conventional combustion modes. One of the primary benefits of a dual-fuel control strategy is the ability to control combustion phasing. This is accomplished by means of two fuels with dissimilar autoignition properties, typically a main fuel with a high-octane number and a secondary fuel with a low octane number. By blending these fuels in different proportions, the autoignition properties of the mixture can be changed, allowing for control over when combustion starts in the LTC engine [1]. Numerous combinations of fuels have been explored for dual-fuel control, including mixtures of ethanol and n-heptane, *n*-heptane and mineral diesel, gasoline and mineral diesel, and gasoline and natural gas. Studies have stated encouraging outcomes with these combinations, showing varying levels of feasibility. This inherent capability to utilize a wide array of fuel combinations, as comprehensively detailed in Table 7.4, is what grants dual-fuel systems 'unparalleled fuel flexibility.' It signifies that engines are not constrained to a single fuel source but can dynamically adapt to diverse market conditions, energy policies, and resource availability. For instance, the feasibility of pairing conventional diesel with gaseous fuels like natural gas or hydrogen, or even renewable options such as biodiesel and ethanol, provides significant operational resilience. This adaptability allows operators to strategize fuel choices based on cost-effectiveness (e.g., opting for cheaper natural gas when diesel prices are high), leverage regionally abundant alternative fuels, and respond effectively to evolving environmental mandates by choosing cleaner-burning combinations. The sheer variety of viable pairs, from those enhancing efficiency to those drastically cutting specific pollutants, underscores the versatility and future-proofing potential of dual-fuel technology. For example, experiments using ethanol and diesel in Premixed Charge Compression Ignition (PCCI) mode, where ethanol was port injected and mineral diesel was direct injected, demonstrated that the concentration of OH radicals, a key parameter governing HCCI combustion, was significantly controlled by the compression ratio [5,6]. Beyond the combustion control and emission benefits within the LTC mode itself, the dual-fuel concept is often integrated into a broader strategy known as DMO. This approach combines the conventional compression ignition (CI) or spark ignition (SI) combustion modes with the LTC mode.

7.3 CLASSIFICATION OF DUAL-FUEL STRATEGIES

The fundamental principles of dual-fuel combustion are applied in diverse ways in IC engines, resulting in distinct strategies. These strategies are classified based on the methods of fuel introduction, the degree of in-cylinder mixing, and the dominant mode of ignition, each exhibiting unique characteristics in terms of performance, emissions, and fuel flexibility. Dual-fuel combustion strategies in IC engines can be categorized based on engine type (CI or SI), fuel delivery method, and the degree of fuel premixing. The classification helps understand how different combinations and configurations influence combustion behavior, emissions, and engine efficiency [4,7] (Table 7.1).

7.3.1 Compression Ignition (CI)-Based Dual-Fuel Combustion

This represents the most common dual-fuel strategy, predominantly applied in diesel engines. In a typical CI-based dual-fuel system, a gaseous or highly volatile primary fuel (such as methane, natural gas, or hydrogen) is introduced through the engine's intake system. Subsequently, a high-cetane number liquid fuel, commonly diesel, is directly injected into the cylinder toward the end of the compression stroke to initiate ignition [14,15]. A key advantage of this dual-fuel approach, particularly beneficial for advanced combustion strategies like LTC, is the precise control over combustion phasing. This is achieved by leveraging the distinct autoignition properties of the two fuels. It typically involves pairing a primary fuel with a lower autoignition propensity (e.g., a high-octane fuel like gasoline or natural gas) with a secondary fuel that ignites readily under compression (e.g., a high-cetane fuel like diesel). By varying the proportions in which these fuels are introduced into the combustion chamber, the overall autoignition characteristics of the mixture can be precisely modulated.

TABLE 7.1
Causes and Mitigation of Emissions and Combustion Instability

Issue	Main Causes	Mitigation Strategies	References
UHC (unburned hydrocarbons)	• Poor mixture formation (fuel-rich or fuel-lean pockets) • Cold boundary layers and crevices (piston head) • Incomplete combustion due to low temperature or insufficient mixing	• Optimize fuel/air mixing (increase injection velocity, reduce droplet size) • Minimize crevice volumes (piston design) • Increase combustion temperature (raise intake/coolant temp, compression ratio) • Exhaust gas recirculation (EGR) to reduce cold regions (with care for CO) • Boost pressure to enhance mixing and chemical rates	[8,9]
CO (carbon monoxide)	• Incomplete oxidation of fuel (often from fuel-rich zones) • Low combustion temperature (due to high EGR or lean operation) • Short combustion duration at high speed	• Improve mixing and combustion temperature (optimize injection timing and boost) • Control EGR rate to avoid excessive oxygen depletion • Increase combustion duration (optimize engine speed/load)	[8,10,11]
Combustion instability	• Rapid or uneven heat release (from high hydrogen content or poor stratification) • Inadequate control of ignition timing • Variations in fuel properties or mixture composition	• Optimize fuel blend ratios (hydrogen/natural gas) • Refine injection timing and duration • Use advanced control strategies (closed-loop feedback, real-time monitoring) • Engine and chamber design modifications to dampen pressure oscillations	[12,13]

This capability allows for unparalleled control over the exact timing of combustion initiation, which is paramount for achieving stringent emission targets and high efficiencies in LTC engines [16,17]. The versatility of CI-based dual-fuel control has led to extensive investigation into various fuel combinations. Research has explored mixtures such as n-heptane and mineral diesel, gasoline and mineral diesel, and gasoline and natural gas. More advanced RCCI applications, a subset of CI-based LTC, have notably included a wide range of pairings. These often involve a low-reactivity fuel (LRF) (like gasoline or specific alcohols) as the dominant primary fuel with a high-reactivity fuel (HRF) (like diesel or n-heptane) as the secondary. Specific examples of successful investigations include DME with diesel blends, natural gas with diesel in heavy-duty RCCI applications, and various alcohols such as methanol, ethanol, and butanol paired with diesel in RCCI setups. Studies consistently report positive results with these diverse combinations, demonstrating varying yet promising levels of feasibility for achieving desired combustion characteristics and emissions benefits.

7.3.2 Spark Ignition (SI)-Based Dual-Fuel Combustion

SI-based dual-fuel combustion systems represent a distinct approach to leveraging multiple fuels for engine optimization. In these engines, a primary fuel (such as gasoline, ethanol, or compressed natural gas (CNG)) is typically mixed with air and subsequently ignited by a spark plug. A secondary fuel, often with different combustion characteristics like hydrogen, ethanol, or LPG, is then introduced to achieve specific performance or emissions benefits, such as improving knock resistance, enhancing combustion efficiency, or allowing for cleaner fuel substitution. These systems have garnered significant attention for their potential to enhance overall engine efficiency and reduce emissions by integrating alternative fuels with conventional gasoline. For instance, the introduction of a secondary fuel with a higher-octane rating, such as ethanol or natural gas, can significantly improve knock resistance, allowing for higher compression ratios or greater turbocharging, which directly contributes to increased thermal efficiency and power output. This "octane-on-demand" capability allows the engine to optimize its performance across varying load conditions. Furthermore, the use of gaseous fuels as the primary or secondary component can lead to reduced PM emissions due to their cleaner burning characteristics compared to liquid fuels. Research indicates that blending hydrogen with gasoline, for instance, can improve brake thermal efficiency and decrease cycle-by-cycle variations, although this may lead to increased NO_x emissions due to elevated combustion temperatures [18,19].

Similarly, studies have demonstrated that incorporating natural gas or methanol into SI engines results in improvements in combustion stability and reductions in hydrocarbon emissions. Furthermore, advanced ignition technologies are crucial enablers for optimizing SI dual-fuel operation. Ultrafast plasma systems, for example, have shown promise in achieving rapid and efficient combustion, particularly in lean-burn scenarios that are often sought for higher efficiency and lower emissions [15,20]. Computational studies also suggest that hydrogen–ethanol blends can enhance power output by up to 20% compared to hydrogen–methanol mixtures, while concurrently lowering CO and HC emissions. These findings collectively underscore the considerable potential of dual-fuel SI combustion strategies in advancing sustainable engine technologies [21].

7.3.3 HCCI Dual-Fuel Combustion

Homogeneous Charge Compression Ignition (HCCI) represents a highly promising LTC concept, characterized by simultaneous autoignition of a well-mixed fuel–air charge, leading to ultra-low NO_x and soot emissions [22]. However, controlling HCCI combustion, particularly its combustion phasing and operating range, presents significant challenges. Dual-fuel strategies offer a robust solution to these control issues by leveraging the distinct autoignition properties of two fuels. This approach enables more accurate supervision of the combustion process, extending the benefits of HCCI across a broader range of engine loads.

Investigations into HCCI dual-fuel operation have demonstrated compelling improvements in engine performance and emissions. For instance, studies on natural gas engines employing a dual-fuel HCCI mode have shown significant enhancements in fuel efficiency, often by 10%–15% compared to conventional SI operation [23]. This improvement is primarily attributed to reduced throttling losses and the leaner combustion inherent to HCCI. Crucially, NO_x emissions in HCCI mode were reported to be over 90% lower than in SI, making this approach highly suitable for meeting stringent emissions regulations. The use of specific fuel pairs, such as natural gas combined with F-T naphtha (enhanced with an additive), has proven effective in achieving controlled autoignition, thereby enabling efficient low-to-moderate load operation without requiring diesel-like compression ratios [24]. This also facilitates smooth transitions between SI and HCCI modes, ensuring high-load power output remains uncompromised. Further research underscores the sophisticated control achievable in HCCI dual-fuel engines through careful fuel selection and blending. For example, studies exploring iso-octane and *n*-heptane as a dual-fuel pair in HCCI engines have confirmed that adjusting the fuel reactivity ratio provides effective control over combustion phasing and heat release [25]. By ensuring a homogeneous charge through premixing and single-port injection, and employing a closed-loop feedback system with cylinder pressure sensors, dynamic optimization of the fuel ratio was achieved to maximize work output. Such optimized approaches have been shown to significantly expand the HCCI engine's operating range, potentially allowing it to achieve approximately 60% of the load range of a boosted SI engine, thus demonstrating clear potential for enhancing efficiency while maintaining stable combustion control. The use of advanced computational models, such as genetic algorithms to fine-tune stochastic reactor models, has further improved the prediction accuracy for multi-cycle HCCI operation [23]. Despite these advancements, practical application of HCCI dual-fuel combustion still faces several hurdles. While NO_x emissions are remarkably low, there is a consistent challenge with increased hydrocarbon (HC) and carbon monoxide (CO) emissions in HCCI mode, often necessitating the integration of oxidation catalysts for posttreatment. Uneven combustion, stemming from air and fuel distribution issues, can lead to misfires and knocking in certain cylinders, impacting stability. Operational complexities, such as the need for SI during startup before transitioning to HCCI, limit seamless low-load performance. Moreover, the availability of certain optimal secondary fuels (like F-T naphtha with specific additives) may pose commercial feasibility challenges. To overcome these limitations, future research must focus on improving fuel options, refining fuel-blending and combustion control strategies (especially to address cylinder-to-cylinder variations), and exploring advanced engine designs like variable compression ratios (VCRs) and optimized injection systems to expand the HCCI load range. Continued efforts are also needed in integrating real-time combustion feedback systems for enhanced operational reliability and developing advanced emission control strategies for CO and HC, as well as low-temperature NO_x control, if applicable to specific fuel sets.

7.3.4 Reactivity Controlled Compression Ignition (RCCI)

RCCI is a progressive dual-fuel combustion technology specifically engineered to achieve highly optimized combustion through precise in-cylinder fuel blending and reactivity management. This strategy utilizes two fuels with significantly dissimilar autoignition properties to shape the combustion duration and magnitude, ultimately leading to high thermal efficiency and ultra-low nitrogen oxides (NO_x) and PM emissions (see Figure 7.2). The core principle of RCCI involves introducing a LRF, such as gasoline or biogas, predominantly into the intake manifold via a port fuel injector, where it premixes with air. This homogeneous charge is then drawn inside the cylinder through the intake stroke. Subsequently, a HRF, typically diesel, biodiesel, or a specific blend, is directly injected into the cylinder during the compression stroke. This direct injection can employ single or multiple strategies, with precise timing crucial for governing the reactivity stratification within the cylinder.

As depicted in Figure 7.3, the mixture reactivity peaks around the bowl rim, leading to a higher concentration of autoignition pockets in that area.

The concept of reactivity stratification is fundamental to RCCI. While global reactivity refers to the overall autoignition propensity determined purely by the total proportions of the fuels, reactivity stratification specifically relates to the spatial variation of fuel reactivity across the combustion chamber. This layering is created by the direct injection of HRF into the LRF/air mixture. For instance, early injected HRF might reach the squish region, while later injections serve as ignition sources,

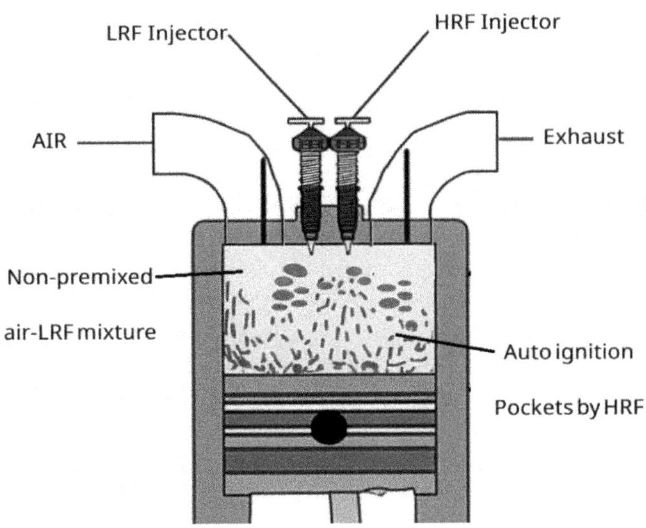

FIGURE 7.2 Schematic diagram of RCCI engine.

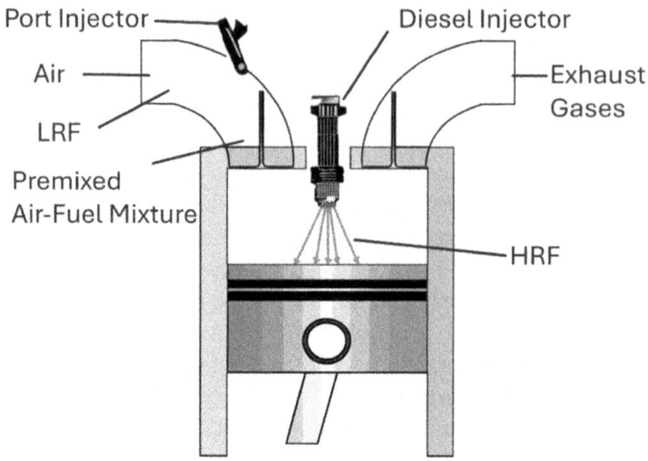

FIGURE 7.3 Autoignition and staged ignition initiated through localized combustion zones.

resulting in a spatially varied reactivity profile. This stratification is significantly influenced by in-cylinder mixing dynamics, engine load, equivalence ratio, fuel blends, and temperature (see Figure 7.3). During the RCCI combustion process, ignition initiates preferentially in areas with a higher concentration of the more reactive HRF and then propagates to the surrounding LRF regions [26]. This controlled, distributed ignition enables a more expansive and gradual combustion of the premixed charge, which yields high thermal efficiency, lower peak pressure rise rates, and consequently, reduced exhaust emissions.

Studies have extensively explored various fuel combinations and their impact on RCCI performance. For example, investigations into direct-injected natural gas as an HRF have shown that gaseous fuel stratification can significantly improve combustion efficiency, particularly at low loads, while simultaneously reducing methane and CO emissions, along with a decrease in NO_x [25]. While these studies note a slight increase in unburnt hydrocarbons (UHC) in some scenarios, the overall potential for cleaner combustion remains high. Autoignition in RCCI effectively combines principles from both conventional diesel and HCCI engines: it involves port injection of a homogeneous low-cetane fuel charge (akin to HCCI) and direct injection of a high-cetane fuel to create charge stratification (similar to conventional diesel). The HRF, injected at high pressure, tends to concentrate around the piston bowl rim area, making the mixture most reactive there, leading to numerous autoignition pockets. The heat release rate (HRR) in RCCI combustion is distinctively shaped by the formation of these fuel-rich zones and the strategic two-stage ignition process. Initially, the HRF drives a controlled low-temperature heat release, which then transitions into a high-temperature heat release from the combustion of the premixed LRF. This staged approach extends combustion duration and lowers peak HRR, thereby minimizing pressure rise rates and mitigating engine knocking. Optimal control of injection timing, fuel reactivity ratios, and thermal management are paramount for shaping the HRR, maximizing efficiency, and minimizing emissions in RCCI engines.

TABLE 7.2
Comparison of Premixed and Stratified Combustion

Feature/Aspect	Premixed Dual-Fuel Combustion	Stratified Dual-Fuel Combustion
Fuel introduction and mixing	A low-reactivity fuel (LRF) (e.g., natural gas and gasoline) is uniformly mixed with air before entering the combustion chamber [27]	LRF is port-injected to form a lean background mixture; high-reactivity fuel (HRF) is directly injected to create rich local zones [25,27]
Ignition mechanism	A small quantity of HRF (e.g., diesel) is injected to initiate combustion within the lean homogeneous mixture [27,28]	Combustion begins in the rich zones formed by the directly injected HRF and propagates through the stratified charge
Combustion characteristics	Rapid and complete combustion due to homogeneity. Can lead to high pressure rise rates	Controlled combustion due to varying fuel-air concentrations. Aims to reduce pressure rise rates and avoid knocking
Primary advantages	High thermal efficiency; Simultaneous reduction in NO_x and particulate matter (PM) emissions (due to lean, homogeneous burn); Potentially simpler in-cylinder flow management [28]	Enables higher load operations; More controlled combustion process; Reliable ignition even under lean overall conditions (due to rich zones near ignition source) [29]
Key challenges	The uniformity of the mixture can result in high pressure rise rates, limiting the engine's maximum load capacity; Potential for combustion instability (e.g., knocking or misfires) [28,30]	Complexity of managing multiple injections and ensuring optimal stratification; Achieving and maintaining optimal stratification across the operating map can be challenging [31,32]
Typical goal	Maximize efficiency and minimize emissions under lean conditions, typically at low to moderate loads	Extend operating range to higher loads, enhance combustion stability, and reduce pressure rise rates

7.3.5 Premixed vs. Stratified Dual-Fuel Approaches

Dual-fuel combustion strategies have emerged as promising solutions for enhancing engine efficiency and reducing emissions. Among these, premixed and stratified dual-fuel approaches stand out, each offering distinct advantages and challenges (Table 7.2).

7.4 VARIABLES IN DUAL-FUEL ENGINES

Dual fuel combustion is influenced directly or indirectly by numerous factors. Key controllable variables significantly impact the intricate combustion behavior and emissions characteristics of dual-fuel engines, parameters frequently investigated by researchers in experimental and computational studies.

7.4.1 Fuel Injection Timing

The precise timing of HRF injection, typically defined by its crank angle degree, is a supremely critical variable for optimizing the performance of RCCI engines. This parameter directly governs the initiation and subsequent phasing of the dual-fuel combustion event, thereby profoundly affecting crucial combustion metrics such as the HRR and its distribution. Strategic adjustment of HRF injection timing, which can involve single or multiple (split) injections (both advanced and retarded), offers a powerful mechanism to control the overall combustion phase [33]. Proper optimization can significantly contribute to the reduction of undesirable unburned hydrocarbons (UHC) and carbon monoxide (CO) emissions, often without notably increasing nitrogen oxides (NO_x) due to the nature of distributed combustion. However, overly advanced HRF injection timing can sometimes lead to a higher peak pressure rise rate, demanding careful calibration. For instance, studies have demonstrated that an optimal HRF injection timing of 53° BTDC, combined with specific injection pressure (580 bar) and premixed ratio (0.76), led to substantial emission reductions (23% for UHC and 39% for CO) alongside marked improvements in combustion efficiency and gross indicated efficiency [33].

7.4.2 Fuel Injection Pulses

The strategic manipulation of the number of HRF injection pulses holds considerable influence over the in-cylinder reactivity profile and the overall dual-fuel combustion process in RCCI engines. Research indicates that employing multiple injection pulses (e.g., double or triple injections) can provide superior control over the heat release and offer benefits for operation at low to mid-engine loads, enabling finer modulation of reactivity gradients [34]. Conversely, a single injection pulse strategy is often found to be more appropriate for high-load conditions, where it helps in mitigating undesirable excessive pressure spikes while maintaining robust combustion. For example, studies utilizing palm oil dimethyl ester have reported improved efficiency and extended load capability when a single injection strategy was carefully employed, showcasing the impact of pulse number on overall engine operability [25].

7.4.3 Fuel Injection Pressure

The pressure at which the pilot fuel is injected into the cylinder is a pivotal parameter in optimizing RCCI engine performance and fuel economy. It directly affects the atomization quality of the HRF spray, its penetration, and ultimately, the degree of fuel–air mixing and reactivity stratification within the dual-fuel charge. Poorghasemi et al. notably found that reducing injection pressure from 450 to 300 bar could yield a 5% increase in gross indicated efficiency and a 4°CA delay in the combustion phase [35]. This adjustment, when coupled with an advanced injection timing, also contributed to reduced UHC, CO, and NO_x emissions. Conversely, findings by Zhang et al. highlight the critical balance required: excessively low injection pressure can lead to poor fuel atomization and incomplete combustion, while overly high pressure might result in fuel misdistribution or spray impingement [36]. Overall, while direct injection pressure itself may have a more limited direct influence on gross combustion phasing, its precise control is instrumental in enabling more controlled combustion, enhancing thermal efficiency, and extending the dual-fuel engine's operational range.

7.4.4 Fuel Spray Angle

The HRF spray angle, which is determined by the atomizer geometry, supply conditions, and fuel properties, represents a critical variable influencing the spatial distribution of fuel droplets and subsequent fuel–air mixing within the dual-fuel charge. A wider spray angle generally promotes broader fuel–air interaction and a more distributed ignition, thereby enhancing overall combustion efficiency and mitigating local hot spots. Conversely, a narrower angle can lead to more concentrated reactivity zones. As demonstrated through a 3D CFD model, reducing the spray angle from 74° to 55° resulted in a significant 27% decrease in UHC and carbon monoxide (CO) [33]. This optimization was also associated with improved combustion characteristics and higher gross indicated efficiency. While the initial, low-temperature stage of heat release might remain relatively unaffected by spray angle changes, its profound influence on the overall combustion phase underscores the paramount importance of carefully optimizing this parameter for desired performance and emissions outcomes in dual-fuel RCCI engines.

7.4.5 Fuels Energy Fraction

The fuels energy fraction, defined as the ratio of energy supplied by one fuel (e.g., the premixed or pilot-injected fuel) to the total energy input to the engine, plays a singularly critical role in dual-fuel RCCI combustion. This ratio directly dictates the overall in-cylinder reactivity of the charge, which is intimately tied to the cetane number characteristics of the fuels involved. By judiciously adjusting this energy proportion, engineers can precisely manipulate key combustion attributes such as ignition delay, peak in-cylinder pressure and temperature development, and the overall shape of the heat release profile. For example, it has been observed that increasing the diesel mass fraction (which represents the HRF energy contribution) at an early injection timing effectively advanced combustion and slightly increased the peak pressure rise rate [37]. Beyond its profound influence on combustion control, optimizing the energy fraction also serves as a vital tool in emissions management, significantly contributing to the reduction of problematic UHC and carbon monoxide (CO) emissions.

7.4.6 Premixed Air to Fuel Ratio

Distinct from the energy fraction, the premixed ratio specifically quantifies the amount of LRF that is thoroughly mixed with air before entering the cylinder. This parameter profoundly influences the homogeneity and reactivity of the main charge, thereby critically affecting the subsequent autoignition and combustion phases in RCCI engines. Studies prominently highlight its central role in achieving improved combustion stability, enhancing overall engine efficiency, and effectively reducing ringing intensity by contributing to lower peak pressure rise rates [33,38]. Furthermore, increasing the premixed ratio generally contributes to the desired reduction of NO_x and soot emissions, characteristic of LTC modes. However, it is an important consideration that excessively high premixed ratios can sometimes lead to increased UHC and carbon monoxide (CO) emissions because of flame quenching or inadequate combustion in lean zones, a challenge actively addressed through continuous advancements in dual-fuel mixture preparation techniques.

7.5 CONTROL STRATEGIES

Dual-fuel combustion, particularly in advanced modes like RCCI, demands sophisticated control to harness its benefits. Given the inherent complexities of dual-fuel combustion, precise control of the combustion phase becomes paramount, fundamentally determining engine efficiency, reliability, and overall operational characteristics.

7.5.1 Split Injection

Split Injection (SPI) strategy is a highly effective technique employed in dual-fuel RCCI engines to precisely manage and optimize the combustion process. By dividing the HRF injection into two or more distinct pulses at carefully timed intervals, this strategy allows for highly refined control over both the ignition delay and the overall combustion phasing. Typically, the initial injection introduces a portion of the HRF (such as diesel) early in the compression stroke. This early pulse forms a strategically premixed region with the LRF, like iso-butanol or gasoline, creating a foundation for controlled autoignition. A subsequent injection, timed closer to TDC, then precisely initiates the main combustion event by creating local zones of even higher temperature and reactivity, facilitating a more controlled and spatially uniform burn throughout the cylinder.

A key benefit of this staged approach is the significant reduction in peak pressure rise rates and associated combustion noise, a critical aspect for engine durability and refinement. Research highlights that relying solely on single injection events often limits load flexibility due to more abrupt and uncontrolled pressure changes [39,40]. Moreover, delaying part of the HRF injection has been shown to notably reduce NO_x emissions, a testament to its ability to spread heat release more evenly and lower peak combustion temperatures. Li et al. also observed that while SPIs might extend the overall ignition delay, this characteristic can be advantageous in spreading the heat release more uniformly over a longer crank angle duration [41]. This distributed heat release can lead to improved thermal efficiency and a reduction in PM, UHC, and carbon monoxide (CO), particularly when other operating parameters like injection pressure and exhaust gas recirculation (EGR) rates are concurrently optimized. Under EGR conditions, SPI strategies achieved up to a 30.0% reduction in PM emissions [42]. This improvement is primarily resulted because of enhanced in-cylinder pressure and temperature, and improved air–fuel mixing. The advanced combustion phasing enabled by SPI leads to shorter combustion durations and higher peak in-cylinder pressures, which in turn elevate the combustion temperature—conditions that significantly promote PM oxidation. Additionally, by dividing the main fuel charge into multiple smaller pulses, SPI minimizes spray overlap and allows each injection to interact with relatively fresh air. This reduces localized rich zones and suppresses soot formation. Together, these thermodynamic and spray-driven effects make SPI an effective strategy for achieving cleaner combustion compared to single-injection approaches. At a high natural gas substitution rate of 93.3%, the implementation of a SPI strategy in a medium-speed dual-fuel marine engine led to a substantial 36.3% reduction in NO_x emissions compared to the optimized single injection approach [43]. This improvement is primarily attributed to the early pilot injection, which promotes low-temperature reactions of *n*-heptane in the squish region, generating reactive radicals like CH_2O and OH. The study explores how post injection strategies can improve combustion and reduce particulate emissions in an EGR-equipped diesel engine during transient operations (rapid load changes). It finds that short loading times (e.g., 3 seconds) significantly worsen particulate, CO, and HC emissions due to intake air lag and poor combustion. Introducing a small post injection (2 mg fuel, 2000 μs after main injection) enhances in-cylinder turbulence and oxygen distribution, which improves combustion and reduces larger particle emissions by up to 50% [44]. However, post injection has minimal effect on NO_x emissions due to already high EGR rates and oxygen deficiency. Overall, post injection is shown to be an effective method for mitigating emission spikes during transient engine operation. In essence, SPI not only offers moderate advancement of combustion phasing but also actively supports cleaner and more efficient dual-fuel engine operation by controlling the reactivity gradients. SPI improves thermal efficiency in light-duty diesel engines by smoothing the in-cylinder pressure rise and distributing heat release more evenly [45]. This reduces peak HRRs, shortens flame penetration, and lowers spray tip velocity, which in turn minimizes heat transfer to the cylinder walls. The study found that an optimized SPI strategy (MSS-7) with an advanced combustion phase reduced fuel consumption by 1.78% compared to a conventional two-pilot injection [42]. This gain in efficiency was attributed to reduced heat transfer loss, lower exhaust heat loss, and decreased combustion loss, especially under low-load and EGR conditions.

7.5.2 Exhaust Gas Recirculation

EGR plays a fundamental role in regulating dual-fuel combustion in RCCI engines, especially crucial under medium to high load conditions where combustion can become excessively rapid. By recirculating a controlled portion of the exhaust gases back into the intake charge, EGR fundamentally alters the in-cylinder mixture by diluting the oxygen concentration and increasing its specific heat capacity. This dilution effectively reduces the global reactivity of the charge, thereby slowing down the combustion rate and significantly mitigating peak pressure rise rates and potentially damaging in-cylinder pressure spikes [46]. Such precise control is particularly vital in RCCI systems, where the stratification of fuel reactivity between the LRF and HRF is already delicately balanced. At higher engine loads, increased EGR rates, when combined with carefully adjusted fuel ratios and advanced diesel injection timing, have consistently demonstrated the ability to suppress premature ignition, consequently extending the viable RCCI operating range to higher power outputs [47]. EGR tends to slightly reduce thermal efficiency in the dual-fuel HCCI engine [17,48]. This is because EGR dilutes the intake charge and lowers the combustion temperature, which leads to incomplete combustion and higher CO and HC emissions—both indicators of unburned fuel and energy loss. The reduced oxygen availability and slower combustion rate also mean less effective conversion of chemical energy into useful work. However, EGR plays a crucial role in enabling stable HCCI operation by moderating ignition timing and preventing knock, especially in the gasoline–diesel dual-fuel setup. While thermal efficiency may drop slightly due to combustion inefficiencies, EGR is essential for achieving controlled, LTC and extending the HCCI operating range.

However, the efficacy of the EGR strategy is heavily dependent on the proper and homogeneous mixing of the recirculated exhaust gases with the fresh intake charge. Inadequate EGR mixing can lead to mixture inhomogeneity and stratification, which may result in uneven ignition timing and a prolonged combustion duration—both highly detrimental to engine stability and emissions performance in a dual-fuel environment. Internally generated EGR, often achieved via throttling or sophisticated valve timing control, can be particularly sensitive to these mixing challenges. While high EGR rates are generally beneficial for reducing NO_x and soot emissions due to lowered combustion temperatures and reduced oxygen availability, excessive use may complicate combustion phasing control and could potentially elevate UHC and CO emissions unless overall combustion temperatures are adequately maintained. Use of EGR in a gasoline–diesel dual-fuel HCCI engine had shown improvement over the combustion control by delaying ignition and lowering peak temperatures, which helps reduce NO_x emissions and supports stable HCCI operation [49]. EGR is especially useful in dual-fuel setups because it moderates the aggressive autoignition behavior of premixed gasoline while allowing diesel to initiate combustion more smoothly. However, this benefit comes with increased CO and HC emissions due to reduced oxygen availability and lower combustion temperatures, which hinder complete fuel oxidation. As reported in various studies, achieving optimal EGR utilization in dual-fuel engines demands a careful and integrated balance between emissions control, combustion efficiency, and mixture uniformity [47,50].

7.5.3 Dual Direct Injection

The dual direct injection (DDI) strategy represents an advanced fuel delivery approach in dual-fuel RCCI engines. This strategy enhances combustion control through precise in-cylinder fuel stratification [46]. DDI achieves this by utilizing distinct injectors—one dedicated to LRF and another for the HRF—allowing for independent regulation over the injection timing, location, and pressure of each fuel. This independent control facilitates the establishment of a highly customized spatial gradient in fuel reactivity across the combustion chamber. This gradient is key to enabling precise modulation of ignition timing and combustion phasing, offering a level of control unparalleled by conventional port-fuel injection for the LRF [46,51]. Furthermore, direct in-cylinder injection of the LRF, as opposed to port injection, significantly mitigates issues related to crevice-induced UHC and carbon monoxide (CO) emissions, as it minimizes the trapping of unburned charge in less-accessible zones of the cylinder.

Extensive experimental studies have robustly demonstrated the effectiveness of DDI in expanding the operating load range of RCCI engines, all while consistently maintaining low emissions and achieving high thermal efficiency. For instance, specific DDI configurations, often employing high EGR rates (e.g., 46%) and optimized injection strategies, have achieved impressive brake thermal efficiencies nearing 49% and indicated mean effective pressure (IMEP) values as high as 21 bar [52].

The inherent cooling effect of early-stage LRF injection, particularly when directed into regions like the squish area, can effectively delay ignition and lower peak pressure rise rates, resulting in smoother and more controlled dual-fuel combustion. Moreover, the dynamic modulation of the LRF's injection pressure and timing offers additional layers of control over the ignition delay and the overall heat release characteristics. Fundamentally, DDI presents a highly promising pathway to achieving cleaner, more efficient dual-fuel combustion by offering superior control over reactivity gradients and intricate combustion dynamics.

7.5.4 Dual-Mode Operation

The DMO strategy offers a dynamic and intelligent pathway to control the combustion process in dual-fuel engines by enabling seamless transitions between conventional combustion modes such as CI or SI and advanced LTC modes like RCCI. This sophisticated strategy allows the engine to initiate combustion under one regime and then fluidly switch to another as dictated by operating conditions, effectively modulating the combustion phasing to optimize both efficiency and emissions across the entire operating cycle. By leveraging the inherent flexibility provided by both HRF and LRF, DMO helps mitigate the inherent limitations of individual combustion modes when applied across a broad load range, offering refined control over ignition timing and tailored heat release characteristics.

Research has consistently demonstrated that while RCCI operation alone may sometimes lead to increased emissions of CO and UHC, particularly under low-load conditions where combustion efficiency can be a challenge, integrating it within a dual-mode framework significantly enhances overall engine performance and emissions compliance. For example, studies revealed that while simple RCCI/conventional diesel switching might result in higher UHC and CO emissions, a more structured dual-mode, dual-fuel strategy can overcome these issues [53]. Conversely, Benajes et al. found that employing DMO with high-ethanol blends, such as E85, led to substantial reductions in nitrogen oxides (NO_x) and soot, alongside marked improvements in thermal efficiency, significant achievements often attained without the need for complex exhaust aftertreatment systems [54]. Furthermore, coupling DMO with complementary strategies like EGR and DDI has been shown to synergistically suppress peak pressure rise rates, improve combustion stability, and further reduce CO and UHC emissions, thereby maximizing the comprehensive benefits of advanced LTC regimes.

7.5.5 Variable Valve Timing

Variable valve timing (VVT) is a pivotal technology in managing dual-fuel combustion, particularly within RCCI engines. By dynamically fine-tuning the precise timing of intake and exhaust valve events, VVT profoundly influences the in-cylinder thermodynamic conditions, thereby enabling precise control over the combustion processes unique to dual-fuel operation. For instance, an early intake valve closing event can effectively lower the effective compression ratio. This reduction in compression ratio can in turn decrease peak combustion temperatures, which is instrumental in mitigating the formation of NO_x, a key challenge in IC engines [55]. Conversely, implementing late intake valve closing can extend the ignition delay period, promoting more thorough fuel–air mixing and consequently contributing to a reduction in soot emissions within the dual-fuel charge. These dynamic adjustments to valve timing allow for a far more flexible and efficient combustion process across the diverse range of engine loads encountered in real-world operation. In the intricate context of dual-fuel RCCI engines, the integration of VVT has demonstrated significant and multifaceted benefits. Numerous studies have shown that combining VVT with RCCI strategies enables engines to not only meet but often surpass stringent emission standards, such as the Euro VI regulations, by consistently achieving ultra-low NO_x and soot emissions, particularly at low and medium loads where controlling these pollutants can be challenging [55]. Beyond emissions, VVT actively contributes to improved thermal efficiency and enhanced fuel economy by optimizing combustion phasing to align with engine load and speed demands, often reducing the necessity for excessively high EGR rates, which can sometimes compromise efficiency. The inherent adaptability of VVT also plays a crucial role in managing combustion stability and reducing pressure rise rates, which are critical parameters for maintaining engine durability and ensuring consistent performance in highly reactive dual-fuel combustion environments [55]. Overall, VVT serves as an indispensable and highly effective tool for enhancing the precision control and overall efficiency of modern dual-fuel combustion systems (Table 7.3).

7.5.6 The Comparative Impact on Emission Reduction and Thermal Efficiency on Various Combustion Control Strategies

TABLE 7.3
Effect of Combustion Strategies on Thermal Efficiency and Emissions in IC Engines

Strategy	Thermal Efficiency	Emissions	Quantitative Impact	References
Split injection	Slight ↓ or stable depending on conditions	PM ↓ up to 30.0%. NO_x ↓ by 36.3%. and HC ↓	• NO_x ↓ 1400 ppm to 200 ppm with 10% EGR at 350 bar injection pressure • Smoke ↓ by 10% vs. single injection. • Thermal efficiency 29%–30% (trade-off with injection type)	[43,56,57]

(Continued)

TABLE 7.3 (*Continued*)
Effect of Combustion Strategies on Thermal Efficiency and Emissions in IC Engines

Strategy	Thermal Efficiency	Emissions	Quantitative Impact	References
Exhaust gas recirculation (EGR)	Slight ↓ in dual-fuel HCCI. ↑ by ~1% (on average)	↓ NO_x, PM, HC and CO ↑	• NO_x ↓ (up to 85% reduction in cited cases) • PM ↓ • HC and CO ↑ • Fuel consumption ↓ by 0.1% • Thermal efficiency: Improved by ~1 percentage point	[17,57,58]
Dual direct injection	Potentially ↑ (e.g., nearing 49% Brake Thermal Efficiency, 21 bar IMEP) with better mixture and combustion	↓ HC, CO, NO_x	• HC and CO ↓ due to improved fuel–air mixing • NO_x ↓ with proper strategy	[52,56]
Variable valve timing	↑ but ↓ pumping losses, better volumetric efficiency, Also contributes to enhanced fuel economy	↓ HC, CO, NO_x	• HC, CO, and NO_x ↓ through optimized valve timing and overlap. • Thermal efficiency ↑ by several percent depending on operating range	[55,59,60]
Dual-mode operation (DMO)	Marked improvements in thermal efficiency (e.g., with E85 blends)	Substantial ↓ NO_x and soot (with E85 blends). Also can ↓ CO and UHC synergistically when coupled with other strategies	Markedly improves thermal efficiency. Can also synergistically suppress peak pressure rise rates, improve combustion stability, and further reduce CO and UHC emissions when coupled with EGR and DDI	[53,54]

7.5.7 Other Potential Strategies

Beyond the primary strategies detailed, several other promising techniques are explored to further optimize dual-fuel combustion. In some RCCI implementations, a single fuel strategy might employ cetane number improvers to enhance the reactivity of a typically LRF, effectively substituting the role of a HR) and simplifying fuel delivery systems. Similarly, other fuel additives can be utilized to fine-tune the reactivity profile of blended fuels, offering another layer of control over the combustion process. The VCR technique also demonstrates considerable potential in improving overall dual-fuel combustion performance. While VCR can sometimes lead to a surge in NO_x releases due to potentially higher in-cylinder temperatures, it has shown clear benefits in reducing CO, HC, and soot emissions, advantages that extend effectively to RCCI operation by optimizing the pressure-temperature history. Furthermore, the implementation of robust feedback and advanced control systems is absolutely essential for achieving the precise reactivity and injection timing control required in dual-fuel engines. These sophisticated control architectures are paramount for optimizing combustion performance not only during steady-state operations but, critically, also during dynamic and transient conditions, ensuring smooth transitions and consistent efficiency across varied demands.

While each of these control strategies offers distinct advantages for dual-fuel engine management, optimal performance, emissions, and operating range extension are often achieved through their synergistic combination. For instance, VVT is frequently utilized to achieve internal EGR, demonstrating how a single mechanism can serve multiple control objectives. Similarly, SPI tactics are commonly implemented within DDI systems to fine-tune in-cylinder reactivity and heat release. DMO then intelligently orchestrates the seamless transition between various combustion regimes, dynamically selecting and optimizing the application of these individual control levers to meet diverse engine operating demands and environmental regulations. This integrated approach underscores the complex, yet powerful, control landscape available in modern dual-fuel engine development.

7.6 ADVANTAGES AND CHALLENGES

Dual-fuel combustion approaches have emerged as a highly capable pathway for the near-to-medium term implementation of LTC in contemporary IC engines. This approach offers a compelling solution for the relatively rapid adaptation of advanced combustion concepts into production-grade solutions [61,62].

7.6.1 Advantages of Dual-Fuel Combustion

The integration of two distinct fuels in engine operation offers a multitude of significant benefits:

- **Exceptional Emissions Reduction:** A primary advantage of dual-fuel combustion, particularly when leveraging LTC modes like RCCI, is its profound potential for achieving ultra-low NO_x and PM emissions. This is mainly attributed to the significantly reduced peak combustion temperatures and the more distributed, homogeneous heat release characteristic of these combustion strategies [63,64].

- **Enhanced Thermal Efficiency:** Under optimized operating conditions, dual-fuel engines consistently demonstrate improved thermal efficiency. This gain often stems from factors such as reduced throttling losses (in premixed systems), more favorable combustion phasing, and improved HRRs, leading to more efficient conversion of fuel energy into useful work.
- **Superior Fuel Flexibility**: Dual-fuel systems inherently offer remarkable fuel flexibility, enabling the concurrent or alternating use of various energy sources. This versatility, evident in the wide range of fuel combinations explored and proven effective (e.g., diesel with natural gas, hydrogen, or various alcohols, as summarized in Table 7.4), allows for the strategic substitution of more expensive or polluting conventional fuels (like diesel) with cleaner, more readily available, or economically advantageous alternatives (such as natural gas, hydrogen, ethanol, or biogas). This versatility is crucial for adapting to evolving fuel markets and environmental regulations. Allows for the strategic substitution of more expensive or polluting conventional fuels (like diesel) with cleaner, more readily available, or economically advantageous alternatives (such as natural gas, hydrogen, ethanol, or biogas). This versatility is crucial for adapting to evolving fuel markets and environmental regulations. Some of the feasible fuel combinations that have been studied are shown in the Table 7.4.

TABLE 7.4
Emission Characteristics of Dual-Fuel Combinations

Primary Fuel	Secondary Fuel	Effect on Emissions	References
Diesel	Natural gas (CNG)	• ↓ NO_x, CO_2, PM • ↑ HC, CO	[65,66]
Diesel	Biodiesel (B20/B100)	• ↓ Smoke, NO_x, PM • ↑ CO, HC (less than CNG) • Improved combustion stability	[67,68]
Diesel	Hydrogen	• ↓ CO_2, PM, HC • ↑ NO_x (at high hydrogen ratios)	[69]
Diesel	Ethanol	• ↓ NO_x, PM, CO_2 • ↑ HC, CO	[66,69]
Diesel	Methanol	• ↓ NO_x, PM, CO_2 • ↑ HC, CO	[70]
Diesel	Propane (LPG)	• ↓ NO_x, PM, CO_2 • ↑ HC, CO	[71]

(*Continued*)

TABLE 7.4 (*Continued*)
Emission Characteristics of Dual-Fuel Combinations

Primary Fuel	Secondary Fuel	Effect on Emissions	References
Diesel	Dimethyl ether (DME)	• ↓ NO_x, PM, CO_2 • ↓ HC, CO	[72]
Diesel	Biogas	• ↓ NO_x, PM, CO_2 • ↑ HC, CO	[73]
Diesel	Ammonia	• ↓ NO, CO, Soot • ↓ Combustion efficiency at high ammonia ratios	[74]
Diesel	Isobutanol	• ↓ Smoke, NO_x, PM, • ↑ CO, HC (less than CNG)	[68]
Diesel	Crude palm oil biodiesel	• ↓ NO_x, PM, CO_2 • ↑ HC, CO. (improved with extra combustion air)	[67]
Gasoline	Ethanol	• ↓ NO_x, PM, CO_2 • ↑ HC, CO	[69]
Natural Gas	Hydrogen	• ↓ CO_2, PM, HC, • ↑ NO_x (at high hydrogen ratios)	[12]
Diesel	Fischer–Tropsch synthetic fuel	• ↓ NO_x, PM, CO_2↓ HC, CO	[75]

- **Reduced Knocking Tendency:** In dual-fuel SI engines, the utilization of a low-reactivity primary fuel (e.g., natural gas) in conjunction with a small pilot ignition source effectively mitigates the knocking tendency associated with conventional SI engines. This allows for higher compression ratios or greater turbocharging, directly contributing to increased power output and efficiency without combustion instability.
- **Extended Operating Range Potential:** Strategies like DDI and DMO have demonstrated significant potential in expanding the viable operating range of LTC engines, pushing the boundaries of what single-fuel LTC can achieve. This allows for the benefits of LTC (high efficiency, low emissions) to be extended to broader engine loads. For instance, configurations employing DDI with high EGR rates have achieved impressive brake thermal efficiencies nearing 49% and IMEP values as high as 21 bar. Similarly, DMO with ethanol blends like E85 has resulted in substantial reductions in NO_x and soot, coupled with improved thermal efficiency, sometimes even eliminating the need for complex exhaust aftertreatment.

7.6.2 Implications of RCCI on Economics Durability and Safety

While the technical advantages of dual-fuel combustion, particularly RCCI, are significant, their widespread real-world deployment necessitates a critical examination of practical implications related to economics, long-term durability, and operational safety. Providing a holistic view, these factors represent crucial trade-offs for successful commercial adoption.

7.6.2.1 Economic Considerations

While dual-fuel systems offer the potential to reduce fuel costs by enabling the use of less expensive or renewable fuels, they inherently demand more complex fuel delivery and control systems [76]. This increased complexity, in turn, leads to higher initial investment and overall maintenance costs. Furthermore, the necessary infrastructure for alternative fuels, such as CNG or hydrogen, may not be universally available, posing additional deployment costs and logistical hurdles [77].

7.6.2.2 Durability and Maintenance

The unique operating conditions and varied fuel properties associated with dual-fuel engines can significantly impact engine component wear, injector life, and compatibility with exhaust aftertreatment systems [78]. For instance, certain alternative fuels like alcohols can be corrosive to fuel lines and seals, while gaseous fuels may contribute to accelerated valve seat recession due to reduced lubrication. Consequently, dual-fuel operation may necessitate more frequent or specialized maintenance, stemming from the increased complexity of the integrated fuel and control systems, which can affect overall operational uptime and lifecycle costs [79].

7.6.2.3 Safety Implications

The handling and storage of multiple fuels, especially gaseous or highly volatile ones, such as hydrogen or natural gas, introduce distinct safety challenges, including inherent risks of leaks or fire hazards [60]. Therefore, system design must rigorously incorporate robust safety features. These include advanced leak detection systems, automatic fuel shut-off mechanisms, and proper ventilation in enclosed spaces. Ensuring comprehensive safety protocols and providing adequate training for personnel involved in the operation and maintenance of these systems are paramount for safe deployment [60].

7.6.3 Challenges in Dual-Fuel Combustion

Despite their numerous advantages, realizing the full potential of dual-fuel LTC engines necessitates addressing several formidable technical and practical challenges:

- **Combustion Control and Stability:** One of the most significant hurdles in dual-fuel LTC, particularly for modes like RCCI, is the precise control and stability of combustion. Unlike conventional engines that rely on direct SI or pilot-diesel injection timing for direct control, LTC ignition timing and combustion rates are largely governed by complex chemical kinetics [16]. This inherent reliance on autoignition chemistry makes it challenging to achieve consistent and repeatable combustion phasing across wide variations in engine loads, speeds, and ambient conditions [6,80]. This often manifests as combustion instability at light loads due to excessively lean mixtures, or severe knocking/ringing at higher loads due to overly rapid heat release. The underlying causes include rapid or uneven heat release (which can stem from high hydrogen content or poor stratification) and inadequate control of ignition timing, coupled with variations in fuel properties or mixture composition [79]. Addressing these issues requires sophisticated control strategies.

Specifically, advanced feedback control systems, utilizing real-time data from in-cylinder pressure sensors, are crucial for dynamically adjusting parameters such as fuel blend ratios, injection timing, and duration [79]. Furthermore, engine and combustion chamber design modifications can help to dampen pressure oscillations and enhance stability. Optimizing the pilot fuel quantity is also critical; too much can lead to increased NO_x, while too little may result in misfire or incomplete combustion.

- **Emissions Profile** and **Aftertreatment Limitations:** While effective in reducing NO_x and soot, dual-fuel LTC engines frequently exhibit elevated emissions of UHC and carbon monoxide (CO), especially at low engine loads. This is often attributed to incomplete combustion stemming from overly lean air-fuel mixtures or difficulties in achieving consistent ignition of the gaseous fuel with minimal pilot injection [48]. More precisely, primary causes for UHC and CO include poor mixture formation (leading to fuel-rich or fuel-lean pockets), flame quenching in cold boundary layers and crevices (e.g., piston head), and incomplete combustion due to low temperatures or insufficient mixing [81]. To mitigate these emissions, strategies focus on optimizing fuel/air mixing (e.g., increasing injection velocity or reducing droplet size), minimizing crevice volumes through piston design, and increasing combustion temperature (e.g., by raising intake or coolant temperature or compression ratio). Careful application of EGR can also reduce cold regions, thereby enhancing chemical rates and overall combustion efficiency. Additionally, controlling EGR rates to avoid excessive oxygen depletion and optimizing injection timing and boost pressure are crucial for improving mixing and combustion temperature [82]. Furthermore, the inherently lower exhaust gas temperatures associated with LTC, while beneficial for engine-out NO_x, complicate the effectiveness of conventional aftertreatment systems designed to oxidize UHC and CO [83,84]. In natural gas-diesel dual-fuel engines, a specific concern is methane slip, where significant amounts of unburned methane can be emitted, particularly at low to medium loads. This undermines the greenhouse gas benefits, as methane has a high global warming potential [64].
- **Cold Start and Load Limitations:** Achieving reliable cold start performance presents a notable concern for dual-fuel LTC engines. At lower ambient temperatures, attaining the necessary in-cylinder conditions for consistent autoignition becomes challenging, often leading to increased misfires and elevated emissions during startup. Moreover, the operating range of dual-fuel LTC engines is typically limited to low-to-medium loads. At higher loads, the rapid and uncontrolled heat release inherent to highly premixed, autoigniting charges can cause excessive pressure rise rates, leading to severe engine knock and a resurgence of NO_x emissions [85].
- **Mode Transition and Transient Operation Complexity:** For dual-fuel engines employing DMO, developing robust control systems for smooth and seamless transitions between different combustion modes (e.g., LTC to conventional CI/SI) is a significant hurdle for commercialization. Achieving combustion robustness throughout these transitions and ensuring reliable mode transition to conventional combustion at high loads are necessary

steps for practical application [43,48]. Strategies are continuously needed to manage these mode changes to ensure satisfactory engine performance while minimizing undesirable spikes in NO_x and smoke emissions during the switch. Transient operation (rapid changes in load or speed) similarly requires advanced control strategies to maintain optimal performance and emissions.

- **Fuel Reactivity Management and System Complexity:** Dual-fuel LTC strategies fundamentally rely on the precise management of fuels with differing reactivities to control combustion phasing and heat release. This necessitates inherently complex fuel delivery systems, often involving multiple injectors (e.g., DDI) and precise control over fuel stratification [86]. Implementing such intricate systems significantly increases the overall complexity and cost of the engine. Additionally, the need for two separate fuel storage systems poses considerable packaging, infrastructure, and logistical challenges, impacting the commercial viability and broad adoption of these technologies.
- **Fuel-Specific Limitations**: Certain dual-fuel combinations introduce unique challenges. For instance, in hydrogen-diesel dual-fuel engines, hydrogen's exceptionally low density can displace a considerable amount of intake air, potentially reducing volumetric efficiency and overall engine power. Furthermore, hydrogen's high flame speed may lead to issues like pre-ignition and backfire, requiring specialized engine designs and control strategies [6,54].

Addressing these multifaceted challenges requires ongoing advancements in fuel injection strategies, combustion chamber design, sophisticated control systems, and innovative aftertreatment solutions. Continued research into areas such as VVT, advanced EGR systems, and integrated hybrid combustion modes is crucial to overcome these hurdles and fully harness the profound benefits of dual-fuel LTC technology [1].

7.7 FUTURE SCOPE AND APPLICATIONS

Dual-fuel combustion strategies have transcended theoretical research to find extensive practical application across diverse sectors, including transportation and stationary power generation. These real-world implementations consistently demonstrate their profound potential for achieving significant reductions in fuel consumption and mitigating harmful emissions. The following outlines various current applications and case studies, highlighting the practical benefits, associated challenges, and observed outcomes of integrating dual-fuel systems.

7.7.1 Automotive Applications

- **Dual-Fuel Diesel-Gasoline Engines (RCCI):** RCCI technology offers a compelling avenue for future automotive diesel engines, utilizing gasoline as the LRF introduced via intake and diesel as the HRF pilot [87]. Such systems could significantly reduce NO_x and PM emissions compared to future conventional diesel powertrains, due to LTC's lower peak combustion temperatures and more distributed heat release, while also enhancing fuel efficiency, especially at medium-to-high loads. However, practical implementation faces hurdles like potential low-load misfire from improper air–fuel mixing, and the necessity for highly complex calibration and control strategies to manage dynamic automotive operation and mode transitions.

- **Methane-Diesel Dual-Fuel Engines (CI Engines):** These systems are anticipated for expanded adoption in heavy-duty trucks and buses, primarily operating on methane (e.g., CNG or LNG) with a small diesel pilot in CI engines [27,88]. Key benefits include lower operational costs from cheaper, cleaner methane, and reduced CO_2 emissions. A notable challenge, however, is methane slip, particularly at low loads, where incomplete combustion could undermine greenhouse gas benefits; additionally, integrating the liquid and gaseous fuel delivery systems requires substantial engine and infrastructure modifications.

7.7.2 Stationary Power Generation

- **Dual-Fuel Power Plants (Natural Gas and Diesel):** These robust systems hold significant potential for future stationary power generation, serving as primary or backup generators for industrial facilities and remote communities by leveraging the flexibility of natural gas and diesel [85,89]. Their main advantage is fuel choice flexibility, ensuring reliable power regardless of fuel availability or price, and allowing substantial NO_x reductions by primarily operating on natural gas. Challenges involve ensuring smooth transitions between fuels, as varying combustion properties demand precise control, potentially leading to increased thermal stresses and engine wear over extended operation.
- **Biomass-Diesel Dual-Fuel Generators:** These innovative generators could find increasing application in off-grid communities, producing electricity by combining readily available renewable biomass (e.g., wood chips) with a diesel pilot [48]. This offers high sustainability and reduced reliance on costly, less environmentally friendly diesel, leading to lower operational costs and greater energy independence. Despite this, biomass variability (moisture and density) could significantly affect combustion stability and efficiency, and handling solid biomass alongside liquid fuels presents considerable technical complexity in system design and operation.

7.7.3 Marine and Locomotive Applications

- **Dual-Fuel Engines in Shipping (LNG-Diesel):** The marine sector is projected to continue its strong shift toward dual-fuel engines using LNG as the primary fuel and diesel as a pilot, envisioned for widespread deployment in modern cargo ships and liners to meet future environmental regulations [48]. These engines are expected to significantly reduce sulfur emissions (meeting IMO 2020), NO_x, and PM compared to conventional marine engines, crucial for operating in ECAs. Challenges include developing robust LNG bunkering infrastructure, especially in smaller ports, and managing LNG boil-off during long voyages, which could lead to fuel loss and necessitates sophisticated tank systems.

- **Diesel-Natural Gas Locomotives:** Dual-fuel technology is anticipated for increasing integration into locomotives for freight and passenger transport, utilizing a combination of diesel and natural gas [90]. This configuration could yield lower fuel costs and reduced emissions, aligning with environmental mandates, while also providing increased fuel flexibility. However, a limited refueling infrastructure for natural gas could hinder widespread adoption in some regions, and the design and integration of complex fuel conversion and storage systems are critical for safe, efficient locomotive operation.

7.7.4 Future Scope and Trends

As global energy demands continue to evolve, driven by population growth and industrialization, and as environmental pressures intensify, the role of dual-fuel combustion strategies is poised for significant expansion across various sectors.

- **Integration of Renewable Fuels:** The future will see a much more widespread adoption of sustainable fuel sources in dual-fuel systems. This includes the increased use of biofuels (e.g., bio-methane and biodiesel), synthetic fuels derived from renewable energy, and critically, hydrogen, as either pilot or primary fuels. This trend aligns directly with the urgent global push for carbon-neutral and truly sustainable energy solutions, moving away from fossil fuel dependence.
- **Advanced Control Technologies:** The burgeoning fields of machine learning, artificial intelligence, and sophisticated real-time feedback systems are set to revolutionize dual-fuel engine management. Future dual-fuel engines will become increasingly adaptable and self-optimizing, capable of adjusting performance parameters based on real-time operating conditions without the need for complex manual tuning. This will enhance efficiency, reduce emissions, and improve robustness across diverse loads and transients.
- **Hybrid Powertrain Integration:** The synergistic integration of dual-fuel combustion engines with hybrid or fully electrified powertrains offers transformative potential, particularly in the transportation sector where energy demand is highly variable. This hybridization allows the IC engine to operate predominantly at its most efficient points while the electric components handle transient loads and low-speed operation, leading to significantly greater overall energy efficiency and reduced emissions.
- **Infrastructure Development:** The continued growth and expansion of refueling infrastructure for alternative fuels such as natural gas (CNG/LNG) and hydrogen are paramount. As these networks mature and become more accessible, the adoption of dual-fuel systems in commercial fleets,

industrial applications, and even passenger vehicles will experience a substantial increase, making these technologies more viable and widespread.

- **Stringent Regulatory Compliance:** Governments worldwide are implementing increasingly stringent emission standards and ambitious carbon reduction goals. Dual-fuel engines, especially those optimized for LTC, will play a critical role in achieving compliance across all sectors, including automotive, marine, and stationary power generation, providing a crucial pathway to cleaner air and a lower carbon footprint.

7.8 CONCLUSION

Dual-fuel combustion strategies stand as a pivotal and highly adaptable approach for significantly enhancing the efficiency, performance, and emissions profile of IC engines. By intelligently combining HRF and LRF, these systems unlock substantial benefits across various operational parameters. A major strength of dual-fuel setups lies in their unmatched operational flexibility, allowing engines to dynamically switch between fuels based on immediate availability, cost fluctuations, and environmental regulations. This inherent fuel flexibility offers crucial resilience against market volatility, making dual-fuel engines particularly well suited for diverse applications from automotive to heavy-duty transport and stationary power generation. These systems also deliver improved thermal efficiency, especially under LTC regimes like RCCI and PCCI, often outperforming conventional single-fuel engines. This efficiency, coupled with the strategic use of cleaner primary fuels, drives significant emission reductions in key pollutants such as NO_x, PM, and CO_2. Achieving these improvements hinges on precise system calibration. Economically, substituting more expensive fuels with cleaner, cheaper alternatives frequently leads to substantial operational cost savings, a powerful incentive for fuel-intensive industries like shipping and heavy-duty transport. However, the increased complexity inherent in dual-fuel systems presents a considerable challenge. Optimizing combustion, managing fuel transitions, and ensuring stable operation across varying load and speed conditions demand advanced engine management and sophisticated control systems. This complexity can also necessitate modifications to existing fuel infrastructure. Despite these hurdles, dual-fuel combustion strategies offer a compelling and practical path forward for IC engine technology. Their capacity to deliver enhanced efficiency, reduced emissions, and impressive fuel flexibility positions them as a critical solution for meeting global energy demands and increasingly stringent environmental mandates in the evolving landscape of sustainable power generation and transportation.

REFERENCES

1. A. K. Agarwal, A. P. Singh, and R. K. Maurya, "Evolution, challenges and path forward for low temperature combustion engines," *Progress in Energy and Combustion Science*, vol. 61, pp. 1–56, 2017, doi: 10.1016/j.pecs.2017.02.001.
2. A. Jaworski, K. Lejda, J. Lubas, and M. Mądziel, "Comparison of exhaust emission from Euro 3 and Euro 6 motor vehicles fueled with petrol and LPG based on real driving conditions," *Combustion Engines*, vol. 178, no. 3, pp. 106–111, 2019.

3. N. N. Mustafi, R. R. Raine, and S. Verhelst, "Combustion and emissions characteristics of a dual fuel engine operated on alternative gaseous fuels," *Fuel*, vol. 109, pp. 669–678, 2013, doi: 10.1016/j.fuel.2013.03.007.
4. S. Mi et al., "Optimization of direct-injection ammonia-diesel dual-fuel combustion under low load and higher ammonia energy ratios," *Fuel*, vol. 375, p. 132611, 2024, doi: 10.1016/j.fuel.2024.132611.
5. Z. Jin, S. Mi, D. Zhou, Y. Qian, and X. Lu, "The characteristics analysis and feasible improvement strategy evaluation for ammonia-diesel dual-fuel engine under heavy load and high ammonia ratio conditions," *Fuel*, vol. 383, p. 133911, 2025, doi: 10.1016/j.fuel.2024.133911.
6. A. Menaa, F. Amrouche, M. S. Lounici, and K. Loubar, "Experimental investigation of hydrogen use in dual fuel and like dual fuel mode," *Fuel*, vol. 393, p. 135031, 2025, doi: 10.1016/j.fuel.2025.135031.
7. F. Wei, Z. Zhang, W. Wei, H. Zhang, and W. Cai, "Multi-objective optimization of the performance for a marine methanol-diesel dual-fuel engine," *Fuel*, vol. 368, p. 131556, 2024, doi: 10.1016/j.fuel.2024.131556.
8. P. Miles, "Sources and mitigation of CO and UHC emissions in low-temperature diesel combustion regimes: Insights obtained via homogeneous reactor modeling," In *13th Diesel Engine Efficiency and Emissions Research Conference*, Detroit, Michigan, 2007.
9. E. Malfi, S. Esposito, and M. De Felice, "Phenomenological model for unburned hydrocarbon emissions and dual-fuel internal combustion engines," *International Journal of Engine Research*, vol. 25, no. 10, pp. 1818–1834, 2024, doi: 10.1177/14680874241255157.
10. M. Ghazikhani, M. R. Kalateh, Y. K. Toroghi, and M. Dehnavi, "An experimental study on the effect of EGR and engine speed on CO and HC emissions of dual fuel HCCI engine," *Proceedings of World Academy of Science, Engineering and Technology*, vol. 40, pp. 136–141, 2009.
11. S. Zeinali, E. Neshat, and S. Z. Mollayousefi, "Numerical investigation on different effects of recirculated species on UHC and CO formation during lean homogeneous combustion," *Scientific Reports*, vol. 15, p. 277449324, 2025.
12. S. J. M. Algayyim, K. Saleh, A. P. Wandel, I. M. R. Fattah, T. Yusaf, and H. A. Alrazen, "Influence of natural gas and hydrogen properties on internal combustion engine performance, combustion, and emissions: A review," *Fuel*, vol. 362, p. 130844, 2024, doi: 10.1016/j.fuel.2023.130844.
13. F. Scrignoli, F. Vecchio, F. Legrottaglie, E. Mattarelli, and C. A. Rinaldini, "Numerical investigation of dual fuel combustion on a compression ignition engine fueled with hydrogen/natural gas blends," *Fuels*, vol. 3, no. 1, pp. 132–151, 2022.
14. A. Hossein, A. Gharehghani, and M. Mahdi, "RCCI combustion of ammonia in dual fuel engine with early injection of diesel fuel," *Fuel*, vol. 365, p. 131182, 2024, doi: 10.1016/j.fuel.2024.131182.
15. Y. Wang, Y. Su, X. Li, F. Xie, and Y. Sun, "Research on the performance of methanol/PODE dual fuel spark-assisted compression ignition combustion with different intake air adjustment," *Applied Thermal Engineering*, vol. 255, p. 123987, 2024, doi: 10.1016/j.applthermaleng.2024.123987.
16. F. Zhang, C. Yang, Z. Wang, and X. Cheng, "Comparison of diesel single/double/triple injection strategies and early/late compression ignition regimes in an ammonia/diesel dual-fuel engine," *Energy*, vol. 322, p. 135753, 2025, doi: 10.1016/j.energy.2025.135753.
17. R. Y. Dahham, H. Wei, R. Zhang, J. Li, G. Shu, and J. Pan, "Numerical study for the comparison between direct dual-fuel stratification and reactivity-controlled compression ignition of ammonia-based engines," *Applied Thermal Engineering*, vol. 258, p. 124624, 2025, doi: 10.1016/j.applthermaleng.2024.124624.

18. X. Li, X. Zhang, P. Ni, Y. Pei, W. Li, and Z. Peng, "Exploring the effects of compression ratio and initial flame kernel radius on combustion characteristics and fuel economy of a dual-fuel spark ignition engine under oxy-fuel combustion mode," *Fuel*, vol. 385, p. 134098, 2025, doi: 10.1016/j.fuel.2024.134098.
19. Z. Wang et al., "Systematic assessment on the ammonia/gasoline combustion performance in a modern dual-fuel spark-ignition engine on different conditions ST," *Fuel*, vol. 392, p. 134806, 2025, doi: 10.1016/j.fuel.2025.134806.
20. Y. Wang et al., "A new criterion and method for classifying combustion modes in a high compression ratio spark ignition engine," *Energy Conversion and Management*, vol. 325, p. 119395, 2025, doi: 10.1016/j.enconman.2024.119395.
21. K. Uddeen, Q. Tang, H. Shi, and J. W. G. Turner, "Ammonia-methanol and ammonia-ethanol dual-fuel combustion in an optical spark-ignition engine: A multiple flame generation approach," *Applied Thermal Engineering*, vol. 265, p. 125544, 2025, doi: 10.1016/j.applthermaleng.2025.125544.
22. F. Königsson, P. Ståhammar, and H. E. Ångström, "Combustion modes in a diesel-CNG dual fuel engine," *SAE Tech. Pap.*, 2011, doi: 10.4271/2011-01-1962.
23. A. Aldawood, S. Mosbach, and M. Kraft, "HCCI combustion control using dual-fuel approach: Experimental and modeling investigations," *SAE Tech. Pap.*, 2012, doi: 10.4271/2012-01-1117.
24. R. H. Stanglmaier, T. W. Ryan, and J. S. Souder, "HCCI operation of a dual-fuel natural gas engine for improved fuel efficiency and ultra-low NO_x emissions at low to moderate engine loads," *SAE Tech. Pap.*, no. 724, 2001, doi: 10.4271/2001-01-1897.
25. H. Wang, Z. Zheng, H. Liu, and M. Yao, "Combustion mode design with high efficiency and low emissions controlled by mixtures stratification and fuel reactivity," *Frontiers of Mechanical Engineering*, vol. 1, pp. 1–15, 2015, doi: 10.3389/fmech.2015.00008.
26. D. Ganesh, P. R. Ayyappan, and R. Murugan, "Experimental investigation of iso-butanol/diesel reactivity controlled compression ignition combustion in a non-road diesel engine," *Applied Energy*, vol. 242, pp. 1307–1319, 2019, doi: 10.1016/j.apenergy.2019.03.166.
27. J. Shin and S. Park, "Numerical analysis and optimization of combustion and emissions in an ammonia-diesel dual-fuel engine using an ammonia direct injection strategy," *Energy*, vol. 289, p. 130014, 2024, doi: 10.1016/j.energy.2023.130014.
28. Y. Yang, W. Long, Y. Qian, P. Dong, Y. Wang, and Z. Cai, "Spatiotemporal regulation of methanol premixed and diffusion co-combustion for marine methanol/diesel dual fuel engine," *Energy*, vol. 315, p. 134349, 2025, doi: 10.1016/j.energy.2024.134349.
29. D. Hariharan et al., "Catalytic partial oxidation reformation of diesel, gasoline, and natural gas for use in low temperature combustion engines," *Fuel*, vol. 246, pp. 295–307, 2019, doi: 10.1016/j.fuel.2019.02.003.
30. J. Cai, K. Mun, R. Rajasegar, A. Srna, X. Bai, and J. H. Walther, "LES of pilot n-heptane ignition and its interaction with the lean premixed methane – air mixture in a dual-fuel combustion engine," *Proceedings of the Combustion Institute*, vol. 40, no. 1–4, p. 105281, 2024, doi: 10.1016/j.proci.2024.105281.
31. Y. Kang, W. Zhou, X. Huang, J. Zhang, H. Wang, and X. Lu, "International journal of hydrogen energy DNS study on turbulent stratified combustion mechanism of ammonia-hydrogen premixtures with varying thermochemical stratifications," *International Journal of Hydrogen Energy*, vol. 115, pp. 157–169, 2025, doi: 10.1016/j.ijhydene.2025.02.450.
32. Z. Li, Z. Zhang, Y. Fan, J. Li, and K. Wu, "ScienceDirect Fuel reactivity stratification assisted jet ignition for low-speed two-stroke ammonia marine engine," *International Journal of Hydrogen Energy*, vol. 49, pp. 570–585, 2023, doi: 10.1016/j.ijhydene.2023.08.256.

33. M. Nazemi and M. Shahbakhti, "Modeling and analysis of fuel injection parameters for combustion and performance of an RCCI engine," *Applied Energy*, vol. 165, pp. 135–150, 2016, doi: 10.1016/j.apenergy.2015.11.093.
34. N. T. Wimer, M. T. Henry de Frahan, and S. Yellapantula, "Deep reinforcement learning to discover multi-fuel injection strategies for compression ignition engines," *International Journal of Engine Research*, vol. 24, no. 9, pp. 3985–4007, 2023, doi: 10.1177/14680874231184002.
35. K. Poorghasemi, R. K. Saray, E. Ansari, B. K. Irdmousa, M. Shahbakhti, and J. D. Naber, "Effect of diesel injection strategies on natural gas/diesel RCCI combustion characteristics in a light duty diesel engine," *Applied Energy*, vol. 199, pp. 430–446, 2017, doi: 10.1016/j.apenergy.2017.05.011.
36. C. Zhang, C. Zhang, L. Xue, and Y. Li, "Combustion characteristics and operation range of a RCCI combustion engine fueled with direct injection *n*-heptane and pipe injection *n*-butanol," *Energy*, vol. 125, pp. 439–448, 2017, doi: 10.1016/j.energy.2017.02.148.
37. A. H. Kakaee, P. Rahnama, and A. Paykani, "CFD study of reactivity controlled compression ignition (RCCI) combustion in a heavy-duty diesel engine," *Periodica Polytechnica Transportation Engineering*, vol. 43, no. 4, pp. 177–183, 2015, doi: 10.3311/PPtr.7756.
38. H. Wang, X. Zhao, L. Tong, and M. Yao, "The effects of DI fuel properties on the combustion and emissions characteristics of RCCI combustion," *Fuel*, vol. 227, pp. 457–468, 2018, doi: 10.1016/j.fuel.2018.04.025.
39. L. Wang et al., "Experimental study on the high load extension of PODE/methanol RCCI combustion mode with optimized injection strategy," *Fuel*, vol. 314, p. 122726, 2022, doi: 10.1016/j.fuel.2021.122726.
40. S. Molina, A. García, J. M. Pastor, E. Belarte, and I. Balloul, "Operating range extension of RCCI combustion concept from low to full load in a heavy-duty engine," *Applied Energy*, vol. 143, pp. 211–227, 2015, doi: 10.1016/j.apenergy.2015.01.035.
41. J. Li, X. Ling, D. Liu, W. Yang, and D. Zhou, "Numerical study on double injection techniques in a gasoline and biodiesel fueled RCCI (reactivity controlled compression ignition) engine," *Applied Energy*, vol. 211, pp. 382–392, 2018, doi: 10.1016/j.apenergy.2017.11.062.
42. S. Kang, S. Lee, and C. Bae, "Effects of multi-stage split injection on efficiency and emissions of light-duty diesel engine," *Energies*, vol. 15, no. 6, p. 15062219, 2022, doi: 10.3390/en15062219.
43. L. Jiang et al., "Effects of split injection strategy on combustion characteristics and NO_x emissions performance in dual-fuel marine engine," *Applied Thermal Engineering*, vol. 248, p. 123153, 2024, doi: 10.1016/j.applthermaleng.2024.123153.
44. S. Feng, W. Hong, Y. Yao, and T. You, "Research of post injection strategy of an egr diesel engine to improve combustion and particulate emissions performance: application on the transient operation," *Symmetry*, vol. 12, no. 12, pp. 1–17, 2020, doi: 10.3390/sym12122002.
45. A. Zehni, N. Balazadeh, M. Hajibabaei, and K. Poorghasemi, "Numerical study of the effects of split injection strategy and swirl ratio for biodiesel PCCI combustion and emissions," *Propulsion and Power Research*, vol. 9, no. 4, pp. 355–371, 2020, doi: 10.1016/j.jppr.2020.11.004.
46. J. Li, W. Yang, and D. Zhou, "Review on the management of RCCI engines," *Renewable and Sustainable Energy Reviews*, vol. 69, pp. 65–79, 2017, doi: 10.1016/j.rser.2016.11.159.
47. V. B. Pedrozo, I. May, T. D. M. Lanzanova, and H. Zhao, "Potential of internal EGR and throttled operation for low load extension of ethanol-diesel dual-fuel reactivity controlled compression ignition combustion on a heavy-duty engine," *Fuel*, vol. 179, pp. 391–405, 2016, doi: 10.1016/j.fuel.2016.03.090.

48. E. Ejder, S. Dinçer, and Y. Arslanoglu, "Decarbonization strategies in the maritime industry : An analysis of dual-fuel engine performance and the carbon intensity indicator," *Renewable and Sustainable Energy Reviews*, vol. 200, p. 114587, 2024, doi: 10.1016/j.rser.2024.114587.
49. M. R. Kalateh and M. Ghazikhani, "an experimental study on the effects of EGR and equivalence ratio on CO and soot emissions of dual fuel HCCI engine," *Biological and Environmental Engineering*, vol. 322, pp. 284–290, 2009, doi: 10.1142/9789814295048_0058.
50. Y. Wang, M. Yao, T. Li, W. Zhang, and Z. Zheng, "A parametric study for enabling reactivity controlled compression ignition (RCCI) operation in diesel engines at various engine loads," *Applied Energy*, vol. 175, pp. 389–402, 2016, doi: 10.1016/j.apenergy.2016.04.095.
51. M. B. Luong, R. Sankaran, G. H. Yu, S. H. Chung, and C. S. Yoo, "On the effect of injection timing on the ignition of lean PRF/air/EGR mixtures under direct dual fuel stratification conditions," *Combustion and Flame*, vol. 183, pp. 309–321, 2017, doi: 10.1016/j.combustflame.2017.05.023.
52. J. H. Lim and R. D. Reitz, "High load (21 Bar IMEP) dual fuel RCCI combustion using dual direct injection," *Journal of Engineering for Gas Turbines and Power*, vol. 136, no. 10, pp. 1–6, 2014, doi: 10.1115/1.4027361.
53. A. García, J. Monsalve-Serrano, V. Rückert Roso, and M. E. Santos Martins, "Evaluating the emissions and performance of two dual-mode RCCI combustion strategies under the World Harmonized Vehicle Cycle (WHVC)," *Energy Conversion and Management*, vol. 149, pp. 263–274, 2017, doi: 10.1016/j.enconman.2017.07.034.
54. J. Benajes, A. García, J. Monsalve-Serrano, and R. Lago Sari, "Fuel consumption and engine-out emissions estimations of a light-duty engine running in dual-mode RCCI/CDC with different fuels and driving cycles," *Energy*, vol. 157, pp. 19–30, 2018, doi: 10.1016/j.energy.2018.05.144.
55. G. Xu, M. Jia, Y. Li, Y. Chang, and T. Wang, "Potential of reactivity controlled compression ignition (RCCI) combustion coupled with variable valve timing (VVT) strategy for meeting Euro 6 emission regulations and high fuel efficiency in a heavy-duty diesel engine," *Energy Conversion and Management*, vol. 171, pp. 683–698, 2018, doi: 10.1016/j.enconman.2018.06.034.
56. W. Guo, M. Xiao, Z. Zhang, Y. Wang, L. Shi, and K. Deng, "Effects of multiple injections on the combustion and hydrocarbon emission characteristics of the start cylinder in direct-start process," *Fuel*, vol. 320, p. 123851, 2022, doi: 10.1016/j.fuel.2022.123851.
57. G. Edara, Y. V. V. Satyanarayana Murthy, P. Srinivas, J. Nayar, and M. Ramesh, "Effect of cooled EGR on modified light duty diesel engine for combustion, performance and emissions under high pressure split injection strategies," *Case Studies in Thermal Engineering*, vol. 12, pp. 188–202, 2018, doi: 10.1016/j.csite.2018.03.004.
58. B. Deng et al., "A comprehensive investigation of EGR (exhaust gas recirculation) effects on energy distribution and emissions of a turbo-charging diesel engine under world harmonized transient cycle," *Energy*, vol. 316, p. 134506, 2025, doi: 10.1016/j.energy.2025.134506.
59. Y. Qiang et al., "Study on the effect of variable valve timing and spark timing on the performance of the hydrogen-fueled engine with passive pre-chamber ignition under partial load conditions," *Energy Conversion and Management*, vol. 302, p. 118104, 2024, doi: 10.1016/j.enconman.2024.118104.
60. G. Fontana and E. Galloni, "Variable valve timing for fuel economy improvement in a small spark-ignition engine," *Applied. Energy*, vol. 86, no. 1, pp. 96–105, 2009, doi: 10.1016/j.apenergy.2008.04.009.

61. Y. Li, L. Huang, Y. Chen, and W. Tang, "Stratified premixed combustion optimization of a natural gas/biodiesel dual direct injection engine," *Energy*, vol. 294, p. 130935, 2024, doi: 10.1016/j.energy.2024.130935.
62. D. Hu, H. Wang, C. Yang, H. Mu, Y. Wang, and E. D. Cnn-gru, "A novel control strategy for DDF/PPCI combustion mode in a diesel micro-pilot ignited dual-fuel engine," *Energy*, vol. 313, p. 133932, 2024.
63. L. Yang, S. Ji, J. Hunicz, R. Wang, A. Zare, and Y. Su, "Analysis of multi-mode combustion and performance in a marine diesel/natural gas dual-fuel engine based on an irreversible equivalent combustion cycle theory," *Energy*, vol. 326, p. 136248, 2025, doi: 10.1016/j.energy.2025.136248.
64. S. Mostasharshahidi, M. Esmaeili, and G. Najafi, "Strategic optimization of dual-fuel diesel/gas engines by numerical approach for environmental and economic benefits," *International Communications in Heat and Mass Transfer*, vol. 159, p. 108156, 2024, doi: 10.1016/j.icheatmasstransfer.2024.108156.
65. M. E. J. Stettler, W. J. B. Midgley, J. J. Swanson, D. Cebon, and A. M. Boies, "Greenhouse gas and noxious emissions from dual fuel diesel and natural gas heavy goods vehicles," *Environmental Science & Technology*, vol. 50, no. 4, pp. 2018–2026, 2016, doi: 10.1021/acs.est.5b04240.
66. H. Liu, Z. Wang, Y. Li, Y. Zheng, T. He, and J. Wang, "Recent progress in the application in compression ignition engines and the synthesis technologies of polyoxymethylene dimethyl ethers," *Applied Energy*, vol. 233–234, pp. 599–611, 2019, doi: 10.1016/j.apenergy.2018.10.064.
67. D. Yuvenda, B. Sudarmanta, A. Wahjudi, and R. Aini, "Effect of adding combustion air on emission in a diesel dual- fuel engine with crude palm oil biodiesel compressed natural gas fuels," *International Journal of Renewable Energy Development*, vol. 11, no. 3, pp. 871–877, 2022.
68. M. G. Krishnan and S. Rajkumar, "Effects of dual fuel combustion on performance, emission and energy- exergy characteristics of diesel engine fuelled with diesel-isobutanol and biodiesel-isobutanol," *Energy*, vol. 252, p. 124022, 2022, doi: 10.1016/j.energy.2022.124022.
69. A. Callinan, "Dual-fuel solutions at low additive percentages for minimizing pollutant emission formation and maximizing energy generation," *Journal of Multidisciplinary Engineering Science and Technology*, vol. 9, no. 9, pp. 15500–15513, 2022.
70. P. Karvounis, G. Theotokatos, C. Patil, L. Xiang, and Y. Ding, "Parametric investigation of diesel–methanol dual-fuel marine engines with port and direct injection," *Fuel*, vol. 381, p. 133441, 2024, doi: 10.1016/j.fuel.2024.133441.
71. A. C. Polk, C. D. Carpenter, K. K. Srinivasan, and S. R. Krishnan, "An investigation of diesel-ignited propane dual fuel combustion in a heavy-duty diesel engine," *Fuel*, vol. 132, pp. 135–148, 2014, doi: 10.1016/j.fuel.2014.04.069.
72. Y. Wang, H. Liu, Z. Huang, and Z. Liu, "Study on combustion and emission of a dimethyl ether-diesel dual-fuel premixed charge compression ignition combustion engine with LPG (liquefied petroleum gas) as ignition inhibitor," *Energy*, vol. 96, pp. 278–285, 2016, doi: 10.1016/j.energy.2015.12.056.
73. A. Henham and M. K. Makkar, "Combustion of simulated biogas in a dual-fuel diesel engine," *Energy Conversion and Management*, vol. 39, no. 16, pp. 2001–2009, 2009.
74. J. Paparao, K. Kumar Pandey, and S. Murugan, "Experimental studies on the effect of TBC piston in a dual-fueled diesel engine," *Fuel*, vol. 306, p. 121700, 2021, doi: 10.1016/j.fuel.2021.121700.
75. P. Cai, C. Zhang, Z. Jing, Y. Peng, J. Jing, and H. Sun, "Effects of Fischer-Tropsch diesel blending in petrochemical diesel on combustion and emissions of a common-rail diesel engine," *Fuel*, vol. 305, p. 121587, 2021, doi: 10.1016/j.fuel.2021.121587.

76. R. D. Reitz and G. Duraisamy, "Review of high efficiency and clean reactivity controlled compression ignition (RCCI) combustion in internal combustion engines," *Progress in Energy and Combustion Science*, vol. 46, pp. 12–71, 2015, doi: 10.1016/j.pecs.2014.05.003.
77. V. B. Pedrozo, X. Wang, W. Guan, and H. Zhao, "The effects of natural gas composition on conventional dual-fuel and reactivity-controlled compression ignition combustion in a heavy-duty diesel engine," *International Journal of Engine Research*, vol. 23, no. 3, pp. 397–415, 2022, doi: 10.1177/1468087420984044.
78. K. Muniappan, S. Sheshadri, and P. P. Duvvuri, "Numerical analysis of the effects of direct dual fuel injection on the compression ignition engine," *ACS Omega*, vol. 5, no. 46, pp. 30047–30058, 2020, doi: 10.1021/acsomega.0c04434.
79. X. Liu, W. Zhong, P. Jiang, L. Dai, Z. He, and Q. Wang, "Optimizing RCCI combustion for improved engine performance under low load conditions: Impact of low-reactivity fuel and direct injection timing," *Fuel*, vol. 351, p. 128871, 2023, doi: 10.1016/j.fuel.2023.128871.
80. J. Hunicz, L. Yang, A. Rybak, S. Ji, and S. Gęca, "Comparison of diesel and hydrotreated vegetable oil as the high-reactivity fuel in reactivity-controlled compression ignition," *Energy Conversion and Management*, vol. 323, p. 119264, 2025, doi: 10.1016/j.enconman.2024.119264.
81. R. Kiplimo, E. Tomita, N. Kawahara, and S. Yokobe, "Effects of spray impingement, injection parameters, and EGR on the combustion and emission characteristics of a PCCI diesel engine," *Applied Thermal Engineering*, vol. 37, pp. 165–175, 2012, doi: 10.1016/j.applthermaleng.2011.11.011.
82. H. Huang, Z. Li, W. Teng, R. Huang, Q. Liu, and Y. Wang, "Effects of EGR rates on combustion and emission characteristics in a diesel engine with *n*-butanol/PODE3–4/diesel blends," *Applied Thermal Engineering*, vol. 146, pp. 212–222, 2019, doi: 10.1016/j.applthermaleng.2018.09.126.
83. K. Ramalingam, S. Vellaiyan, and G. M. Satyanarayana, "Case studies in thermal engineering an experimental and ANN analysis of ammonia energy integration in biofuel powered low-temperature combustion engine to enhance cleaner combustion," *Case Studies in Thermal Engineering*, vol. 63, p. 105284, 2024, doi: 10.1016/j.csite.2024.105284.
84. M. De Felice et al., "A novel phenomenological model for stratification-controlled low temperature combustion applied to a dual-fuel marine internal combustion engine supplied with natural gas and light fuel oil end of injection," *Fuel*, vol. 386, p. 134218, 2025, doi: 10.1016/j.fuel.2024.134218.
85. P. V Rameez and M. M. Ibrahim, "Maximizing hydrogen utilization in CI engines: An investigation of dual fuel and RCCI combustion approaches under low and mid-load conditions in a medium duty automotive engine," *Energy Conversion and Management*, vol. 321, p. 119100, 2024, doi: 10.1016/j.enconman.2024.119100.
86. X. Yin et al., "Investigation into the operating range of a dual-direct injection engine fueled with methanol and diesel," *Energy*, vol. 267, no. 28, p. 126625, 2023.
87. S. Pan, X. Liu, K. Cai, X. Li, W. Han, and B. Li, "Experimental study on combustion and emission characteristics of iso-butanol/diesel and gasoline/diesel RCCI in a heavy-duty engine under low loads," *Fuel*, vol. 261, p. 116434, 2020, doi: 10.1016/j.fuel.2019.116434.
88. P. A. Harari, N. R. Banapurmath, V. S. Yaliwal, T. M. Y. Khan, M. E. M. Soudagar, and A. M. Sajjan, "Experimental studies on performance and emission characteristics of reactivity controlled compression ignition (RCCI) engine operated with gasoline and Thevetia Peruviana biodiesel," *Renewable Energy*, vol. 160, pp. 865–875, 2020, doi: 10.1016/j.renene.2020.07.009.

89. Z. Wang, C. Yang, F. Zhang, and X. Cheng, "Effects of diesel injector nozzle angle and split diesel injection strategy on combustion and emission characteristics of an ammonia/diesel dual-fuel engine," *Energy*, vol. 307, p. 132686, 2024, doi: 10.1016/j.energy.2024.132686.
90. W. Li, Z. Liu, and Z. Wang, "Experimental and theoretical analysis of the combustion process at low loads of a diesel natural gas dual-fuel engine," *Energy*, vol. 94, pp. 728–741, 2016, doi: 10.1016/j.energy.2015.11.052.

8 Combustion Control and Optimization Techniques

Nidhi, Bulbul Ongar, and Zhannat Bekbolatova

8.1 INTRODUCTION

Combustion in internal combustion (IC) engines refers to the rapid exothermic oxidation of a hydrocarbon-based fuel with an oxidizer, typically air, within a confined space, converting chemical energy into thermal energy and subsequently mechanical work (Heywood, 1988). This process can be categorized into premixed, diffusion, or partially premixed modes, each exhibiting distinct flame propagation and thermal characteristics. In conventional compression ignition (CI) engines, combustion is typically characterized by nonuniform spray-driven fuel–air mixing and rapid heat release, often resulting in localized stoichiometric and rich mixtures. These zones promote high-temperature regions (>2000 K), leading to excessive formation of nitrogen oxides (NO_x), while fuel-rich pockets favor incomplete combustion and particulate matter (PM) formation (Krishnamoorthi et al., 2019; Reitz & Duraisamy, 2015). As a result, traditional combustion modes face inherent thermochemical inefficiencies and emission challenges.

The environmental and regulatory pressures have intensified the necessity to address these limitations. High levels of NO_x and PM not only contribute to smog and adverse health outcomes but also violate increasingly stringent global emission norms such as Euro VI and Bharat Stage VI standards. Conventional aftertreatment solutions including diesel oxidation catalysts, diesel particulate filters, and selective catalytic reduction while effective, incur additional costs, complexity, and parasitic losses in engine performance due to increased backpressure and fuel penalty (Kalghatgi, 2014). Moreover, the stochastic nature of ignition timing, heat release rate (HRR), and in-cylinder turbulence in uncontrolled combustion impairs the optimization of brake thermal efficiency (BTE) and may result in knock or misfire, particularly under varying load conditions (Zhao, 2007). Therefore, improving the in-cylinder combustion process itself is considered a more thermodynamically rational and sustainable pathway for enhancing engine performance and reducing emissions.

Consequently, the control of combustion has become a cornerstone of modern IC engine research, particularly through the development of advanced low-temperature combustion (LTC) strategies such as Homogeneous Charge Compression Ignition (HCCI), Premixed Charge Compression Ignition (PCCI), and Reactivity Controlled Compression Ignition (RCCI). These modes aim to decouple the tight trade-off between NO_x and soot emissions by creating a lean, homogeneous, and well-premixed air–fuel mixture that undergoes autoignition at lower temperatures, thus

 DOI: 10.1201/9781003633631-8

suppressing thermal NO_x pathways while minimizing fuel-rich soot-forming zones (Musculus et al., 2013; Krishnamoorthi et al., 2019). However, achieving stable combustion in LTC regimes requires precise control over ignition delay, fuel reactivity, and charge stratification, typically managed via high EGR rates, split injections, variable valve timing (VVT), and multiple fuels with differential reactivity (Miers et al., 2012; Kokjohn et al., 2011). Thus, advanced combustion control not only addresses emission compliance but also lays the foundation for improved fuel efficiency and fuel-flexibility in next-generation engines.

LTC has emerged as a transformative strategy in internal IC engine research, offering a pathway to decouple the long-standing trade-off between NO_x and PM emissions inherent in conventional combustion. In traditional CI engines, high in-cylinder temperatures (>2000 K) in stoichiometric and rich zones lead to intense NO_x formation and soot production, respectively (Musculus et al., 2013; Krishnamoorthi et al., 2019). LTC aims to address this by promoting combustion at lower peak temperatures and with more homogeneous air–fuel mixtures, thereby simultaneously suppressing thermal NO_x generation and soot precursors (Ju et al., 2021). RCCI, an advanced LTC mode, leverages dual-fuel stratification to control ignition phasing and reactivity gradients, achieving over 85% NO_x and 95% PM reductions under certain conditions (Splitter et al., 2013). These innovations align with growing global pressure to reduce tailpipe pollutants and achieve sustainable transport.

The urgency for widespread LTC adoption in IC engines is underscored by the Intergovernmental Panel on Climate Change (IPCC), which highlights the need for rapid reductions in anthropogenic greenhouse gas emissions to limit global warming to 1.5°C above pre-industrial levels (IPCC, 2018). Transportation, being a major contributor to CO_2 and other harmful pollutants, must transition toward cleaner technologies. LTC supports this shift not only by reducing combustion-generated NO_x and PM but also by enabling higher thermal efficiencies (Krishnamoorthi et al., 2019; Splitter et al., 2013). For example, RCCI engines have demonstrated BTEs approaching 60%, compared to 40%–45% for conventional diesel engines, thereby reducing fuel consumption and associated CO_2 emissions (Splitter et al., 2013). Furthermore, cool flame kinetics intrinsic to LTC enhance the thermodynamic efficiency by distributing heat release more evenly, reducing exergy destruction and combustion noise (Ju, 2021).

Another critical advantage of LTC combustion is its compatibility with a wide range of fuels, including low-carbon biofuels, alcohols, and synthetic blends. These fuels often exhibit unique reactivity and volatility characteristics that are favorable for LTC operation but problematic for conventional combustion modes due to ignition delay or incomplete oxidation (Kohse-Höinghaus et al., 2010; Agarwal, 2007). LTC allows the use of low-cetane or high-octane fuels by decoupling ignition from fuel injection timing (IT), thereby enabling precise control over combustion phasing (Kokjohn et al., 2011; Krishnamoorthi et al., 2019). For instance, bio-alcohols with high latent heat and oxygen content reduce in-cylinder peak temperatures and promote cleaner combustion when operated under HCCI or PCCI modes. Moreover, the ability to operate under ultra-lean conditions and high exhaust gas recirculation (EGR) levels in LTC further expands the fuel flexibility and robustness of next-generation IC engines (Pitz & Mueller, 2011; Wang et al., 2016). LTC is characterized by the presence of cool and hot flames. Cool flame converts the fuel into alkenes and aldehydes below 800 K and is typically a dull blue color front. The warm flame converts the intermediate products into CO and water and temperature lies between 800 K and 1100 K with typical two flame fronts. LTC is not merely a combustion but a technology for enabling low-emission, high-efficiency, and fuel-diverse powertrains for the future. Table 8.1 shows a general comparison between different LTC strategies.

8.2 FACTORS AFFECTING LTC

LTC combustion is affected by charge characteristics such as intake temperature, air–fuel mixture, and engine operating conditions including IT, combustion chamber geometry, and valve timings (Krishnamoorthi et al., 2019; Jia et al., 2022). The effect of various parameters is discussed below.

8.2.1 Effect of Charge

8.2.1.1 Premixed Ratio

The premixed ratio indicates the fraction of energy from premixed fuel (low reactive) relative to the total fuel energy from both low reactive and high reactive fuels. Increasing the premixed ratio improves combustion, leading to longer ignition delay and improved phasing (Zheng et al., 2018). Zheng et al. observed that a premixed ratio of 100% leads to typical HCCI combustion characteristics, whereas 0% means the conventional combustion. Fully premixed fuel results in the highest average in-cylinder pressure and HRR, with a relatively lower average gas temperature. With the increase of premixed portion from 0% to 70%, diffusion-dominated combustion diminishes due to extended ignition delay, and availability of more time enhances air–fuel mixing and thus, promotes premixed combustion. At elevated EGR levels (~45%), diffusion combustion weakens regardless of premixed ratio. Lower premixed ratios or EGR lead to greater dominance of diffusion combustion. Combustion efficiency drops when increasing the premixed fuel share from 0% to 50%, primarily due to ultra-lean mixtures created by low-reactivity fuel (LRF) alongside locally rich zones from high-reactivity fuel (HRF), concentrating combustion in core areas rather than uniformly in the chamber. Beyond 50%, equivalence ratio (φ) improves, enabling uniform ignition both centrally and near chamber walls, increasing temperatures and combustion efficiency (Zheng et al., 2018; Jia et al., 2022).

Jia et al. investigated the effect of premixing methanol with diesel fuel at a fixed diesel IT of 27° bTDC in RCCI combustion. Retarding or advancing IT from 7° to 37° bTDC changed the peak HRR due to varying LRF and HRF distribution. A longer ignition delay accumulated more fuel before ignition, shifting combustion toward the expansion stroke. The highest in-cylinder temperatures were recorded at 60% premixed ratio and the peak temperature was the lowest with conventional diesel. Substitution by Methanol lowered gas temperature through evaporative cooling owing to its high latent heat of vaporization. Methanol promoted premixed combustion and reduced mixing-controlled combustion near the end of combustion.

An increase of premixed ratio from 0% to 70% reduced ringing intensity (RI) due to lower peak pressure and HRR. Soot emissions peaked at 12° bTDC IT, likely from a higher local equivalence ratio at crank angle of 10% heat release (CA10). Advancing start of main injection to 22° bTDC reduced soot to 0.001 g/kWh but increased combustion ringing, which can be mitigated by volatile fuels and delayed injections (Jia et al., 2022).

It is reported that advancing the premixed combustion increases peak pressure and reduces mean gas temperature. NO_x emission reduction of 21% was observed in RCCI mode versus conventional combustion at peak loads. In addition to this, HC and CO emissions increased with lean mixtures at high premixed ratio due to crevice volumes and wall quenching (Krishnamoorthi et al., 2019; Gao et al., 2018; Hosseini et al., 2020). Higher premixed ratios smooth and advance HRR, reducing peak temperatures. Wing et al. showed that dimethyl ether (DME) as premixed fuel increased cylinder temperature and reduced ignition delay by advancing the start of combustion. It is observed that optimal charge preparation at 45% premixed ratio and 50% EGR yielded thermal efficiency of 47.5% (Wing et al., 2017; Liu et al., 2019).

It is reported that BTE of 32.65% was obtained at 51% premixed ratio and intake temperature of 323 K; however, thermal efficiency declined due to volumetric effi ciency loss above intake temperature of 60°C. Higher premixed ratios reduce NO_x via lower peak temperatures and homogeneous charge, however, can increase HC and CO emissions due to incomplete oxidation and wall quenching. Hydrogen addition up to 30% reduced HC by 29.8% and CO by 35.5%, but NO_x increased by 23.1% due to increased combustion temperatures. Methanol substitution in RCCI mode enhanced premixed combustion phase due to its high latent heat of vaporization reducing NO_x emission but increased the unburned HC in some cases (Gross et al., 2018; Zhang et al., 2020).

Fuel properties, including fuel type and characteristics such as cetane number (CN), latent heat, and volatility, significantly influence premixed LTC combustion behavior. Fuels like ethanol, methanol, hydrogen, and DME modify ignition delay, combustion temperature, and emissions. LTC combustion modes such as RCCI and PCCI are notably affected by fuel properties including octane number, CN, chemical composition, and molecular structure (Kokjohn et al., 2011).

Fuel reactivity, often characterized by CN or octane number, controls ignition delay in LTC systems. High-cetane fuels such as diesel ignite early, causing challenges due to advanced combustion and knock risks. Conversely, low-cetane fuels like gasoline and methanol allow longer ignition delays for better timing control (Kokjohn et al., 2011). Authors reported that oxygenated biodiesel with higher CN accelerates premixed combustion. At 50% engine load, compressed natural gas (CNG) combined with biodiesel in RCCI mode yielded the highest cylinder pressure and HRR due to prolonged ignition delay, whereas neat biodiesel had a shorter ignition delay in conventional engines. CNG/biodiesel RCCI produced approximately 32.5% lower hydrocarbon (HC) emissions compared to CNG/diesel, attributed to biodiesel's better ignitability (Hosseini et al., 2019).

Fuel blends such as diesel-gasoline or alcohol-diesel are commonly used to moderate fuel reactivity and enable tailored combustion phasing in RCCI engines (Reitz et al., 2014). Highly volatile fuels like gasoline and ethanol promote homogeneous air–fuel mixing, leading to cleaner combustion and lower soot emissions. Fuels with low volatility may cause uneven mixing and localized rich zones, raising soot formation risk under cold-start or stratified conditions (Zhao et al., 2024).

It is reported that a 10% biodiesel-diesel blend (B10) produced a peak pressure of 58 bar, about 8% higher than conventional combustion. Biodiesel/diesel RCCI combustion achieved lower in-cylinder temperatures and a BTE of 31%, compared to 23% in conventional mode, due to enhanced premixed combustion. NO_x emissions reduced by 24% at full load with 30% biodiesel content due to leaner mixtures and higher latent heat of vaporization lowering combustion temperatures. Smoke emissions also decreased with increasing biodiesel across loads, attributed to oxygen content present in biodiesel facilitating better oxidation and faster burn rates. Low boiling points and high-octane ratings extended ignition delay. Advanced main IT increased physical ignition delay; at 50% premixed ratio, IT of 38° bTDC resulted in an ignition delay of 28°, while 18° bTDC reduced it to 14°. Increasing intake oxygen concentration also influenced ignition timing through improved fuel reactivity and oxidation (Thirumalini et al., 2023). Figure 8.1 indicates the variation of in-cylinder temperature, pressure, HRR, and emissions (NO_x, HC, CO, and soot) with premixed ratio in LTC.

It is reported that CO_2 emissions decreased in dual-fuel PCCI mode at low loads due to lower in-cylinder gas temperatures, with further decreases with diesel/15% ethanol blends (E15D85). At medium loads, CO_2 emissions were 31% and 30% lower in diesel/E15D85 compared to conventional dual-fuel and conventional compression modes, respectively. NO_x emissions decreased by 28% in diesel/diesel dual-fuel

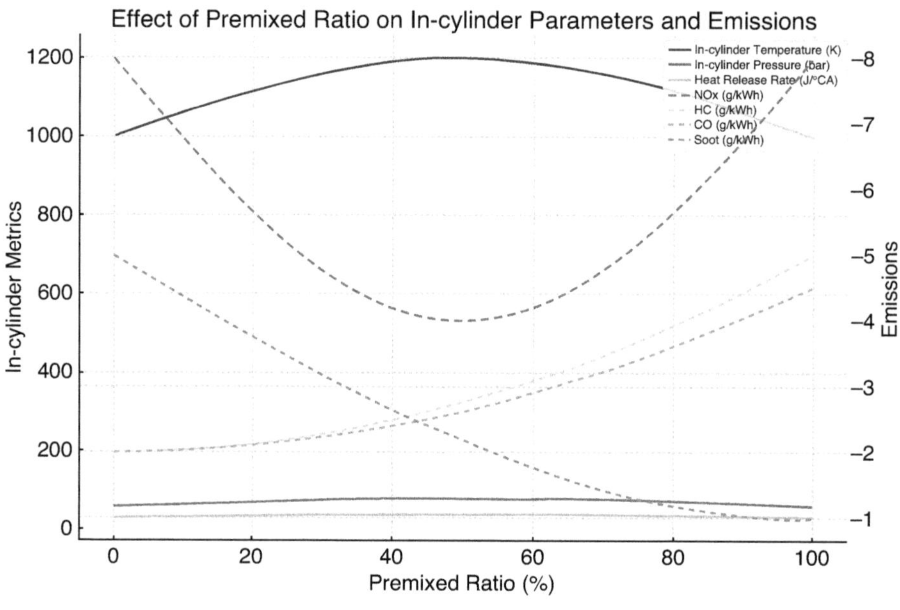

FIGURE 8.1 Variation of in-cylinder temperature, pressure, HRR, and emissions with premixed ratio in RCCI mode

compared to conventional modes, since NO_x strongly depends on temperature. At low loads, NO_x emissions were reduced by 85% and 81% in E15D85/E15D85 and diesel/E15D85 PCCI modes, respectively (Srihari et al., 2021).

Authors investigated the effect of alcohol blends such as methanol and ethanol in LTC systems finding improvements in combustion efficiency and reductions in NO_x and soot. However, they noted increased HC and CO emissions, highlighting the need for optimized combustion chamber design and air–fuel ratio (AFR) control (Virt & Zöldy, 2024). It is reported that fuel spray characteristics at ultra-low temperatures relevant for Gasoline Compression Ignition applications, observing that fuel volatility and ambient conditions critically influence spray vaporization and mixing, which are key for optimal combustion efficiency in LTC systems (Zhao et al., 2024).

8.2.1.2 Air–Fuel Ratio

It is reported that the combustion efficiency decreases as AFR increases. In dual-fuel combustion systems, a combustion efficiency of 99% was achieved at an AFR of 25:1, dropping to 97% at an AFR of 50:1. The study also noted that EGR further reduced combustion efficiency by lowering in-cylinder temperatures due to the presence of stable molecules with high specific heat in the recirculated gas (Bae et al., 2023). The authors studied variations in AFR and intake charge temperature (ICT) affecting combustion dynamics in HCCI engines. Their findings revealed that leaner mixtures (higher AFR) increased combustion instability, whereas richer mixtures improved stability but elevated NO_x emissions, emphasizing the importance of AFR optimization to balance stability and emissions (Singh & Maurya, 2023). Figure 8.2 shows the influence of AFR on combustion efficiency and NO_x emission. On the leaner side, the combustion efficiency decreases with a reduction in NO_x emission due to a drop in in-cylinder temperature.

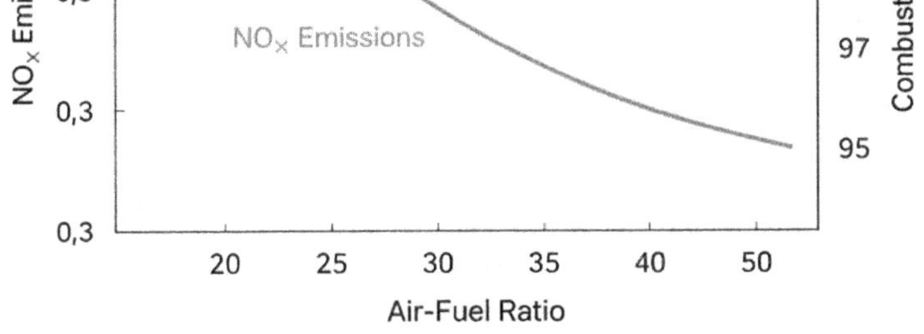

FIGURE 8.2 Trends of Combustion Efficiency and NO_x Emissions with air fuel ratio in LTC

Authors investigated the effects of high EGR rates (up to 64%) on flame temperature and emissions in diesel LTC. They observed that increasing EGR reduced peak flame temperatures from approximately 2580 K to 1400 K, significantly decreasing NO_x and soot emissions but increasing unburned hydrocarbons (HC) and carbon monoxide (CO) by 30%–50%, along with a 5%–7% increase in fuel consumption. Charge dilution was more influential on combustion than IT, although advancing IT helped maintain stability under high dilution (Sarangi et al., 2023).

Authors studied the impact of diesel-gasoline blending ratios on particulate emissions in LTC. A dual-fuel blend of 15% diesel and 85% gasoline (D15G85) yielded the most significant PM reduction. Particle size and number concentration varied with fuel ratio, affecting aftertreatment strategies (Barman & Deshmukh, 2023).

8.2.1.3 Intake Charge Temperature

ICT and pressure are critical parameters in LTC systems such as HCCI, PCCI, and RCCI, influencing ignition timing, stability, emissions, and efficiency. Boost pressure and intake temperature synergistically enable flexible combustion control in RCCI by adjusting phasing and fuel reactivity. Higher boost pressures suppress knock and permit earlier IT (Singh & Maurya, 2023; Virt & Zöldy, 2024).

Increased ICT shortens ignition delay by accelerating combustion reactions, advancing combustion timing. Moderate ICT elevation improves timing control in HCCI and RCCI engines, especially for LRFs like gasoline or ethanol, which otherwise struggle to auto-ignite at low temperatures (Yao et al., 2009; Singh & Maurya, 2023). Low ICT results in misfires or incomplete combustion due to insufficient energy, critical during cold starts or low loads (Zhao et al., 2024). Excessively high ICT can cause advanced combustion and knock-like behavior, increasing pressure rise rates and engine noise (Sarangi et al., 2023). Generally, increasing ICT lowers soot emissions by enhancing oxidation but increases NO_x due to higher combustion temperatures (Kokjohn et al., 2011). Lower ICT reduces HC and CO emissions by improving combustion completeness. Boosting intake pressure increases charge density, allowing more EGR or lean mixtures and enabling LTC operation without misfire (Barman & Deshmukh, 2023). Boosting extends LTC load limits by ensuring sufficient oxygen availability for complete combustion (Reitz et al., 2014).

8.2.2 Effect of Operating Parameters on LTC

8.2.2.1 Injection Timing (IT)

IT is a critical parameter in LTC strategies including HCCI, PCCI, and RCCI. It directly influences mixture formation, ignition phasing, combustion stability, and emissions.

IT determines the start of combustion by influencing the timing and extent of low-temperature heat release. In HCCI and PCCI, advanced IT promotes better premixing, which helps control ignition phasing, while retarded timing negatively affects combustion (Kokjohn et al., 2011; Soudagar et al., 2023). In RCCI, precise control of dual-fuel IT (low-reactivity port fuel vs. high-reactivity direct injection) is essential to manage reactivity stratification and phasing (Barman & Deshmukh, 2023; Singh & Maurya, 2023). Advanced IT may result in higher CO and HC emissions due to over-lean mixtures and lower combustion temperatures (Sarangi et al., 2023). Lower combustion temperature poses difficulty in oxidation of these emissions. Retarded IT increases soot risk due to poorer premixing and local fuel-rich zones but can reduce NO_x owing to lower peak temperatures (Reitz et al., 2014). Generally, advanced IT yields higher peak pressures, while excessively delayed timing can cause incomplete combustion (Yao et al., 2009; Virt & Zöldy, 2024). Optimized timing maximizes thermal efficiency by aligning peak pressure with the optimal crank angle after top dead center (TDC) (Kokjohn et al., 2011). Split injection strategies further balance mixing and stratification in RCCI and PCCI modes, improving efficiency and emissions control (Singh & Maurya, 2023).

8.2.2.2 Spray Geometry and Injection Pressure

Spray geometry parameters such as spray tip penetration, spray angle, and spray cone angle play an important role in combustion. It is reported that reducing spray angle increased cylinder gas pressure by directing more fuel toward the high-temperature piston bowl instead of the squish region. For example, reducing the spray angle from 74° to 50° caused a pressure rise of ~20 bar. Narrow spray angles (25° and 35°) can lead to fuel droplet impingement on the piston bowl surface, especially when the spray is perpendicular, causing poor air-fuel mixing and reduced peak HRR. Maximum cylinder temperature was observed at 35°, and the lowest at 74°. At 55°, in the early expansion stroke, ignition of some fuel pockets led to rapid temperature rise. Reactivity index remained below the critical 5 MW/m^2 limit for all angles except 25°. The highest thermal efficiency (~43%) was obtained at 55°, while it was the lowest at 35° (~38%). Narrow spray angles with partial premixed fuel injection resulted in higher HC and CO emissions due to incomplete combustion (Nazemi et al., 2023). Figure 8.3 illustrates the process of spray impingement and air–fuel mixing with varying spray cone angles in a piston bowl.

Higher fuel injection pressure reduces droplet size, enhancing evaporation and mixture formation, shortening physical ignition delay. However, higher pressure can increase spray–wall impingement, reducing combustion efficiency. Engine speed affects indicated mean effective pressure (IMEP) more in EGR mode than without,

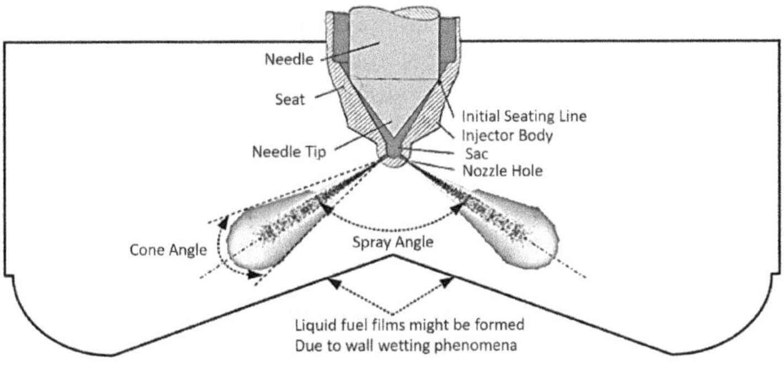

FIGURE 8.3 Illustration of spray impingement and air–fuel mixing with varying spray cone angles in a piston bowl

due to smoother combustion and reduced specific fuel consumption. Singh et al. noted smoke reduction at high fuel injection pressures. Advancing IT raised HC and NO_x emissions, while increasing fuel injection pressure increased HC but reduced NO_x emissions (Singh et al., 2023). Optimal combustion was found at 400 bar due to better diffusion combustion and higher cylinder temperature which is the characteristic of PCCI mode. At 700 bar, wall wetting lowered peak temperature and NO_x emission. Changing pressure from 400 to 1000 bar improved air–fuel mixing, and reduced the PM concentration (Singh et al., 2023).

Split or multiple injection strategies shortened combustion duration compared to single injections, further improved by increased injection pressure. For example, 10% fuel injected early at 54° bTDC and the rest at 10° bTDC allowed pilot ignition to enhance vaporization and oxidation of the main fuel. At low load and 200 bar, single injection produced higher cylinder temperatures (1340 K) than split injection (1150 K), whereas at higher pressures, split injection yielded higher temperatures (1500 K vs. 1250 K). Low injection pressure single injections prolong stoichiometric combustion, while split injections promote lean combustion and avoid wall wetting at high pressures (Singh & Maurya, 2023).

8.2.2.3 Combustion Chamber

The geometry of the combustion chamber plays a crucial role in the LTC modes. It significantly affects in-cylinder air motion, fuel-air mixing, combustion phasing, and emission characteristics. Piston bowl geometry, such as Toroidal Re-entrant Combustion Chamber (TRCC), enhances turbulence and fuel–air mixing, leading to better combustion efficiency and reduced NO_x and PM (Mishra et al., 2023;). TRCCs, in particular, have demonstrated a 3%–5% improvement in BTE over traditional designs. However, flat or hemispherical bowls offer less effective mixing, leading to increased emissions (Christensen et al., 2002). The geometry also governs the development of swirl and tumble motion, which are essential for uniform charge preparation and low-temperature ignition (Lin et al., 2000). Improved in-cylinder turbulence supports more homogeneous combustion, reducing soot and NO_x formation (Vressner et al., 2008). However, excessive surface area or poorly directed spray impingement due to chamber design may result in higher unburned hydrocarbons and CO emissions Thus, optimizing combustion chamber geometry remains a key strategy in achieving cleaner and more efficient LTC engine performance. In conclusion, the combustion chamber geometry strongly affects the movement and mixing of charge. Better mixing of charge causes efficient combustion and lowering of emissions.

8.3 CONTROL AND OPTIMIZATION OF LOW-TEMPERATURE COMBUSTION

LTC strategies offer significant benefits in thermal efficiency and emissions reduction. However, the narrow controllability window of LTC combustion makes optimization essential to achieve stable, efficient, and low-emission engine operation across various loads and speeds. LTC modes operate at lower peak combustion temperatures (~1500 K) compared to conventional diesel combustion, reducing NO_x and soot simultaneously. However, challenges persist such as controlling combustion phasing without direct ignition control, increased hydrocarbon (HC) and carbon monoxide (CO) emissions due to incomplete oxidation, and maintaining high BTE with low NO_x emissions. The techniques to control and optimize combustion are based on the parameters that affect the LTC. The methods and techniques for control and optimization of LTC are discussed below.

8.3.1 Variable Valve Timing (VVT)

VVT plays a critical role in controlling combustion in LTC modes. VVT adjusts the effective compression ratio and residual gas fraction, significantly influencing the ignition delay. It enables better control of in-cylinder temperature and composition, aiding combustion phasing in HCCI. By varying intake and exhaust valve timing, VVT controls temperature, pressure, residual fraction of gases, and ignition delay, parameters essential for precise LTC combustion management (Murata et al., 2006; Guan et al., 2020). Residual gases decrease the in-cylinder temperature by dilution effect. It affects the ignition delay time.

Early Intake Valve Closing reduces the effective compression ratio by closing the intake valve before bottom dead center (BDC), limiting inducted air and increasing residual gases. This lowers peak combustion temperatures, suppressing NO_x and facilitating autoignition in LTC (Murata et al., 2006). Late Intake Valve Closing, where the intake valve closes after BDC during compression, causes intake backflow, reducing effective compression ratio and temperature rise, thus better controlling ignition timing and mitigating knock and pressure rise rates (Ashkezari, 2016).

Adjusting valve overlap allows internal EGR by recirculating exhaust gases into the cylinder, diluting fresh charge, lowering oxygen availability, and cooling combustion—all favorable for LTC (Ojeda, 2010). VVT facilitates advanced cycles like the Miller cycle, applying early or late intake valve closing to reduce compression work while maintaining expansion, supporting lower combustion temperatures, NO_x reduction, and LTC stability (Luján et al., 2020). Advanced VVT strategies demonstrate effective combustion phasing control across various operating conditions in light- and heavy-duty engines, enabling stable LTC operation with reduced CO, HC, and NO_x emissions (Guan et al., 2020).

8.3.2 Exhaust Gas Recirculation

By recirculating a portion of the exhaust gas back into the intake manifold, EGR effectively dilutes the intake charge, lowers peak combustion temperatures, and delays the onset of autoignition. This makes EGR a pivotal technique in LTC regimes for suppressing NO_x and controlling combustion phasing (Zhang et al., 2022). In LTC modes such as HCCI, PCCI, and RCCI, EGR is extensively used to modulate in-cylinder conditions and achieve stable, low-emission combustion. However, a high percentage of EGR drops the combustion quality and thus decreases BTE of the engine.

The recirculated exhaust gas primarily comprises inert species like carbon dioxide (CO_2) and water vapor (H_2O), which possess high specific heat capacities. These components absorb significant amounts of combustion heat, thereby lowering the adiabatic flame temperature and leading to a thermal buffering effect. This thermal moderation smoothens the pressure rise and facilitates a more uniform temperature distribution, both of which are essential for achieving combustion stability in LTC (Modabberian, 2024).

Moreover, EGR reduces the concentration of oxygen in the intake charge, which in turn suppresses the formation of thermally activated NO_x via the Zeldovich mechanism. This oxygen dilution prolongs the ignition delay, enabling more time for air–fuel mixing and promoting charge homogeneity, which significantly reduces the likelihood of localized high-temperature regions responsible for soot and NO_x formation (Dec, 2009). Enhanced ignition delay also assists in decoupling the end of fuel injection and start of combustion, enabling better control over combustion phasing in RCCI and PCCI modes (Singh et al., 2023). Figure 8.4 shows the effect of EGR on in-cylinder temperature and emissions.

Experimental studies have shown that both high-pressure and low-pressure EGR systems can be optimized for different LTC strategies. For example, in HCCI engines, external EGR helps control combustion timing and prevents misfire at part loads, while internal EGR (achieved through negative valve overlap) can be more effective in light-duty applications (Kokjohn et al., 2011). EGR also mitigates the high pressure rise rate and combustion knock phenomena often encountered in LTC regimes, improving engine durability and acoustic performance (Ghazimirsaied et al., 2020).

In modern control systems, EGR is dynamically coupled with other LTC control parameters such as intake boosting, VVT, and fuel reactivity to maintain

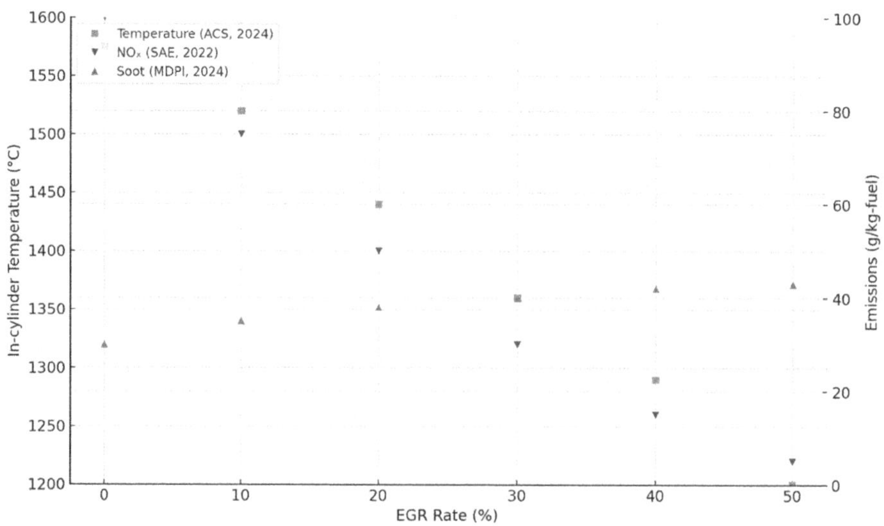

FIGURE 8.4 Effect of EGR on cylinder Temperature and Emissions

combustion stability across a wide load range. The effectiveness of EGR is also influenced by the temperature and composition of the recirculated gases; cooled EGR systems have been found more effective in reducing combustion temperature and NO_x, while hot EGR can assist with ignition under cold-start conditions (Wang et al., 2018). As LTC strategies become more advanced, especially with the use of alternative and LRFs, EGR will remain a cornerstone of emission control and combustion management.

8.3.3 Fuel Injection Strategies

Fuel injection strategies are critical for managing combustion phasing, ignition delay, and emissions in advanced LTC modes such as HCCI, Partially Premixed Compression Ignition (PPCI/PCCI), and RCCI. In these regimes, combustion is governed by chemical kinetics and mixture preparation rather than flame propagation, making precise control of injection parameters essential (Reitz & Duraisamy, 2015). Tailoring IT, pressure, duration, and pulse count enables fine-tuning of combustion to achieve high efficiency with ultra-low emissions. Split or multistage injection techniques are commonly employed to control the degree of premixing and stratification. These strategies improve combustion control by manipulating ignition timing and the spatial distribution of the air–fuel mixture, thereby reducing the rate of pressure rise and lowering peak in-cylinder temperatures (Ghazikhani et al., 2017; Zhang et al., 2022). Dividing fuel injection into pilot, main, and postinjection pulses offers several advantages. Pilot injection initiates early heat release, stabilizing ignition and allowing smoother pressure rise. Main injection delivers the bulk of the fuel, while postinjection can aid in the oxidation of soot. This approach not only enhances air–fuel mixing but also facilitates more complete combustion, reducing emissions of unburned hydrocarbons (HC), carbon monoxide (CO), and PM (Ciatti, 2010; Musculus et al., 2013). Early IT favors better premixing and reduces soot and NO_x formation by promoting low-temperature heat release. However, excessively early injection can lead to overly lean mixtures and wall impingement, increasing CO and HC emissions due to incomplete combustion in over-diluted zones (Zhao et al., 2007; Benajes et al., 2016). High injection pressures enhance atomization and spray penetration, promoting uniform mixture formation. Nevertheless, very high pressures may reduce ignition delay excessively, leading to advanced combustion phasing and higher NO_x due to increased peak temperatures (Zhang et al., 2022). Multiple injections, such as triple or quadruple injection strategies, are gaining attention for their ability to distribute heat release over a longer duration, resulting in smoother pressure traces and improved combustion stability. These strategies are particularly effective in RCCI engines where precise control of reactivity gradients and stratification is essential (Kokjohn et al., 2011; Reitz & Duraisamy, 2015). Furthermore, modern diesel and dual-fuel engines increasingly rely on adaptive injection systems integrated with closed-loop feedback for optimal performance under variable operating conditions (Miers et al., 2012; Malan et al., 2023).

8.3.4 Reactivity Gradient Control

Reactivity Gradient Control (RGC) has emerged as a pivotal strategy in LTC, particularly within RCCI systems. RGC enables precise modulation of combustion phasing, HRR, and emission characteristics by manipulating the spatial distribution of fuel reactivity within the combustion chamber (Wissink et al., 2018; Wissink & Reitz, 2016; Soloiu et al., 2019). This is typically accomplished through the use of two fuels with markedly different reactivities—commonly an LRF such as gasoline, iso-octane, or ethanol delivered via port fuel injection (PFI), and an HRF like diesel, *n*-heptane, or DME injected directly (DI) into the cylinder (Dempsey et al., 2014; Kokjohn et al., 2011).

The resulting stratified reactivity profile allows for tailored ignition behavior: low-reactivity regions delay autoignition, while high-reactivity zones initiate combustion, yielding a spatially distributed and temporally extended heat release. This stratification smooths pressure rise rates and suppresses combustion knock, thus enhancing both thermal efficiency and mechanical durability (Reitz & Duraisamy, 2015; Hanson et al., 2010). The gradual energy release also reduces combustion noise—an important factor in LTC acceptance for commercial applications.

Through dynamic control of IT, fuel quantity, and charge mixing, RGC facilitates real-time combustion phasing control across varying loads and speeds. Notably, studies have demonstrated that adjusting the proportion and timing of PFI and DI fuels allows precise tuning of ignition delay, combustion duration, and emissions trade-offs (Kokjohn et al., 2011; Dempsey et al., 2014). This level of control is critical for maintaining low emissions of NO_x and PM)under different operating conditions. The avoidance of localized high-temperature regions limits NO_x formation through the Zeldovich mechanism, while the premixed nature of RCCI combustion curtails soot by reducing fuel-rich zones (Soloiu et al., 2019; Wissink et al., 2018).

Advanced engine simulation and optical diagnostics have provided deeper insights into RGC dynamics. Three-dimensional computational fluid dynamics (CFD) models incorporating detailed chemical kinetics have been instrumental in evaluating fuel stratification, ignition timing, and pollutant formation under various RGC strategies (). Additionally, experimental studies using chemiluminescence and laser-based diagnostics confirm that reactivity gradients correlate with improved combustion uniformity and reduced combustion variability (Splitter & Reitz, 2014).

As LTC technology evolves, RGC is being further refined through machine learning (ML) models and closed-loop control strategies that optimize fuel reactivity distribution in real time (Duan et al., 2022). The compatibility of RGC with a wide array of alternative fuels—including biofuels and synthetic fuels—also underscores its potential to support decarbonization objectives in future powertrain systems.

8.3.5 Temperature and Pressure of Intake Charge

Intake charge heating involves increasing the temperature of the air–fuel mixture before it enters the combustion chamber. This technique directly impacts the chemical kinetics of autoignition and combustion phasing in LTC modes. Warmer intake air enhances mixture reactivity by promoting low-temperature oxidation pathways, thus reducing ignition delay and aiding cold-start conditions (Zhao, 2007). For fuels with higher octane number or low CN such as gasoline or ethanol, intake heating helps initiate and sustain combustion by compensating for slow chemical kinetics (Ganesh & Subramanian, 2014). Moreover, intake charge heating improves fuel–air mixture homogeneity, crucial for uniform combustion in HCCI engines. However, excessive heating may lead to early ignition or knocking, especially at higher loads, necessitating precise temperature control (Yao et al., 2009).

Intake boosting, via turbocharging or supercharging, increases intake pressure and air density, enhancing mixture preparation and in-cylinder pressure and temperature during compression. Boosting expands LTC engine load range by compensating for the low energy density of dilute or lean mixtures (Dempsey et al., 2013). By increasing pressure and improving mixing, boosting advances ignition timing to optimal phasing while enhancing combustion completeness and thermal efficiency. Higher intake pressures promote turbulence and stratification, beneficial in partially premixed combustion, allowing better control over HRR and mitigating overly rapid combustion typical in pure HCCI (Saxena & Bedoya, 2013). Figure 8.5 depicts a 3D plot showing BTE as a function of intake temperature and boost pressure. This model assumes Peak BTE (~50%) occurs around 130°C intake temperature and 2.2 bar boost pressure. Both too low and too high intake conditions reduce efficiency due to incomplete combustion or engine knock.

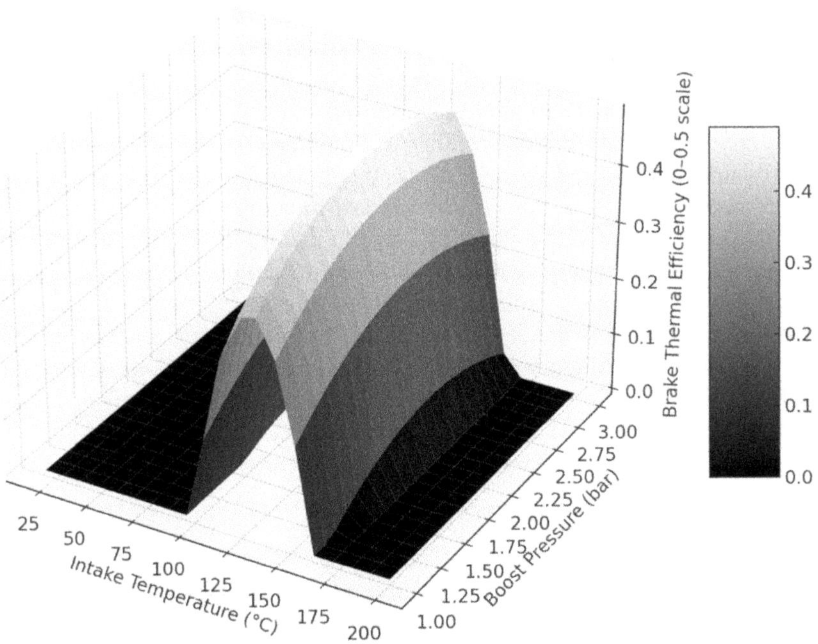

FIGURE 8.5 3D surface plot illustrating the impact of intake temperature and boost pressure on Brake Thermal Efficiency (BTE) in LTC engines

Combining intake heating and boosting enables effective LTC control. Heating is more beneficial at low loads or during cold start, while boosting is essential at medium to high loads to prevent misfire and maintain efficiency. Trade-offs exist: heating increases NO_x at higher temperatures, and excessive boosting may raise soot due to localized rich zones (Yao et al., 2009). Thus, advanced control systems and sensors dynamically adjust intake temperature and pressure to optimize LTC across operating conditions.

8.3.6 Advanced Control Systems

Advanced control systems are critical for successful LTC implementation and optimization. Unlike conventional diesel or spark-ignition engines, LTC lacks direct ignition control, making it highly sensitive to temperature, pressure, and reactivity variations. Advanced electronic controls ensure stable, efficient, and clean operation (Carlucci et al., 2014; Ou et al., 2022; Malan et al., 2023). The Closed-Loop Combustion Control (CLCC), Model Predictive Control (MPC), and virtual sensors enhance the monitoring and control of LTC.

CLCC uses real-time in-cylinder pressure feedback, such as crank angle at 50% heat release (CA50), to dynamically adjust fuel IT, quantity, and intake conditions, maintaining optimal combustion timing over varied loads and speeds. Carlucci et al. (2014) showed CLCC improves LTC combustion stability, reduces cycle-to-cycle variation, and facilitates cold starts and transient operation. MPC forecasts engine behavior to optimize control inputs in LTC engines with nonlinear and variable combustion characteristics. MPC tunes injection strategies, valve timings, and EGR rates to maintain target combustion phasing and emissions. Ou et al. (2022) reported MPC significantly enhances LTC responsiveness and accuracy under rapid load changes. Due to the challenges and cost of physical sensors in harsh in-cylinder environments, virtual sensors estimate critical combustion parameters like in-cylinder temperature, pressure profiles, and ignition delay. Integrated into the ECU and combined with real sensor data, these models provide real-time high-fidelity estimates. Malan et al. (2023) demonstrated that virtual sensors in diesel LTC engines achieve precise combustion control comparable to physical instrumentation. Advanced LTC engines often employ hybrid control architectures integrating closed-loop feedback with predictive and estimation strategies, adapting to fuel variability, ambient conditions, and driver demands while maintaining low NO_x and soot emissions (Carlucci et al., 2014; Ou et al., 2022). Such systems support flexible operation across wide loads and speeds, addressing major LTC commercialization challenges.

8.3.7 Modeling and Simulation Techniques

Modeling and simulation are pivotal in advancing LTC technologies. Accurate modeling facilitates understanding combustion complexities, optimizing engine performance, and reducing reliance on extensive experimental testing.

Zero-dimensional (0D) and quasi-dimensional models simplify the combustion chamber into single or multiple zones, enabling rapid simulations of pressure, temperature, and HRRs, useful for preliminary design and control development.

Kong and Reitz (2003) used detailed chemical kinetics in a 0D framework to study fuel composition effects on HCCI combustion. Yao et al. (2020) provided insights into HCCI engine trends using such models.

Multi-zone models improve fidelity by dividing the chamber into zones with distinct thermodynamics, capturing spatial variations critical for LTC. Sadabadi et al. (2015) developed a physics-based control-oriented model for RCCI combustion validated against detailed multi-zone thermo-kinetic simulations, effectively predicting combustion phasing, and IMEP.

CFD is an essential tool for studying turbulent flow, fuel–air mixing, spray-wall interaction, evaporation, ignition timing, and heat transfer in LTC strategies (Patil et al., 2014; Yao et al., 2009). Advanced CFD packages like CONVERGE and AVL FIRE M use detailed spray, combustion chemistry, and pollutant formation submodels under real engine conditions AVL, 2021). High-resolution turbulence models such as Large Eddy Simulation and Partially Averaged Navier-Stokes resolve turbulent eddies affecting autoignition and heat release (Kokjohn et al., 2011). These simulations optimize combustion chamber design, injector orientation, swirl/tumble, and injection strategies, improving combustion phasing, minimizing RI, and controlling NO_x and soot via optimized air–fuel interactions and EGR (Saxena & Bedoya, 2013; Reitz et al., 2010).

Sophisticated modeling tools capture complex physical and chemical interactions such as autoignition, stratification, heat release, and emissions. The Stochastic Reactor Model Engine Suite is particularly effective for LTC modes like HCCI and PPCI, using stochastic particle-based modeling to represent spatial in-cylinder variations critical for ignition delay, combustion phasing, and emissions prediction (Andrae et al., 2007).

8.4 DIAGNOSTIC TOOLS FOR LTC CONTROL AND OPTIMIZATION

Diagnostic tools enable detailed characterization of combustion processes, emissions, and thermal dynamics in LTC engines. It is important for developing and refining control strategies to enhance efficiency, reduce emissions, and ensure reliable operation across different operating conditions. This section discusses the tools helpful in the study of LTC control and optimization. In-cylinder pressure measurement via piezoelectric transducers provides high-resolution real-time data to derive HRR, ignition delay, and combustion phasing (CA50) parameters crucial for evaluating combustion quality. Figure 8.6 shows the chart of various diagnostic tools required for diagnosis of parameters in LTC.

Optical access engines equipped with quartz windows or Bowditch configurations enable high-fidelity imaging of combustion phenomena. Techniques like high-speed imaging, laser-induced fluorescence (LIF), and chemiluminescence imaging visualize flame development, ignition kernels, fuel–air mixing, and pollutant formation. LIF measures species such as formaldehyde (HCHO), OH radicals, and fuel distribution, critical markers of LTC behavior (Sick & Stojkovic, 2001).

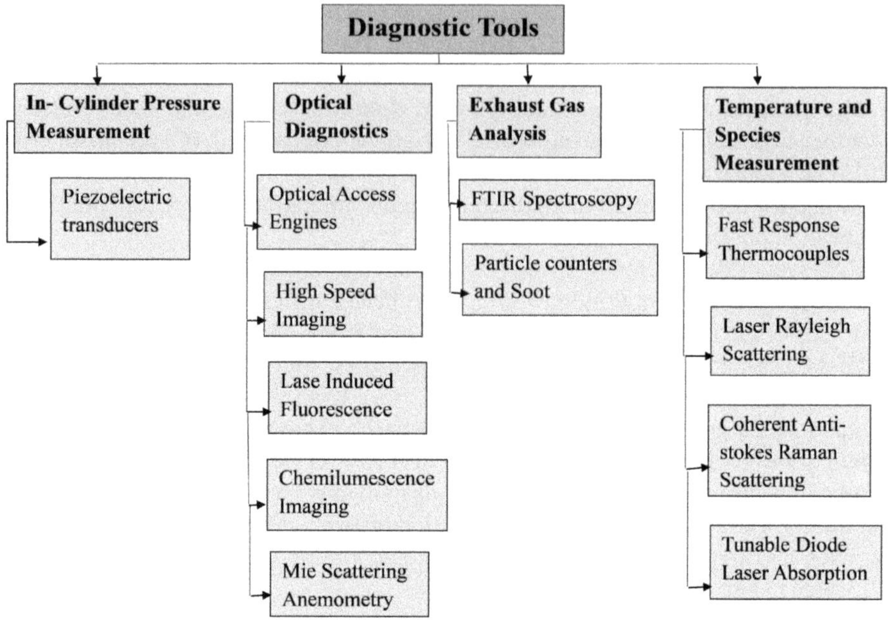

FIGURE 8.6 Schematic representation of diagnostic tools used for LTC engine analysis

Chemiluminescence imaging of radicals such as CH*, OH*, and CO* reveals low-temperature heat release and ignition evolution.

Exhaust gas analyzers measure NO_x, CO, HC, and PM using non dispersive infrared radiation (NDIR) principle, free induction decay (FID) spectroscopy, and chemiluminescence detectors for steady-state and transient analysis. Ultra-low soot emissions in LTC require advanced particle counters and soot photometers like Engine Exhaust Particle Sizer and Scanning Mobility Particle Sizer to characterize particle size and number concentrations (Saxena & Bedoya, 2013). Fourier transform infra red (FTIR) spectroscopy is a powerful tool to examine engine exhaust emissions by simultaneously measuring both regulated and unregulated substances. Its utility extends to analyzing the impacts of alternative fuels, NO_x reduction systems, and catalyst effectiveness.

Spray dynamics, atomization, and fuel–air mixing in LTC are studied using optical diagnostics such as Mie scattering, Schlieren imaging, and Phase Doppler Anemometry, correlating injection parameters with combustion characteristics. Schlieren imaging visualizes air–fuel stratification critical for RCCI ignition control (Bai et al., 2019).

Fast-response thermocouples, laser Rayleigh scattering, and Coherent Anti-Stokes Raman Scattering measure local gas temperatures, validating CFD models and understanding heat release. Species-specific probes like Tunable Diode Laser Absorption Spectroscopy provide nonintrusive, high-resolution sensing of temperature and species such as H_2O and CO_2 under engine conditions (Goldenstein et al., 2014).

8.5 CASE STUDIES IN LTC

The practical implementation of LTC in modern IC engines has advanced significantly in recent years through experimental validation and simulation-driven design. Several case studies illustrate the viability of LTC modes in reducing NO_x and PM emissions while maintaining acceptable thermal efficiency. These studies highlight the challenges, control strategies, and application potentials of LTC in passenger cars, commercial vehicles, and marine engines.

An early demonstration of HCCI combustion was conducted by the Massachusetts Institute of Technology (MIT), where a gasoline engine retrofitted with HCCI control demonstrated ultra-low NO_x emissions and improved part-load efficiency. HCCI operated successfully under variable load and speed conditions using precise control of intake temperature, valve timing, and EGR. However, misfire and combustion instability limited high-load operation, necessitating mode switching to spark-ignition at higher speeds (Zhao et al., 2009.

RCCI was extensively investigated at the University of Wisconsin–Madison. Kokjohn et al. (2011) demonstrated that dual-fuel RCCI using gasoline and diesel achieved over 50% BTE with near-zero NO_x and soot emissions. In a multicylinder heavy-duty engine, RCCI extended the operating range compared to HCCI, showing promise for long-haul transport. Caterpillar Inc. incorporated RCCI into their heavy-duty platform with direct diesel injection and port-injected gasoline, controlling combustion phasing via fuel reactivity gradients and EGR (Splitter et al., 2012). PPC was explored by Lund University using a Volvo D5 engine running PPC with gasoline and ethanol, exhibiting low NO_x and soot while maintaining high efficiency with late injection strategies. PPC provided flexible combustion control via IT and pressure, improving stability and cold-start performance relative to HCCI (Kalghatgi et al., 2013).

MAN Energy Solutions demonstrated LTC in large-bore marine engines using methanol and natural gas in RCCI and dual-fuel modes. These configurations leveraged fuel flexibility and autoignition control to reduce SO_x and NO_x emissions, meeting IMO Tier III compliance. The marine sector offers fertile ground for large-scale LTC deployment, especially when combined with clean fuels and hybridization (Kjarstad & Johnsson, 2019).

8.6 FUTURE DIRECTIONS IN LTC

LTC technologies in IC engines represent a crucial pathway to achieving ultra-low emissions and high thermal efficiency. This section explores key research directions and technological opportunities for advancing LTC combustion optimization. By leveraging hybridization, intelligent control, advanced fuels, and high-fidelity simulations, LTC combustion technologies can transition from laboratory-scale demonstrations to real-world deployment, supporting the automotive industry's shift toward cleaner, more efficient IC engines.

Integrating LTC combustion into hybrid-electric vehicles offers the opportunity to circumvent LTC limitations during cold starts and high-load operation. The electric motor can compensate for LTC inefficiencies in off-design conditions

(Reitz et al., 2020). Emerging sensor technologies, such as ion current sensors and fiber-optic-based pressure transducers, offer the potential for more affordable and durable real-time combustion monitoring. Additionally, artificial intelligence (AI) and ML algorithms are being developed to predict combustion behavior and adapt control strategies dynamically (Vohra et al., 2017). Furthermore, hybrid control systems can dynamically switch between LTC and conventional combustion modes to ensure drivability and efficiency.

The nonlinearity and multivariable nature of LTC combustion processes present a significant challenge to traditional control algorithms. In response, researchers are increasingly applying AI and ML techniques to model, predict, and optimize LTC behavior. Data-driven models, such as neural networks and reinforcement learning algorithms, can learn from historical combustion data to identify patterns, predict combustion phasing, and adapt control strategies in real time (Vohra et al., 2017). These intelligent systems offer the potential to manage complex engine conditions without explicit physical modeling, improving robustness and adaptability across varying fuel types and operational scenarios.

8.7 CONCLUSIONS

LTC strategies represent a transformative pathway toward achieving high thermal efficiency and ultra-low emissions in IC engines. However, the inherently nonlinear and multiparameter nature of LTC necessitates a detailed understanding of the coupled physical and chemical processes involved. This chapter has provided an in-depth evaluation of the critical factors affecting LTC performance, including IT, spray geometry, combustion chamber architecture, premixed charge properties, and AFR. The role of intake temperature and pressure was also explored as a means of modulating in-cylinder reactivity and combustion phasing.

Control methodologies such as VVT, EGR, and RGC were shown to be effective in shaping heat release profiles and stabilizing combustion. In particular, advanced fuel injection strategies such as split and multi-pulse injection enable precise regulation of ignition timing and mixture stratification. These strategies are further augmented by the integration of adaptive and model-based control systems, which leverage real-time combustion feedback to dynamically optimize performance within stringent operational constraints.

Simulation techniques, spanning 0D chemical kinetics models to full three-dimensional CFD simulations, were highlighted for their ability to capture the complex thermochemical behavior inherent to LTC modes. These tools serve not only as platforms for mechanistic understanding but also as enablers for the predictive design of robust control strategies. Diagnostic tools, including high-resolution in-cylinder pressure transducers, laser-based diagnostics, and advanced spectroscopic methods, offer high-fidelity insights into ignition dynamics, pollutant formation, and combustion stability.

The inclusion of case studies has underscored the practical applicability of LTC strategies across diverse engine architectures, validating simulation models against experimental data and demonstrating the synergistic benefits of integrated control, diagnostics, and modeling. The collective evidence supports the feasibility of optimized LTC operation as a pathway to next-generation engine technologies.

Parameter	HCCI (Homogeneous Charge Compression Ignition)	PCCI (Premixed Charge Compression Ignition)	RCCI (Reactivity Controlled Compression Ignition)
Mixture preparation	Fully homogeneous (premixed air–fuel)	Early direct (DI) or port fuel injection (PFI) (partially mixed)	Dual-fuel: low-reactivity (PFI) + high-reactivity (DI)
Ignition control	Autoignition; kinetically controlled	Controlled by injection timing (IT) and exhaust gas recirculation (EGR)	Controlled by fuel reactivity and injection strategies
Fuel type	Gasoline, alcohols, and light HC fuels	Diesel, biodiesel, and gasoline blends	Dual fuels (e.g., gasoline + diesel and natural gas + diesel)
Combustion temperature	Low (850–1200 K)	Low to moderate	Low (high dilution with EGR and stratification)
Emissions (NO_x/PM)	Very low/near zero	Low/low	Ultra-low/ultra-low
Thermal efficiency	High (≈45–50%)	Moderate to high (40–48%)	Very high (up to 60%)
Load range	Narrow (struggles at high load)	Moderate	Wide (light to medium-heavy load)
Control complexity	High (due to phasing unpredictability)	Moderate	High (requires dual-fuel system and control strategy)
Cycle-to-cycle variability	High	Moderate	Low (better phasing control)
Transient operation	Poor	Fair	Good
Need for EGR	High	Moderate to high	High
Commercial feasibility	Limited (research stage)	Emerging	Promising (pilot trials ongoing)

REFERENCES

A. Barman and P. Deshmukh, "Particulate emissions from diesel-gasoline blending ratios in LTC engines," *Appl. Energy*, vol. 322, p. 119416, 2023.

A. B. Dempsey, N. R. Walker, E. Gingrich, and R. D. Reitz, "Comparison of Low Temperature Combustion Strategies for Advanced Compression Ignition Engines with a Focus on Controllability," *Combust. Sci. Technol.*, vol. 186, no. 2, pp. 210–241, 2014.

A. Dempsey, R. D. Reitz, and Y. Yang, "Comparison of low temperature combustion strategies for advanced compression-ignition engines with a focus on controllability," *Int. J. Engine Res.*, vol. 15, no. 1, pp. 97–114, 2014.

A. Ghazimirsaied, A. Jaichandar, and K. Kannan, "Effect of EGR on combustion noise and pressure rise rate in LTC engines fueled with biofuels," *Appl. Therm. Eng.*, vol. 165, p. 114556, 2020.

A. Hosseini, M. Karimi, and A. Tabejamaat, "Effects of cetane number on RCCI combustion and emissions characteristics using biodiesel and compressed natural gas," *Fuel*, vol. 242, pp. 102–110, 2019.

A. K. Agarwal, "Biofuels (alcohols and biodiesel) applications as fuels for internal combustion engines," *Prog. Energy Combust. Sci.*, vol. 33, no. 3, pp. 233–271, 2007.

A. Mishra, S. Kulshrestha, F. M. Patel, N. Tiwari, and A. Sharma, "Effect of piston bowl geometry and spray angle on engine performance and emissions in HCCI engine using multi-stage injection strategy," *Env Prog and Sustain Energy*, vol. 43, no. 2, p. e14203, 2024.

A. Modabberian et al., "Low Temperature Combustion Modeling and Predictive Control of Marine Engines," *Applied Sciences*, vol. 14, no. 5, p. 2033, 2024.

A. P. Carlucci, D. Laforgia, S. Motz, R. Saracino, and S. P. Wenzel, "Advanced closed loop combustion control of a LTC diesel engine based on in-cylinder pressure signals," *Energy Conversion and Management*, vol. 77, pp. 193–207, 2014.

A. Soloiu, S. Wissink, and R. Reitz, "Noise and emission reduction in RCCI engines via reactivity gradient control," *Int. J. Engine Res.*, vol. 20, no. 2, pp. 186–200, 2019.

A. Vressner, A. Hultqvist, and B. Johansson, "Study on Combustion Chamber Geometry Effects in an HCCI Engine Using High-Speed Cycle-Resolved Chemiluminescence Imaging," SAE Paper 2007-01–0217, 2007.

A. Z. Ashkezari, A. H. Nezhad, and S. Farahat, "Reduction of pollutant emissions by developing a variable valve timing system in a direct injection diesel engine using computational fluid dynamics modeling," *Env Prog and Sustain Energy*, vol. 35, no. 5, pp. 1430–1440, 2016.

B. Bae, J. Kim, and S. Lee, "Effect of air-fuel ratio on combustion efficiency in dual-fuel systems," *Int. J. Engine Res.*, vol. 24, no. 5, pp. 1234–1245, 2023.

C. Dec, "Advanced compression-ignition engines—understanding the in-cylinder processes," *Proc. Combust. Inst.*, vol. 32, no. 2, pp. 2727–2742, 2009.

C. Ojeda, "Internal EGR and combustion control using variable valve timing in diesel engines," *J. Eng. Gas Turbines Power*, vol. 132, no. 1, p. 012801, 2010.

D. A. Splitter and R. D. Reitz, "Fuel reactivity effects on the efficiency and operational window of dual-fuel compression ignition engines," *Fuel*, vol. 118, pp. 163–175, 2014.

D. L. Kokjohn, R. M. Hanson, D. A. Splitter, and R. D. Reitz, "Fuel reactivity controlled compression ignition (RCCI): A pathway to controlled high-efficiency clean combustion," *Int. J. Engine Res.*, vol. 12, no. 3, pp. 209–226, 2011.

D. Malan, R. Anderson, and M. Johannes, "Virtual sensor-based combustion control in LTC diesel engines," *IEEE Trans. Veh. Technol.*, vol. 72, no. 2, pp. 1345–1355, 2023.

D. Splitter, M. Wissink, D. DelVescovo, and R. D. Reitz, "RCCI Engine Operation Towards 60% Thermal Efficiency," SAE Paper 2013-01–0279, 2013.

D. Splitter, M. Wissink, S. Kokjohn, and R. D. Reitz, "Effect of Compression Ratio and Piston Geometry on RCCI Load Limits and Efficiency," SAE Paper 2012-01–0383, 2012.

G. Duan, J. Han, Z. Pan, and H. Liu, "Machine learning-based predictive control of RCCI combustion with reactivity gradient control," *Appl. Energy*, vol. 306, p. 117991, 2022.

G. Kalghatgi, A. A. Ericsson, and P. Holmen, "Partially premixed combustion: A path towards low NOx and soot emissions in diesel engines," *Fuel*, vol. 111, pp. 169–178, 2013.

G. T. Kalghatgi, "Fuel/engine interactions," *Int. J. Engine Res.*, vol. 15, no. 3, pp. 383–398, 2014.

Gao Z, Scott Curran J, Stuart Daw, Robert Wagner M., "Light duty drive cycle simulations of diesel engine out exhaust properties for an RCCI enabled vehicle," 8th US National Combustion Meeting, 2013.

H. Liu et al., "Effects of charge concentration and reactivity stratification on combustion and emission characteristics of a PFI-DI dual injection engine under low load condition," *Fuel*, vol. 231, pp. 26–36, 2018.

H. Zhao, *HCCI and CAI Engines for the Automotive Industry*, Woodhead Publishing, 2007.

H. Zhao et al., "Spray characteristics of gasoline fuel at ultra-low temperatures for GCI engines," *Fuel*, vol. 343, p. 127819, 2024.

IPCC, "Global warming of 1.5°C: An IPCC special report," 2018.

J. B. Heywood, *Internal Combustion Engine Fundamentals*, McGraw-Hill, 1988.

J. Benajes et al., "An investigation on the influence of injection pressure and orifice diameter on RCCI combustion using gasoline and diesel," *Fuel*, vol. 180, pp. 327–335, 2016.

J. C. G. Andrae, "Kinetic modeling of the influence of NO on the combustion phasing of gasoline surrogate fuels in an HCCI engine," *Energy Fuels*, vol. 27, no. 11, pp. 7098–7107, 2013.

J. Goldenstein et al., "TDLAS measurements of temperature and species in internal combustion engines," *Appl. Phys.* B, vol. 114, no. 3, pp. 371–384, 2014.

J. M. Luján et al., "High-pressure exhaust gas recirculation line condensation model," *Int. J. Engine Res.*, vol. 22, no. 2, pp. 407–416, 2021.

K. Khodadadi Sadabadi, "Modelling and Control of Combustion Phasing of an RCCI Engine," M.S. thesis, 2015.

L. Liu et al., "Optimization of premixed ratio and EGR on thermal efficiency in LTC," *Appl. Therm. Eng.*, vol. 156, pp. 120–128, 2019.

L. Zhang et al., "Hydrogen addition effects on emissions in RCCI engines," *Int. J. Hydrogen Energy*, vol. 45, no. 25, pp. 13348–13360, 2020.

M. Bai et al., "Optical diagnostics of spray dynamics in RCCI combustion," *Fuel*, vol. 242, pp. 749–758, 2019.

M. Ghazikhani et al., "Effect of injection strategy on performance and emissions in HCCI engines," *Energy Convers. Manag.*, vol. 136, pp. 404–416, 2017.

M. Kjarstad and J. Johnsson, "LTC combustion strategies for marine engines using alternative fuels," *Energy Procedia*, vol. 158, pp. 3332–3338, 2019.

M. Krishnamoorthi et al., "A review on low temperature combustion engines," *Renew. Sustain. Energy Rev.*, vol. 116, p. 109404, 2019.

M. L. Wissink et al., "Spray–Wall Interactions in a Small-Bore, Multicylinder Engine Operating With RCCI," *J. Eng. Gas Turbines Power*, vol. 140, no. 9, p. 092808, 2018.

M. Mohebbi et al., "Performance and emissions of a reactivity controlled light-duty diesel engine," *Applied Thermal Engineering*, vol. 134, pp. 214–228, 2018.

M. Musculus et al., "Conceptual models for partially premixed low-temperature diesel combustion," *Prog. Energy Combust. Sci.*, vol. 39, no. 2–3, pp. 246–283, 2013.

M. Virt and B. Zöldy, "Alcohol fuel blends for low temperature combustion engines: A review," *Fuel*, vol. 338, p. 127111, 2024.

M. Vohra et al., "Design analysis for optimal calibration of diffusivity in reactive multilayers," *Combust. Theory Modelling*, vol. 21, no. 6, pp. 1023–1049, 2017.

M. Wissink and R. Reitz, "The role of the diffusion-limited injection in direct dual fuel stratification," *Int. J. Engine Res.*, vol. 18, no. 4, pp. 351–365, 2017.

N. Kohse-Höinghaus et al., "An experimental and kinetic modeling study of the oxidation of alcohols," *Prog. Energy Combust. Sci.*, vol. 36, pp. 404–469, 2010.

N. Nazemi et al., "Effect of spray angle on combustion and emissions in RCCI engines," *Int. J. Engine Res.*, vol. 24, no. 8, pp. 1743–1756, 2023.

R. D. Reitz and G. Duraisamy, "Review of high-efficiency and clean RCCI combustion," *Prog. Energy Combust. Sci.*, vol. 46, pp. 12–71, 2015.

R. D. Reitz et al., "Multi-fuel combustion and fuel blending strategies," *Annu. Rev. Chem. Biomol. Eng.*, vol. 5, pp. 119–142, 2014.

R. D. Reitz et al., "Modeling advanced combustion modes," *Prog. Energy Combust. Sci.*, vol. 36, no. 6, pp. 677–688, 2010.

R. D. Reitz et al., "Integration of LTC combustion in hybrid-electric vehicles," *Prog. Energy Combust. Sci.*, vol. 76, p. 100808, 2020.

R. Ganesh and S. Subramanian, "Effect of intake charge heating on combustion phasing in LTC engines," *Appl. Therm. Eng.*, vol. 75, pp. 324–331, 2014.

R. Kruth, "AVL's Perspective on Future Powertrains," SAE Paper 2021-26–0047, 2021.

R. M. Hanson et al., "An experimental investigation of fuel reactivity controlled PCCI combustion," *SAE Int. J. Engines*, vol. 03, no. 1, pp. 700–716, 2010.

S. Ciatti, "Advanced low temperature combustion concepts for diesel engines," SAE Paper 2010-01–1088, 2010.

S. Gross et al., "Brake thermal efficiency and emissions at varying intake temperatures," *Energy Rep.*, vol. 4, pp. 43–51, 2018.

S. Kong and R. Reitz, "Numerical study of premixed HCCI engine combustion," *Combust. Theory Model*, vol. 7, no. 2, pp. 417–433, 2003.

S. Miers et al., "Experimental investigation of high EGR low temperature combustion," SAE Paper 2012-01–1096, 2012.

S. Sarangi et al., "Impact of high EGR rates on flame temperature," *Fuel*, vol. 334, p. 126832, 2023.

S. Saxena and I. Bedoya, "Low temperature combustion strategies," *Prog. Energy Combust. Sci.*, vol. 39, no. 5, pp. 457–488, 2013.

S. Sick and D. Stojkovic, "Laser diagnostics of low temperature combustion," *Prog. Energy Combust. Sci.*, vol. 27, no. 3, pp. 243–296, 2001.

S. Singh and R. Maurya, "Combustion dynamics in HCCI engines," *Energy Convers. Manag.*, vol. 275, p. 116491, 2023.

S. Singh et al., "Influence of injection pressure and split injection," *Energy Rep.*, vol. 9, pp. 358–369, 2023.

S. Srihari and S. Thirumalini, "Investigation on reduction of emission in PCCI-DI engine," *Renewable Energy*, vol. 114, pp. 1232–1237, 2017.

V. Charitha et al., "Investigation on performance and emissions of RCCI dual fuel combustion," *Renewable Energy*, vol. 134, pp. 1081–1088, 2019.

V. Patil et al., "CFD simulation of low temperature combustion in diesel engines," *Int. J. Engine Res.*, vol. 15, no. 4, pp. 555–570, 2014.

Y. Cui et al., "Optical diagnostics on HCCI to CDC via PPC," *Fuel*, vol. 269, p. 117441, 2020.

Y. Jia et al., "Influence of methanol-diesel premixing on RCCI combustion," *Energy Convers. Manag.*, vol. 245, p. 114532, 2022.

Y. Ju, "Cool flame assisted LTC combustion," *Combust. Flame*, vol. 230, p. 111456, 2021.

Y. Ju et al., "Recent advances and prospects in low temperature combustion," *Renew. Sustain. Energy Rev.*, vol. 145, p. 111043, 2021.

Y. Murata et al., "Achievement of Medium Engine Speed and Load Premixed Diesel Combustion with VVT," SAE Paper 2006-01–0203, 2006.

Y. Ou et al., "Model predictive control for advanced combustion engines," *Energy Rep.*, vol. 8, pp. 792–811, 2022.

Y. Wang et al., "Effects of cooled and hot EGR on combustion and emissions," *Fuel*, vol. 216, pp. 819–829, 2018.

Y. Zhang et al., "Effects of DI-RCCI on GHG and NOx emissions," *Fuel*, vol. 322, p. 124270, 2022.

Y. Zhao et al., "Combustion and emission characteristics of a DME-diesel dual fuel PCCI engine," *Energy*, vol. 72, pp. 608–617, 2014.

Zhang, H. Xie, and H. Zhao, "Investigation of SI-HCCI Hybrid Combustion and Control Strategies," *Combustion Science and Technology*, vol. 181, no. 5, pp. 782–799, 2009.

W. Guan et al., "Variable valve actuation–based combustion control strategies," *Int. J. Engine Res.*, vol. 21, no. 4, pp. 578–591, 2020.

9 Emission Formation and Control in Low-Temperature Combustion

Pratibha Mahajan, Bhanu Priya, and Nancy Gulati

9.1 INTRODUCTION TO EMISSIONS IN LOW-TEMPERATURE COMBUSTION

Global primary energy consumption increased by 2% to 620 exajoules (EJ) in 2023, marking the second consecutive record year. Its growth rate was more than 5% higher than its pre-COVID level in 2019 and 0.6% more than its 10-year average [1]. Oil continues to be the most widely used fuel in the world, accounting for little more than one-third of all energy used. Currently, the transportation sector uses petroleum products, of which petrol and diesel are two key fuels (approximately 65%). Fuel combustion in internal combustion (IC) engines results in a variety of pollutants and greenhouse gases. The primary components of the exhaust gas include particulate matter (PM), carbon dioxide (CO_2), smoke opacity, unburned hydrocarbon (UHC), carbon monoxide (CO), and nitrogen oxides (NO_x) [2–4]. Ground-level ozone, which is hazardous to the living environment, is created when HC and CO react with NO_x and sunlight [5–7].

In 2023, global primary energy consumption reached 620 EJ, marking a 2% increase over the previous year and surpassing pre-pandemic levels. As shown in Figure 9.1, oil remains the largest global energy source in 2023, accounting for over one-third of total consumption, followed by natural gas and coal, based on data from the Energy

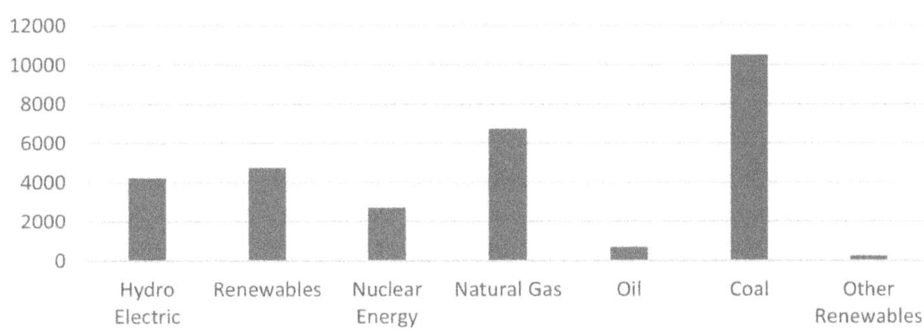

FIGURE 9.1 Consumption of energy from different sources.

DOI: 10.1201/9781003633631-9

Institute's Statistical Review of World Energy 2024 [8]. Although renewable energy continues to grow, its share still lags behind fossil fuels. The transportation sector is particularly reliant on petroleum products, primarily petrol and diesel, which are significant contributors to internal combustion engine (ICE) emissions. This energy mix underscores the urgent need for cleaner combustion strategies such as low-temperature combustion (LTC) to reduce environmental impact while maintaining energy demand.

9.2 OVERVIEW OF EMISSIONS IN INTERNAL COMBUSTION ENGINES

ICEs are recognized for producing various pollutants during their operation. These emissions generally consist of:

- **Nitrogen Oxides (NO_x):** Generated at elevated combustion temperatures when nitrogen reacts with oxygen in the air.
- PM**:** Tiny particles formed from the incomplete combustion of fuel, notably in diesel engines.
- **Carbon Monoxide (CO**): Released when fuel burns incompletely, particularly when oxygen is scarce.
- **Hydrocarbons (HCs):** The unburned fuel that escapes combustion and is discharged as petrol.
- **Unburned Fuel:** This is the result of incomplete combustion, which releases too much fuel.

There are noticeable differences in emissions between LTC and conventional combustion techniques like spark ignition (SI) and compression ignition (CI) engines. Conventional combustion typically produces more NO_x and PM emissions because it operates at higher temperatures and with richer fuel–air mixes. LTC technology, on the other hand, offers a revolutionary approach to ICE design, improving thermal efficiency while also significantly lowering emissions of nitrogen oxides (NO_x) and soot (PM). This cutting-edge combustion technique uses exact temperature control and fuel–air mixing to produce a more effective and clean burn.

9.2.1 Key Characteristics of LTC Emissions

LTC is an innovative method designed to enhance engine efficiency while minimizing harmful emissions. Below are essential features of emissions associated with LTC:

- **Reduced NO_x and PM:** LTC lowers these emissions by maintaining cooler combustion temperatures and utilizing leaner fuel–air mixtures. This approach limits NO_x formation, which typically occurs in high-temperature environments. Similarly, the cooler temperatures contribute to decreased PM formation.
- **Increased CO and UHC:** Although LTC lowers NO_x and PM, it tends to elevate CO and UHCs. This increase stems from incomplete combustion

resulting from lower temperatures and lean mixtures, which can leave more fuel unreleased. These emissions are particularly significant in LTC, especially under higher loads or transient conditions.

- **Trade-Offs:** Transitioning to LTC provides substantial advantages in reducing NO_x and PM emissions; however, it often results in elevated CO and UHC emissions. Balancing the trade-offs among efficiency, power output, and emissions is essential. In some scenarios, pursuing the lowest emissions may compromise engine power or fuel efficiency, while prioritizing maximum power output may increase pollutant emissions.

In conclusion, LTC shows potential for lowering harmful emissions, yet it's vital to navigate the trade-offs between emissions, efficiency, and power output. Ongoing research and engineering efforts aim to improve LTC strategies for optimal performance in real-world settings.

9.2.2 Formation of NO_x (Nitrogen Oxides)

Nitrogen oxides (NO_x), primarily consisting of nitric oxide (NO) and nitrogen dioxide (NO_2), are major pollutants in ICEs. In conventional diesel and SI engines, NO_x formation follows the Zeldovich mechanism, which is highly temperature dependent. However, LTC strategies such as Homogeneous Charge Compression Ignition (HCCI), Partially Premixed Compression Ignition (PPCI), and Reactivity-Controlled Compression Ignition (RCCI) significantly modify NO_x formation mechanisms due to their low-temperature and lean combustion characteristics.

9.2.2.1 Thermochemical Basis of NO_x Formation in LTC

In LTC regimes, the characteristic low peak combustion temperatures (<1800 K) suppress the formation of thermal NO_x, which primarily occurs via the temperature-sensitive Zeldovich mechanism above 2000 K in conventional diesel engines.

Moreover, LTC combustion modes generally operate under lean or near-stoichiometric equivalence ratios, reducing the presence of high-temperature, stoichiometric or fuel-rich zones where NO_x typically forms.

The reaction kinetics in LTC favor low-temperature oxidation pathways, where intermediate species like RO_2 and QOOH dominate rather than the high-temperature chain branching seen in conventional combustion. This shift slows the rate of NO-forming reactions and reduces the residence time of hot combustion products, further inhibiting NO_x buildup.

In HCCI and RCCI combustion modes, the use of premixed or partially premixed mixtures ensures that temperature and mixture homogeneity prevent localized thermal spikes—a key differentiator from conventional CI modes that produce stratified combustion zones conducive to NO_x formation.

9.2.2.2 Suppression of Thermal NO_x at Lower Temperatures

Thermal NO_x formation, governed by the Zeldovich mechanism, is prominent in high-temperature (>1800 K) flames due to the activation of the reaction:

$$N_2 + O \rightarrow NO + N$$

$$N + O_2 \rightarrow NO + O$$

LTC operates at peak combustion temperatures ranging from approximately 1000 to 1500 K, which is below the threshold necessary for substantial thermal NO_x production. Several factors contribute to NO_x suppression in LTC:

- The use of homogeneous lean mixtures prevents localized high-temperature regions, thereby limiting NO formation.
- High exhaust gas recirculation (EGR) rates, a common feature in LTC, decrease oxygen availability and further reduce peak combustion temperatures.

Studies show that HCCI and RCCI reduce NO_x emissions by over 90% compared to conventional diesel combustion due to the elimination of diffusion flames and lower combustion temperatures [1,9]. This suppression of thermal NO_x due to reduced flame temperature and a lean premixed charge has been further supported by [10], who demonstrated that reactivity-controlled stratification in RCCI combustion could achieve ultra-low NO_x emissions under varying load conditions.

9.2.2.3 Role of Prompt NO_x and N_2O Pathways

While thermal NO_x formation is minimized in LTC, other pathways contribute to residual NO_x emissions:

- **Prompt NO_x (Fenimore Mechanism):**
  - Occurs in fuel-rich flames where hydrocarbons (HCs) react with atmospheric nitrogen (N_2) to form NO.
  - More relevant in PPCI and RCCI modes with locally rich zones but remains lower than in conventional diesel combustion.
- **N_2O Pathway:**
  - At low temperatures (~1000 K), NO formation via the N_2O-intermediate mechanism becomes significant:

$$N_2 + O \rightarrow N_2O$$

$$N_2O + O \rightarrow 2NO$$

  - This pathway is especially relevant in high-EGR LTC modes, where increased residence time allows N_2O accumulation [11].
  - In RCCI, leaner mixtures reduce NO formation via this route.

While prompt NO_x is generally negligible in LTC due to lean mixtures, the N_2O pathway may contribute significantly in certain high-EGR cases [12].

9.2.2.4 Effect of Local Hotspots and Stratification

Even in LTC, localized temperature inhomogeneities can cause NO_x spikes due to:

- **Local Hotspots:**
  - Occur due to incomplete mixing, leading to higher temperature zones in PPCI and RCCI combustion.
  - These hotspots can trigger NO formation in localized regions, even when bulk temperatures remain low.
  - Advanced computational fluid dynamics modeling helps predict and mitigate such temperature excursions [2].
- **Charge Stratification:**
  - Intentional fuel stratification produces areas with different equivalency ratios in RCCI and PPCI.
  - Richer areas may produce more NO_x, especially if combustion phasing occurs later.
  - Research on optical engines shows that fuel stratification levels are correlated with NO generation, and excessive stratification should be avoided [13].
- **Injection Timing and Spray Dynamics:**
  - Local equivalence ratios and temperatures are greatly impacted by the fuel injection approach.
  - While late injection can result in concentrated rich pockets, which increase NO generation, early injection in LTC encourages greater mixing and decreased NO creation.
  - By spreading gasoline more uniformly, multiple injections help avoid hotspots.

Advanced fuel injection techniques, fuel selection, and combustion phasing modifications are all necessary to efficiently manage NO_x emissions in LTC systems. The following strategies preserve effectiveness and performance while reducing NO_x formation:

- NO_x is reduced via combustion phasing control through injection time optimization.
- To avoid very rich areas, stratified combustion must be carefully adjusted.
- By using fuels with a lesser propensity for autoignition, such as gasoline in RCCI, hotspots are reduced and better mixing is maintained [4].

9.2.3 Particulate Matter (PM) Formation

ICEs, especially diesel engines running in typical combustion modes, are a major source of PM emissions. However, the mechanics of PM production have changed as a result of the introduction of LTC techniques. By improving fuel–air premixing and lowering peak combustion temperatures, LTC seeks to minimize the production of soot and nitrogen oxides (NO_x). However, the complex interplay between temperature, fuel composition, and oxygen availability in LTC conditions leads to unique PM formation pathways that distinguish it from conventional diesel and SI engines.

9.2.3.1 Thermochemical Basis of PM Formation in LTC

Unlike conventional diesel combustion where diffusion flames and fuel-rich regions drive soot PM formation via pyrolysis and acetylene-based mechanisms, LTC regimes rely on premixed or partially premixed combustion (PPC). These conditions significantly reduce the occurrence of fuel-rich cores, thereby lowering soot precursor formation.

However, the influence of equivalence ratio stratification remains critical in LTC. If rich pockets exist (e.g., in RCCI stratification), incomplete oxidation due to low flame temperatures and slower reaction rates may still yield some PM. Moreover, the in-cylinder pressure and temperature affect the soot oxidation rates; under low-temperature operation, this oxidation is often incomplete, especially near cylinder walls or crevice volumes.

In summary, while fuel–air homogeneity and reduced temperature suppress soot nucleation, inefficient late-stage oxidation may still lead to residual PM emissions under some LTC conditions.

9.2.3.2 Influence of Low-Temperature Mixing-Controlled Combustion

Unlike traditional diesel combustion, which relies on a diffusion-controlled process where fuel burns as it mixes with air, LTC promotes premixed combustion with lower in-cylinder temperatures. The primary effects of this combustion mode on PM formation include:

- **Reduced Soot Formation:** LTC operates at lower combustion temperatures, which inhibits the thermal breakdown of hydrocarbons, thereby limiting the formation of soot precursors such as acetylene (C_2H_2) and polycyclic aromatic hydrocarbons (PAHs) [14].
- **Extended Ignition Delay:** A longer ignition delay in LTC allows for improved fuel–air premixing, reducing local fuel-rich pockets that contribute to soot nucleation [15].
- **Effects of Stratification:** While overall PM formation is lower in LTC, fuel stratification due to incomplete mixing can result in localized regions with incomplete combustion, potentially increasing ultrafine PM emissions [5].

9.2.3.3 Role of Fuel Composition, Oxygen Availability, and Injection Strategies

PM formation in LTC engines is influenced by fuel properties, oxygen availability, and injection techniques. Oxygenated fuels (e.g., biodiesel and ethanol blends) reduce PM by enhancing oxidation, while high-aromatic fuels promote soot precursor formation. Higher volatility fuels improve atomization and mixing, reducing local fuel-rich zones that lead to PM [14].

TABLE 9.1
Comparison of PM Formation in LTC, Diesel, and SI Engines

Engine Type	PM Formation Characteristics
Low-Temperature Combustion engines	• Lower soot emissions due to enhanced premixing and lower combustion temperatures • Possible increase in ultrafine particles from partial oxidation [15]
Diesel engines	• High soot emissions due to fuel-rich regions in diffusion combustion • Large accumulation-mode particles with lower number concentration [14]
SI engines	• Reduced particulate matter (PM) emissions as a result of homogeneous charge combustion • PM is mostly composed of nucleation-mode particles, which are frequently impacted by interactions between lubricants and fuel additives [5]

Injection strategies such as split injections and pilot injections help reduce local soot formation, while late injection increases premixing, minimizing soot but potentially increasing UHC and CO [15]. EGR can suppress NO_x but excessive use may increase PM due to incomplete oxidation [5].

PM emissions in LTC can be considerably decreased by combining improved injection techniques, oxygen control, and fuel selection optimization. Recent CFD and optical diagnostic work by Ma et al. [16] shows that reducing spray wall impingement and optimizing in-cylinder mixture formation significantly reduces PM emissions in PPCI modes, further validating LTC's clean combustion potential.

9.2.3.4 Comparison of PM Formation in LTC vs. Diesel and SI Engines

Table 9.1 presents a comparative study of PM production mechanisms in SI, LTC, and conventional diesel engines.

9.2.4 Unburned Hydrocarbon (UHC) and Carbon Monoxide (CO) Emissions

Because of incomplete oxidation, flame quenching, and charge stratification, UHC and CO emissions from LTC engines are often higher than those from traditional diesel combustion. Due to reduced combustion temperatures and difficulties with fuel–air mixing, LTC frequently results in higher UHC and CO emissions even if it lowers NO_x and PM emissions [14].

9.2.4.1 Thermochemical Basis of UHC and CO Formation in LTC

LTC conditions suppress complete fuel oxidation due to low combustion temperatures and extended ignition delays. As a result, the formation of UHC and CO increases. The slower reaction kinetics—especially those related to low-temperature oxidation pathways such as alkylperoxy (RO_2) and hydroperoxyalkyl (QOOH) intermediates—play a pivotal role in incomplete oxidation.

In conventional CI combustion, high temperatures quickly convert CO to CO_2; however, in LTC, the lower post-flame temperatures result in reduced conversion of CO. Similarly, UHC formation is intensified by:

- Incomplete fuel oxidation in low-temperature zones.
- Wall impingement and crevice quenching in HCCI.
- Fuel trapped in cold zones during stratified charge combustion.

The oxygen availability (air–fuel ratio) and residence time at combustion temperature are both limited, which directly affects the rate of hydrocarbon oxidation.

9.2.4.2 Effect of Incomplete Oxidation and Flame Quenching

LTC suppresses NO_x generation by operating at lower temperatures, but this may result in incomplete fuel oxidation, which raises UHC and CO emissions. The primary reasons include:

- **Lower Peak Temperatures:** LTC combustion temperatures often fall below the activation energy required for complete oxidation of hydrocarbons, increasing UHC levels [1].
- **Quenching Near Cold Surfaces:** In direct-injection (DI) LTC engines, fuel spray impingement on cylinder walls and piston surfaces lowers local temperatures, preventing complete oxidation and leading to UHC emissions [17].
- **Overmixing of Lean Regions:** In some LTC strategies, over-lean air–fuel mixtures may not reach the flammability limits, resulting in fuel escaping unburned as UHC [18].

Kalghatgi et al. [19] reported that optimization of ignition delay and stratification level in partially premixed gasoline combustion reduces UHC while maintaining low soot and NO_x levels.

9.2.4.3 Wall Impingement and Fuel Overmixing Issues

Unlike conventional diesel combustion, which relies on diffusion flames, LTC promotes premixed and mixing-controlled combustion, which can create fuel-rich or fuel-lean pockets that lead to incomplete combustion.

- **Wall Wetting:** In high-EGR or late-injection strategies, fuel droplets may not fully vaporize before reaching cooler cylinder surfaces, leading to pool fires and UHC emissions [1].
- **Charge Stratification Issues:** Excessive fuel–air stratification can result in regions where combustion is incomplete, producing higher CO emissions due to insufficient oxidation [14].

9.2.4.4 Strategies to Mitigate CO and UHC Emissions

To address these challenges, several strategies have been explored:

- **Optimized Injection Timing:** Using early pilot injections or multiple split injections allows better air–fuel mixing and prevents wall wetting [17].
- **Higher Intake Temperatures:** Increasing intake air temperatures can promote fuel vaporization and enhance oxidation, reducing UHC and CO [18].
- **Variable Valve Timing:** Adjusting intake and exhaust valve timings can enhance in-cylinder turbulence, improving fuel–air homogenization [17].
- **Aftertreatment Systems:** Oxidation catalysts and advanced catalytic converters can help oxidize residual UHC and CO before they are released into the atmosphere [1].

9.2.5 Other Pollutants (Soot Precursors, Aldehydes, and PAHs)

Apart from NO_x, PM, CO, and UHC emissions, LTC also produces other harmful pollutants, including PAHs, aldehydes, and soot precursors. The formation of these emissions depends on fuel chemistry, combustion conditions, and operating strategies [17].

9.2.5.1 Formation of Polycyclic Aromatic Hydrocarbons (PAHs) and Aldehydes

- **Polycyclic Aromatic Hydrocarbons (PAHs):** PAHs are formed due to incomplete combustion of hydrocarbons, especially in fuel-rich regions. They serve as precursors to soot and are associated with carcinogenic effects [18].
- **Aldehyde Emissions:** Aldehydes such as formaldehyde (CH_2O) and acetaldehyde (C_2H_4O) are produced due to low-temperature fuel oxidation and incomplete breakdown of hydrocarbon chains [1].

9.2.5.2 Impact of Fuel Chemistry on Soot Precursors

- Higher aromatic content fuels tend to increase PAH and soot formation. Low-aromatic fuels such as Fischer–Tropsch synthetic fuels can significantly reduce these emissions [17].
- Oxygenated fuels (e.g., biodiesel and ethanol blends) promote complete oxidation, thereby reducing soot precursors and PAH formation [18].

While Table 9.2 provides an overview of LTC combustion emissions, a more detailed comparison of the three key LTC strategies—HCCI, RCCI, and PPCI—is presented in Table 9.3 to highlight their operational principles, control requirements, fuel characteristics, and practical considerations.

9.2.5.3 Measurement Techniques for Advanced Emission Characterization

To analyze and control these emissions, several advanced diagnostic tools are employed:

- **Laser-Induced Incandescence (LII):** Used for real-time soot measurement in combustion environments.
- **Fourier Transform Infrared Spectroscopy (FTIR):** Helps in identifying aldehydes and UHCs.
- **Gas Chromatography–Mass Spectrometry (GC–MS):** Used to characterize PAHs and other hydrocarbon species in exhaust emissions [1].

TABLE 9.2
Comparison of Emissions between Conventional CI, SI, and LTC Combustion

Emission Type	Conventional CI (Diesel)	Conventional SI (Gasoline)	LTC (HCCI, RCCI, PCCI, etc.)
NO_x (nitrogen oxides)	High due to high-temperature combustion	Moderate, controlled by Three-Way Catalysts (TWCs)	Low due to lower combustion temperatures
PM (particulate matter)	High, requires diesel particulate filter for control	Low, but higher in direct-injection gasoline engines	Lower than compression ignition but can vary based on fuel and injection strategy
CO (carbon monoxide)	Low due to excess oxygen in combustion	Moderate, reduced by TWC	High due to incomplete oxidation in LTC conditions
UHC (unburned hydrocarbons)	Low, as diesel combustion is diffusion-controlled	Moderate, oxidized in TWC	High due to lower in-cylinder temperatures and flame quenching
SO_2 (sulfur dioxide)	Present, depends on fuel sulfur content	Low due to low-sulfur gasoline	Minimal, as low-temperature combustion (LTC) often uses low-sulfur or alternative fuels
CO_2 (carbon dioxide)	Lower per unit of fuel energy due to high efficiency	High due to lower engine efficiency	Potentially lower due to high efficiency and alternative fuels
Aldehydes and polycyclic aromatic hydrocarbons	Moderate, formed from incomplete combustion	Low, generally oxidized in TWC	High, especially in HCCI/PCCI due to low combustion temperatures

TABLE 9.3
Comparison of HCCI, RCCI, and PPCI Combustion Modes

Aspect	HCCI	RCCI	PPCI
Operating range	Low to mid load	Medium to high load	Medium load range
Control parameters	Intake temp, exhaust gas recirculation (EGR), equivalence ratio, compression ratio	Fuel reactivity, timing, ratio between fuels	Injection timing, pressure, and EGR
Fuel requirement	Single fuel (high octane/low cetane)	Dual fuel (low- and high-reactivity fuels)	Single fuel with stratified injection (e.g., diesel)
Advantages	Ultra-low NO_x and soot, high efficiency	Improved control, high load capability	Balance between HCCI and CI, better controllability
Limitations	Difficult combustion phasing control, knocking	Complex injection system, fuel logistics	Limited operation range, NO_x control under high load

9.3 STRATEGIES FOR EMISSION REDUCTION IN LTC

The efficiency of LTC for emission reduction heavily relies on improving combustion processes, the characteristics of fuel, and engine technologies. To further minimize emissions of NO_x, soot, CO, and UHC, several advanced techniques have been proposed.

9.3.1 In-Cylinder Combustion Control for Emission Reduction

The in-cylinder combustion process must be precisely controlled to limit emissions from LTC engines. This entails putting EGR into practice, controlling air–fuel mixing, and optimizing fuel injection.

9.3.1.1 Fuel Injection Strategies

Injection Timing: Modifying fuel injection timing impacts combustion phasing, thus affecting emissions and efficiency.

Injection Pressure: Increased injection pressure enhances atomization and air–fuel mixing, leading to more complete combustion.

Multiple Injections: Dividing fuel injection into multiple pulses can lower peak temperatures, which in turn reduces NO_x while enhancing combustion efficiency.

In LTC modes, injection timing, split injection, and injector design are not just fueling parameters but vital levers for controlling ignition delay, stratification, and heat release rate. For instance, advanced single injections increase premixing and homogeneity, which may benefit soot reduction but can lead to excessive pressure rise. On the other hand, split injections, such as a pilot-main-post strategy, allow finer control over ignition timing and the shape of the heat release curve.

In systems like RCCI, the injector configuration may involve dual injection systems—one for a high-reactivity fuel (DI) and another for a low-reactivity fuel (port injection). This supports reactivity stratification, which is crucial for stable LTC and for suppressing NO_x and soot simultaneously. The number of nozzle holes, spray angle, and droplet atomization characteristics also determine the mixture formation quality, directly influencing both charge stratification and local combustion temperatures.

9.3.1.2 Exhaust Gas Recirculation (EGR) for NO_x Reduction

EGR reintroduces a portion of exhaust gases into the intake, which reduces combustion temperatures and significantly diminishes NO_x production.

Nevertheless, excessive EGR may result in higher CO and UHC emissions due to incomplete combustion.

9.3.2 Alternative Fuels and Fuel Additives

Alternative fuels and fuel additives significantly influence reactivity and charge preparation in LTC modes, thereby affecting emissions. Figure 9.2 presents simulated

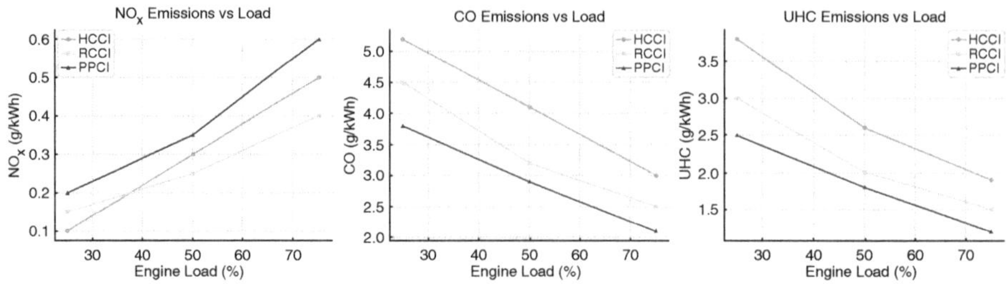

FIGURE 9.2 Simulated emission trends (NO-, CO, and UHC) for HCCI, RCCI, and PPCI across varying engine loads.

emission trends (NO_x, CO, and UHC) for HCCI, RCCI, and PPCI across varying engine loads, highlighting how combustion mode and fuel-reactivity tailoring impact emission trade-offs.

9.3.2.1 Low-Carbon Fuels

Fuels like methanol, ethanol, natural gas, and biofuels have increased oxygen content, which supports cleaner combustion with less soot formation.

Using natural gas and hydrogen facilitates low-temperature ignition and reduces both NO_x and PM emissions.

9.3.2.2 Oxygenated Fuels for Soot Reduction

Fuels enriched with oxygen, such as biodiesel and ethanol blends, promote complete combustion and greatly reduce particulate emissions.

Dual-Fuel Strategies, Ignition Delay, and Fuel Reactivity

In LTC performance, fuel reactivity is crucial, particularly for emissions control and combustion phasing. Alcohols like ethanol and methanol prolong ignition delay because of their high-octane rating, whereas oxygenated fuels (like biodiesel) typically decrease it because of their higher cetane number. Dual-fuel LTC systems like RCCI, which premix a low-reactivity fuel (like petrol or ethanol) and inject a high-reactivity fuel (like diesel) directly, rely heavily on this interaction.

In comparison to conventional diesel CI modes, RCCI operation with diesel-ethanol blends produced a 65%–80% reduction in NO_x and PM while retaining comparable thermal efficiency, according to a case study by Kalghatgi et al. [2]. However, in low-load circumstances, ethanol's extended ignition delay resulted in higher CO and UHC emissions.

9.3.2.3 Fuel Volatility and Chemical Composition

Fuels with higher volatility (e.g., gasoline and alcohols) enhance vaporization and mixing, resulting in fewer UHCs. The molecular makeup of fuels also affects ignition characteristics and emissions; for example, paraffins burn cleaner compared to aromatic-rich fuels that typically generate more soot.

- **Left:** NO_x emissions—consistently low across LTC modes but rise slightly with load.

- **Center:** CO emissions—highest in HCCI, lower in PPCI and RCCI.
- **Right:** UHC emissions—also higher in HCCI due to incomplete combustion at lower temps.

Data indicates that while all LTC modes suppress NO_x effectively, HCCI tends to produce higher CO and UHC due to delayed and incomplete oxidation, especially under low-load conditions. RCCI offers a more balanced emission profile by utilizing dual-fuel strategies.

9.3.3 Hybrid Combustion Modes and Advanced Engine Technologies

Incorporating various combustion strategies alongside advanced boosting technologies further enhances LTC engines for lower emissions and greater efficiency.

9.3.3.1 Hybrid Combustion Modes for Emission Control

HCCI: Enables low-temperature, lean combustion producing minimal NO_x and soot emissions.

RCCI: Utilizes dual fuels with varying reactivity to manage ignition timing, thus reducing emissions while boosting efficiency.

PPC: Balances between conventional diesel and HCCI to minimize NO_x and soot while ensuring combustion stability.

9.3.3.2 Turbocharging and Supercharging on Emissions

Turbocharging amplifies air supply, enhancing combustion efficiency and minimizing CO and UHC emissions. Supercharging increases cylinder pressure to improve air–fuel mixing, leading to lower emissions.

9.4 EXHAUST AFTERTREATMENT TECHNOLOGIES FOR LTC

LTC strategies significantly reduce NO_x and PM emissions compared to conventional diesel and SI combustion. However, they often lead to increased UHC, CO, and specific NO_x and PM fractions that require advanced aftertreatment solutions for effective emission control. This section explores catalytic converters, NO_x reduction strategies, particulate filters, and emerging catalyst technologies for LTC engines. The various exhaust aftertreatment technologies employed to mitigate emissions in LTC engines are summarized in Figure 9.3, offering a structured overview of current and emerging solutions for effective emission control.

9.4.1 Three-Way Catalysts (TWCs) and Lean NO_x Traps (LNTs)

Three-Way Catalysts (TWCs) and Lean NO_x Traps (LNTs) are two well-known aftertreatment technologies that are necessary for the efficient control of emissions in LTC engines. Although TWCs are well-established for stoichiometric SI engines, lower exhaust temperatures and extra oxygen pose a challenge to their performance in LTC circumstances. Similarly, LNTs present a viable strategy for lowering NO_x in lean-burn applications; however, questions still surround their long-term durability and regeneration.

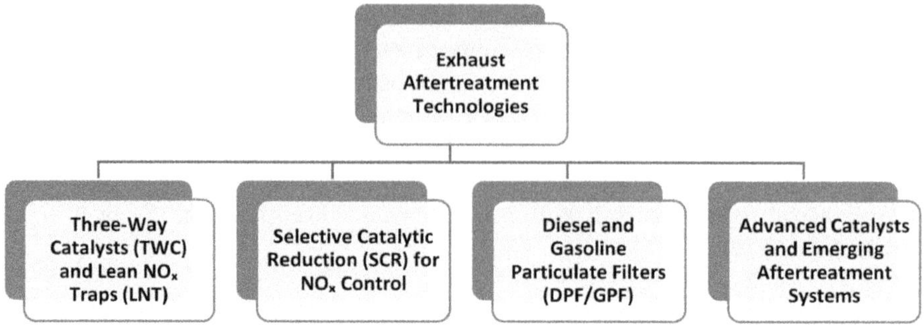

FIGURE 9.3 The various exhaust aftertreatment technologies employed to mitigate emissions in LTC engines.

9.4.1.1 Effectiveness of TWC in LTC-SI Engines

To reduce NO_x, CO, and UHC emissions, stoichiometric SI engines frequently employ TWCs. However, their effectiveness in LTC–SI configurations is mostly dependent on the control of the air–fuel ratio and exhaust gas temperature.

- **Catalytic Performance:** TWC operates efficiently in stoichiometric combustion but struggles in lean LTC conditions, where oxygen-rich exhaust gases hinder NO_x reduction [20].
- **Temperature Sensitivity:** Lower exhaust temperatures from LTC–SI may postpone catalyst light-off, lowering conversion efficiency overall [18].
- **Precious Metal Optimization:** In low-temperature settings, rhodium (Rh) and palladium (Pd) coatings are frequently employed to improve NO_x and CO oxidation.

9.4.1.2 Lean NO_x Trap (LNT) Applications and Limitations

LNTs store NO_x emissions in lean conditions, while they periodically renew in rich environments. Its use in LTC applications is encouraging despite the challenges.

- **NO_x Adsorption Mechanism:** During lean phases, NO_x is absorbed by the barium oxide (BaO) and platinum (Pt) catalysts present in LNTs, and during rich operation, it is released [21].
- **Regeneration Challenges:** Because LTC functions at low exhaust temperatures, effective LNT regeneration necessitates hydrocarbon injection or rich-mode transitions.
- **Durability Concerns:** Over time, LNT performance is diminished by sulfur poisoning and heat deterioration, which renders them less appropriate for long-term LTC use.

9.4.2 Selective Catalytic Reduction (SCR) for NO_x Control

A popular technique for reducing NO_x in lean-burn and diesel engines is called selective catalytic reduction (SCR), which uses ammonia as a reductant to change NO_x

into innocuous nitrogen and water. To improve efficiency in low-temperature situations, however, better catalysts and other strategies like plasma-assisted SCR are required due to the lower exhaust temperatures in LTC engines, which present difficulties for traditional SCR systems.

9.4.2.1 Urea-Based SCR Systems and Ammonia Slip Concerns

One of the best methods for reducing NO_x is SCR, which uses urea-based ammonia infusion to change NO_x into innocuous nitrogen (N_2) and water (H_2O).

- **Temperature Dependency:** SCR catalysts require exhaust temperatures of 250°C–400°C, whereas LTC exhaust temperatures are often lower, affecting conversion efficiency [3].
- **Ammonia Slip Issues:** Ammonia slip catalysts are necessary to prevent environmental pollution in LTC because of the uneven NO_x generation and low temperature conditions that can result in excess ammonia (NH_3) emissions [22].

9.4.2.2 Emerging Plasma-Assisted SCR Technologies

To improve SCR performance in low-temperature LTC applications, plasma-assisted SCR is gaining attention.

- **Plasma Activation:** Nonthermal plasma enhances NH_3 decomposition, enabling efficient NO_x reduction at lower temperatures [7].
- **Catalyst Enhancement:** Combining plasma with copper-exchanged zeolite (Cu-Zeolite) catalysts improves low-temperature NO_x conversion efficiency.
- **Energy Constraints:** Despite its effectiveness, high energy consumption and system complexity remain challenges for large-scale automotive applications.

9.4.3 Diesel and Gasoline Particulate Filters (DPF/GPF)

Diesel and Gasoline Particulate Filters (DPF/GPF) serve a significant role in controlling PM emissions from ICEs. Even while LTC techniques greatly lessen the production of soot when compared to traditional diesel combustion, they do produce fine PM that needs to be effectively filtered. Particulate size distribution, exhaust temperature, and regeneration techniques are some of the variables that affect particulate filter efficiency in long-term care applications.

9.4.3.1 Suitability of DPF and GPF for LTC Applications

LTC strategies produce finer PM than traditional diesel engines, necessitating particulate filter systems for emission control.

- **DPF:** Required for LTC–diesel engines, as they capture nanometer-sized soot particles, preventing their release into the atmosphere [23].
- **GPF:** Essential for LTC–SI configurations, particularly for DI gasoline engines that produce high PM levels [24].

9.4.3.2 Passive vs. Active Regeneration Strategies

- **Passive Regeneration:** Relies on oxygen and NO_2 reactions to continuously oxidize accumulated PM, which is effective at higher temperatures but may be insufficient in LTC due to lower exhaust heat.
- **Active Regeneration:** Uses fuel postinjection or electric heating to elevate temperatures and burn off soot deposits, but increases fuel consumption and operational costs [22].

9.4.4 Advanced Catalysts and Emerging Aftertreatment Systems

As LTC strategies continue to evolve, conventional aftertreatment systems face challenges in effectively reducing emissions at lower exhaust temperatures. To address these limitations, researchers are developing advanced catalysts and emerging aftertreatment technologies that enhance NO_x and PM reduction while maintaining efficiency in LTC conditions. Innovations such as novel catalyst formulations, metal–organic frameworks (MOFs), and nanocatalysts are being explored to improve emission control in next-generation LTC engines. Table 9.4 provides an overview of the performance of conventional aftertreatment technologies under LTC conditions, highlighting the motivation for developing such advanced alternatives.

9.4.4.1 Development of Low-Temperature NO_x Reduction Catalysts

Since conventional SCR and LNT systems struggle at LTC's low exhaust temperatures, research has focused on next-generation NO_x reduction catalysts.

TABLE 9.4
Performance of Aftertreatment Technologies Under LTC Conditions

Aftertreatment Technology	Primary Pollutants Targeted	Effectiveness in LTC	Challenges in LTC Applications	Possible Improvements
Three-way catalysts	NO_x, CO, UHC	Limited effectiveness due to lean-burn conditions	Oxygen-rich exhaust prevents NO_x reduction; low exhaust temperatures delay catalyst activation	Use of advanced catalyst formulations and passive NO_x adsorbers
Lean NO_x traps	NO_x	Moderate effectiveness at mid-range temperatures	Low-temperature operation reduces NO_x adsorption and regeneration efficiency; sulfur poisoning affects durability	Improved catalyst materials and periodic rich-mode operation

(Continued)

TABLE 9.4 (*Continued*)
Performance of Aftertreatment Technologies Under LTC Conditions

Aftertreatment Technology	Primary Pollutants Targeted	Effectiveness in LTC	Challenges in LTC Applications	Possible Improvements
Selective catalytic reduction (SCR)	NO_x	Reduced efficiency at low exhaust temperatures	Ammonia slip due to incomplete reaction; requires external urea injection	Plasma-assisted SCR, low-temperature catalysts (Fe-Zeolite, Cu-Zeolite)
Diesel particulate filter	Particulate matter (PM), soot	Effective for soot control but regeneration is challenging	Lower LTC temperatures hinder passive regeneration, increasing soot accumulation	Active regeneration strategies, catalyzed filters, and optimized exhaust heat management
Gasoline particulate filter	PM from DI gasoline engines	Effective but with increased fuel consumption in regeneration	Lower PM oxidation rates at low-temperature combustion (LTC) temperatures; potential backpressure increase	Advanced filter coatings and low-temperature oxidation catalysts
Hydrocarbon-SCR (HC-SCR)	NO_x	Potential alternative to urea-based SCR for LTC	Low hydrocarbon reactivity; formation of unwanted by-products	Optimization of fuel-based reductants and novel catalyst formulations
Metal–organic frameworks and nanocatalysts	NO_x, PM, and CO	Emerging technology with potential for LTC	Stability under real-world conditions; scalability challenges	High-surface-area catalysts with improved durability and selectivity

- **Low-Temperature SCR Catalysts:** Iron-exchanged zeolites (Fe-Zeolite) and perovskite-based catalysts exhibit promising NO_x conversion at 150°C–250°C, ideal for LTC applications [6].
- **Hydrocarbon-SCR (HC-SCR):** Uses fuel-derived hydrocarbons as NO_x reductants, offering an alternative to urea-based SCR in LTC engines [22].

9.4.4.2 Role of Metal–Organic Frameworks (MOFs) and Nanocatalysts

- **MOF-Based Catalysts:** MOFs with high surface area and tuneable porosity are being explored for NO_x and PM reduction [25].

- **Nano Catalysts:** Platinum (Pt), palladium (Pd), and ceria (CeO_2)–based nanoparticles demonstrate enhanced catalytic activity at low temperatures, improving aftertreatment system efficiency for LTC engines.

9.5 FUTURE DIRECTIONS AND CHALLENGES IN LTC EMISSION CONTROL

As LTC technology is being worked on and developed, a number of challenges and opportunities arise as the industry searches for cleaner and more efficient engines. Future development will largely be determined by the regulations of the industry, hybrids, and the progress of combustion research.

9.5.1 Challenges in Regulations and Compliance

With increased stringency in emission norms and regulations for fuels, the implementation of LTC has to be made.

9.5.1.1 Current and Future Emission Norms

EURO 7 (Europe): Will further lower NO_x, PM, and CO_2 emissions, therefore affecting future LTC strategies.

Environmental Protection Agency (EPA) (United States): EPA mandates lower greenhouse gas emissions and higher fuel efficiency that has caused the need for the development of cleaner engines like LTC.

Bharat Stage VI (India): Like EURO 6 standards, it demands low NO_x and PM emissions with the use of cleaner combustion technologies.

9.5.1.2 Impact of Fuel Regulations on LTC Strategies

Strict sulfur content specifications of fuels can support the application of LTC by minimizing soot generation.

Blended biofuels policies stimulate the use of more oxidized fuels, which improve combustion efficiency and reduce emissions.

9.5.2 Integration of LTC with Hybrid and Electrified Powertrains

The ongoing automotive electrification and hybridization developments could see LTC assisting very effectively with the enhancement of efficiency in not just hybrid electric but specific plug-in hybrid electric vehicle (PHEV) systems too.

9.5.2.1 Hybrid and PHEV Systems

In hybrid systems, LTC is vital as it greatly aids in the battery range extension through its high thermal efficiency. The optimized engine start–stop strategies aid in reducing cold-start emissions, a common hybrid issue.

9.5.2.2 AI-Driven Combustion Optimization Actions

Machine learning algorithms can predict and modify numerous combustion parameters according to real-time information from the operations of the engine to provide enhanced emission control. AI-driven adaptive control systems for fuel injection rate, EGR rate withheld, air–fuel ratio structure itself will help optimization depending on changing driving conditions.

Recent research highlights the potential of AI-driven control strategies to manage the complex dynamics of LTC combustion. Techniques such as model predictive control (MPC) and reinforcement learning (RL) have shown promise in regulating combustion phasing, ignition timing, and emissions under varying load and fuel conditions.

MPC enables proactive control by predicting future engine states and adjusting injection or EGR parameters accordingly. Studies such as [26] demonstrated MPC's effectiveness in controlling HCCI combustion timing while maintaining NO_x and PM limits. Similarly, RL has been explored for real-time adaptive control in dual-fuel LTC engines due to its ability to learn from engine–environment interactions [27].

Hybrid data-driven models, combining physical modeling with neural networks or Gaussian process regression, are also being developed to predict CA50, pressure rise rates, and emissions trends, facilitating fast and intelligent control [28]. These approaches are expected to enhance combustion stability and emissions compliance in next-generation LTC engines.

9.5.3 Research Gaps and Emerging Technologies

Even though this technology has a big positive note in terms of implementation, the real application of this technology has still to be done, and there is a lot of research still needed. Real emission measurement needed: The laboratory results differ from real-world emissions due to various changeable driving scenarios and load/road conditions, hence advanced portable emissions measurement systems (PEMS) to really understand how LTC performs in real time.

9.5.3.1 Future Potential of Low-Carbon Combustion

Emerging hydrogen and e-fuels could enable an even more significant reduction in emissions while still achieving the benefits of high thermal efficiency from the LTC. Advanced catalytic converters and plasma-assisted combustion could enhance emission control.

9.6 SUMMARY AND CONCLUSION

An overview of the main conclusions regarding emissions in LTC engines is given in this part, with a focus on the advantages, difficulties, and emission control techniques. The conversation also emphasizes the importance of cutting-edge technologies and potential avenues for future research to make LTC a viable option for next-generation clean combustion engines.

9.6.1 Recap of Emission Trends in LTC

Because of their lower peak combustion temperatures and leaner fuel–air mixtures, LTC technologies like PPC, RCCI, and HCCI have shown notable reductions in NO_x

and PM emissions [1]. But there are drawbacks to these technologies as well, such as higher emissions of UHC and CO, which necessitate sophisticated oxidation methods and post-combustion treatment [9].

Additionally, LTC engines have trade-offs between power output, emissions, and efficiency; therefore, to get the required performance, fuel injection techniques, charge stratification control, and alternate fuel integration must be optimized. According to recent research, EGR and better fuel–air mixing strategies are essential for balancing these trade-offs and guaranteeing adherence to strict emission standards [11].

9.6.2 Summary of Key Control Strategies

To tackle the emission issues in LTC engines, numerous control strategies have been researched and developed. A particularly effective method involves managing combustion within the cylinder, utilizing techniques like precise fuel injection timing, multiple injections, and adjustments to EGR, which aid in controlling temperature and emissions in the combustion chamber [2]. These alterations have demonstrated a notable reduction in NO_x and PM production, enhancing overall efficiency.

Another key approach involves utilizing alternative fuels and additives, which have shown encouraging outcomes in reducing emissions. Oxygenated fuels, including biodiesel and ethanol blends, improve oxidation and encourage complete combustion, resulting in decreased PM emissions [7]. Furthermore, fuels with greater volatility, such as methanol and natural gas, have been found to enhance vaporization and mixing, thus lessening the generation of UHCs [18].

In addition to fuel modifications, hybrid combustion techniques have proven to be an effective approach for controlling emissions. Methods such as RCCI, HCCI, and PPC provide enhanced control over fuel reactivity and ignition timing, optimizing combustion to achieve lower emissions. These techniques are often supported by exhaust aftertreatment technologies, such as SCR, LNTs, and advanced catalytic converters, which further reduce NO_x, CO, and PM emissions [9].

9.6.3 Future Outlook on Sustainable LTC Technologies

The progression of LTC technologies will remain influenced by changing industry standards and regulatory guidelines. Upcoming research will concentrate on AI-enhanced combustion optimization, in which machine learning algorithms will be able to adjust fuel injection rates, EGR levels, and air–fuel ratios dynamically according to real-time operating conditions [1]. This incorporation of artificial intelligence into engine control systems is anticipated to improve both efficiency and adherence to emissions standards.

Furthermore, LTC technology is expected to be essential in hybrid and electrified powertrains, aiding in improved thermal efficiency and prolonged battery life in hybrid-electric vehicles. By enhancing engine start–stop strategies and minimizing cold-start emissions, LTC can support PHEV systems in attaining greater overall energy efficiency [18].

Research on low-carbon and carbon-neutral fuels will be crucial in determining the direction of LTC as sustainability becomes more and more of a worldwide concern. Potential substitutes that can further cut emissions while preserving high thermal efficiency include hydrogen, synthetic e-fuels, and biofuels. Furthermore, improvements in real-time emissions monitoring and plasma-assisted combustion will improve fuel efficiency and regulatory compliance, making LTC a more appealing option for upcoming transportation requirements [7].

REFERENCES

1. Dec, J.E. (2009). Advanced compression-ignition engines: Understanding the in-cylinder processes. *Proceedings of the Combustion Institute*, 32(2), 2727–2742.
2. Kalghatgi, G.T. (2010). Fuel and engine interactions in HCCI. *SAE International Journal of Engines*, 2(2), 484–497.
3. Koebel, M., Elsener, M., & Kleemann, M. (2019). Urea-SCR: A promising technique to reduce NO_x emissions from diesel engines. *Catalysis Today*, 59(3–4), 335–345.
4. Kook, S. (2012). Fuel spray-wall interactions in LTC engines. *Combustion and Flame*, 159(9), 2569–2582.
5. Kumar, P., & Joshi, A. (2018). Characteristics of particulates emitted by IC engines using advanced combustion strategies. In: Agarwal, A., Gupta, J., Sharma, N., Singh, A. (eds) *Advanced Engine Diagnostics* (pp. 57–71). Springer, Cham.
6. Li, J., Zhao, Y., & Chen, H. (2021). Advances in low-temperature SCR catalysts for NO_x reduction. *Environmental Science & Technology*, 55(5), 1234–1247.
7. Matsumoto, T., Kimura, T., & Inoue, H. (2020). Plasma-assisted SCR for NO_x removal: Experimental insights. *Applied Catalysis B: Environmental*, 264, 118–124.
8. https://www.energyinst.org/statistical-review#:~:text=2023%20saw%20a%20second%20consecutive,its%202019%20pre%2DCOVID%20level.
9. Reitz, R.D., & Duraisamy, G. (2015). Review of high-efficiency low-emission LTC strategies. *Progress in Energy and Combustion Science*, 46, 12–71.
10. Yin, J., Liu, H., & Reitz, R.D. (2021). Optimizing RCCI combustion via in-cylinder stratification and dual-fuel chemistry. *Fuel*, 289, 119833.
11. Zhao, H. (2007). *HCCI and CAI Engines for the Automotive Industry*. Woodhead Publishing, New York.
12. Musculus, M.P.B. (2008). Understanding fuel effects on low-temperature diesel combustion. SAE Technical Paper 2008-01-1074.
13. Johansson, B., & Reuss, D.L. (2018). HCCI and PPC combustion—Concepts and control. SAE Technical Paper 2018-01-0245.
14. Jung, Y., Hwang, J., & Bae, C. (2016). Assessment of particulate matter in exhaust gas for biodiesel and diesel under conventional and low-temperature combustion in a compression ignition engine. *Fuel*, 165, 413–424.
15. Su, J., Zhu, H., & Bohac, S.V. (2013). Particulate matter emission comparison from conventional and premixed low-temperature combustion with diesel, biodiesel, and ethanol fuels. *Fuel*, 113, 221–227.
16. Ma, Z., Wang, Z., & Chen, Z. (2022). CFD and optical diagnostics analysis of soot and UHC formation in PPCI combustion mode. *Energy Conversion and Management*, 252, 115088.
17. Singh, A., Kumar, R., & Zhao, H. (2016). Effects of fuel stratification and injection strategies on emissions in LTC engines. *Applied Energy*, 183, 1–10.
18. Wang, X., Lee, C.F., & Longman, D.E. (2018). Low-temperature combustion strategies and their impact on engine emissions. *Energy Conversion and Management*, 156, 520–531.
19. Kalghatgi, G., Risberg, P., & Ångström, H.E. (2020). Partially premixed auto-ignition of gasoline to attain low smoke and low NO_x at high efficiency in a compression ignition engine. *SAE International Journal of Engines*, 13(3), 3211–3224.

20. Gandhi, H.S., & Graham, G.W. (2018). Automotive emissions control catalysis. *Journal of Catalysis*, 360, 1–19.
21. Zheng, Z., Liu, T., & Huang, J. (2019). Impact of fuel composition on soot precursors and particulate emissions in LTC engines. *Combustion and Flame*, 207, 287–298.
22. Zhang, H., Zhao, X., & Liu, W. (2021). Development of hydrocarbon-SCR systems for lean NO_x reduction. *Fuel Processing Technology*, 223, 106–115.
23. Yuan, C., Xu, H., & Wang, J. (2018). Advances in diesel particulate filter (DPF) technology for modern engines. *Fuel*, 221, 345–358.
24. Zhao, Y., Tang, D., & Chen, L. (2019). Gasoline particulate filters (GPFs) and their role in emission control. *Applied Catalysis B: Environmental*, 243, 789–799.
25. Wang, Y., Zhang, X., & Liu, J. (2022). Advances in particulate matter control for low-temperature combustion engines. *Fuel*, 320, 124053.
26. Zhao, X., et al. (2020). Model predictive control of HCCI combustion phasing using fast-response combustion sensors. SAE Technical Paper 2020-01-0134.
27. Wang, Y., et al. (2021). Reinforcement learning-based adaptive control for dual-fuel LTC engines. *Applied Energy*, 298, 117271.
28. Patel, R., et al. (2022). Data-driven combustion model for predictive control of LTC modes. *Energy*, 240, 122557.

10 Future Directions and Emerging Trends in Low-Temperature Combustion

Aashish Sharma, Sudhanshu Dogra, Amit Thakur, Almagul Mergalimova, Yashvardhan, and Hardik

10.1 INTRODUCTION

10.1.1 Overview of Combustion Technologies

Combustion is the main way that internal combustion engines (ICEs), gas turbines, and other industrial applications change energy. When fossil fuels are burned at high temperatures in traditional high-temperature combustion (HTC) processes, a lot of energy is released, but a lot of pollutants are also released, such as nitrogen oxides (NO_x), particulate matter (PM), and unburned hydrocarbons (UHCs). These pollutants make problems with the environment worse, such as bad air quality, climate change, and health problems [1].

Researchers have looked into different ways of burning that produce fewer emissions while keeping or improving efficiency to deal with these problems. Low-temperature combustion (LTC) is one such promising method. LTC is a new way of burning that is meant to work at much lower peak temperatures than traditional methods [2]. This reduction in combustion temperature results in a decrease in NO_x emissions and soot formation, making it an environmentally friendly alternative to the conventional IC engines.

10.1.2 Fundamentals of Low-Temperature Combustion

LTC strategies aim to optimize the trade-off between thermal efficiency and pollutant formation by precisely controlling in-cylinder conditions such as pressure, temperature, fuel composition, and air–fuel mixing. The fundamental principles of LTC revolve around achieving auto-ignition of a well-mixed charge at lower temperatures to suppress NO_x formation while maintaining efficient combustion strategies try to find the best balance between thermal efficiency and pollutant formation by carefully controlling things like pressure, temperature, fuel composition, and air–fuel mixing inside the cylinder [3]. The main ideas behind LTC are to get a well-mixed charge to ignite on its own at lower temperatures to stop NO_x from forming while keeping combustion efficient (Figure 10.1).

DOI: 10.1201/9781003633631-10

1 **Homogeneous Charge Compression Ignition (HCCI):**

•A combustion mode where a premixed air-fuel mixture auto-ignites due to compression heating, eliminating the need for spark ignition or direct fuel injection timing control.

2 **Partially Premixed Compression Ignition (PPCI):**

•A hybrid combustion mode that combines elements of both diffusion and premixed combustion, offering better control over ignition timing and combustion stability.

3 **Reactivity Controlled Compression Ignition (RCCI):**

•A dual-fuel strategy that leverages fuels with different reactivity levels to precisely control ignition timing and combustion phasing.

FIGURE 10.1 Key LTC strategies.

Each of these LTC methods has its own pros and cons. HCCI, for example, has almost no NO_x and PM emissions, but it is hard to control ignition and has a limited load range. PPCI and RCCI get around some of these problems by using fuel reactivity and injection strategies to better control how combustion works.

10.1.3 Advantages of LTC over Conventional Combustion

LTC strategies aim to optimize the trade-off between thermal efficiency and pollutant formation by precisely controlling in-cylinder conditions such as pressure, temperature, fuel composition, and air–fuel mixing. The fundamental principles of LTC revolve around achieving auto-ignition of a well-mixed charge at lower temperatures to suppress NO_x formation while maintaining efficient combustion [4,5]. There are a number of important benefits to using LTC technologies instead of traditional combustion methods:

1. **A Big Drop in the Amount of Pollutants Released:** LTC lowers combustion temperatures, which stops the formation of NO_x. Also, mixing the fuel and air correctly reduces the amount of soot that forms, which makes the combustion cleaner.
2. **Better Thermal Efficiency:** LTC engines can get higher brake thermal efficiency than regular diesel and spark-ignition (SI) engines by optimizing the pressure and temperature inside the cylinder.
3. **Fuel Flexibility:** LTC can use a wide range of fuels, such as biofuels, synthetic fuels, and regular petroleum-based fuels. This makes it possible to use renewable energy sources.
4. **Less Need for Cooling:** LTC engines have lower peak combustion temperatures, which means they reject less heat. This means that the cooling system doesn't have to work as hard, which makes the engine more efficient overall.
5. **Improvement in Engine Durability:** The lower combustion temperatures reduce thermal stresses on engine components, potentially increasing engine longevity and reducing maintenance costs.

10.1.4 Challenges in Implementing LTC

Despite its advantages, LTC also faces several challenges that must be addressed for widespread adoption in commercial and industrial applications. These challenges include:

1. **Ignition Control and Stability:** Because LTC relies on internal circumstances for ignition rather than external sources like fuel injectors or spark plugs, it is difficult to control auto-ignition timing.
2. **Limited Operating Range:** LTC performs well at mid-to-low engine loads but has trouble with combustion stability at high loads because of misfire and knock problems.
3. **Cold-Start Performance:** Because incomplete combustion can result from lower in-cylinder temperatures, achieving consistent ignition under cold-start conditions is a major challenge.
4. **Fuel Sensitivity:** The effectiveness of LTC is significantly influenced by the characteristics of fuel, such as octane/cetane ratings, volatility, and chemical reactivity.
5. **Emissions of UHCs and Carbon Monoxide (CO):** Although LTC effectively diminishes NO_x and soot emissions, it may result in heightened UHC and CO emissions stemming from the incomplete oxidation of fuel.

10.1.5 Technological Innovations Driving LTC Development

Researchers are actively working to overcome LTC limitations through innovative solutions like:

- **Advanced Fuel Injection Systems:** Multiple injection methods and high-pressure common rail fuel injection enable precision control over ignition timing and fuel delivery.
- **Exhaust Gas Recirculation (EGR):** EGR lowers NO_x emissions and modifies combustion temperature by adjusting in-cylinder oxygen concentration.
- **Variable Compression Ratio (VCR) Engines:** VCR technology optimizes performance under a variety of load scenarios by enabling dynamic compression ratio change.
- **Hybrid Powertrain Integration:** By making up for LTC's shortcomings under high load circumstances, hybrid-electric powertrains and LTC allow for flexible operation.
- **Machine Learning (ML) and** Artificial Intelligence **(AI) Optimization:** To improve efficiency and control techniques, AI is being employed more and more to create prediction models for LTC combustion behaviour.

10.1.6 Future Prospects of LTC

The future of LTC depends on advancements in engine design, fuel development, and regulatory frameworks [6]. As the world focuses more on reducing carbon emissions and improving energy efficiency, LTC technology will be crucial to next-generation combustion systems. Among the potential directions for the future are (Figure 10.2):

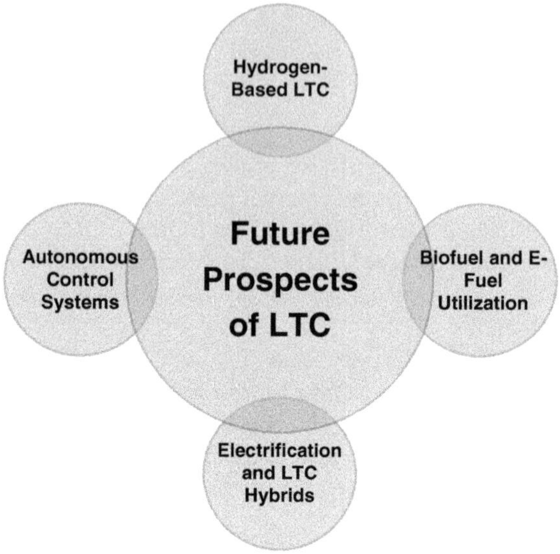

FIGURE 10.2 Prospects of LTC.

- Hydrogen-Based LTC: Research on hydrogen combustion in LTC regimes is expanding, and it may result in carbon emissions that are zero.
- Research is being done on sustainable fuels derived from biomass and synthetic production methods to increase LTC viability.
- Electric drivetrains and LTC engines: Work together to reduce dependency on fossil fuels and increase efficiency.
- Autonomous Control Systems: By enabling real-time adjustments in LTC engines, AI-based control systems will increase adaptability and efficiency.

Put it all up, LTC presents a groundbreaking advancement in combustion technology that achieves a balance between high efficiency and lowering emissions. Even though there are still many challenges to be solved, LTC is becoming a feasible alternative for combustion applications in the future thanks to current research and advancements in technology.

10.2 ADVANCES IN FUEL FORMULATIONS FOR LTC

10.2.1 Importance of Fuel Formulation in LTC

The formulation of fuel is essential for the performance and viability of LTC strategies. In contrast to traditional combustion processes, which function at elevated temperatures with swift oxidation, LTC depends on meticulously regulated ignition timing and gradual combustion reactions. The properties of fuels, such as their chemical makeup, reactivity, volatility, and oxygen levels, have a profound effect on combustion efficiency, emissions, and stability within LTC systems [7,8].

As the global shift towards cleaner and more sustainable energy solutions accelerates, the development of fuels for LTC must comply with rigorous environmental standards, resource availability, and economic viability. Researchers are investigating various fuel modifications and alternative fuels to enhance LTC, reduce pollutant emissions, and improve engine performance.

10.2.2 Alternative Fuels for LTC

The advancement of alternative fuels, including biofuels, synthetic fuels, and hydrogen, is vital for the effective implementation of LTC. These fuels possess characteristics that promote low-temperature reactions and reduce emissions. Studies suggest that fuels with elevated oxygen content improve combustion efficiency at lower temperatures [9].

10.2.2.1 Biofuels

Biofuels, which are sourced from renewable biological materials such as plant biomass, algae, and waste products, represent a promising alternative for LTC. They provide numerous benefits, including lower carbon emissions, increased oxygen content, and compatibility with current ICE technologies [10].

Types of biofuels used in LTC include:

- **Biodiesel:** Sourced from either vegetable oils or animal fats, biodiesel possesses a higher cetane number, which aids in enhancing ignition within LTC environments. Nevertheless, its elevated viscosity may necessitate modifications to the fuel system.
- **Ethanol and Methanol:** Fuels based on alcohol demonstrate remarkable reactivity at low temperatures, rendering them appropriate for LTC uses. Additionally, their significant latent heat of vaporization plays a role in charge cooling effects, thereby lowering peak combustion temperatures.
- **Hydrotreated Vegetable Oils (HVOs):** These types of fuels provide superior oxidation stability, improved cold flow characteristics, and greater compatibility with engines when compared to traditional biodiesel.

10.2.2.2 Synthetic Fuels

Synthetic fuels, commonly referred to as e-fuels, are generated through sophisticated chemical methodologies such as Fischer–Tropsch (FT) synthesis, power-to-liquid (PtL) technology, and biomass-to-liquid conversion. These fuels can be customized to possess favourable characteristics for LTC operation, which include optimized ignition delay, elevated hydrogen-to-carbon ratios, and reduced aromatic content.

Key synthetic fuels for LTC include:

- **Gas-to-Liquid (GTL) Fuels:** Derived from natural gas, GTL fuels are characterized by low sulphur content and high cetane numbers, rendering them suitable candidates for LTC applications.
- **PtL Fuels:** These fuels are produced utilizing captured carbon dioxide (CO_2) and hydrogen sourced from renewable origins, offering a carbon-neutral energy alternative for LTC.

- **FT Fuels:** These liquid hydrocarbons are designed to enhance combustion efficiency and minimize particulate emissions within LTC systems.

10.2.2.3 Hydrogen and Ammonia

Hydrogen is increasingly recognized as a viable zero-emission fuel for LTC. It possesses high energy density, rapid flame propagation, and remarkable reactivity under controlled combustion conditions. Ammonia, serving as a hydrogen carrier, is also under investigation due to its carbon-free combustion properties [11].

Challenges linked to the combustion of hydrogen and ammonia in LTC include:

- **Storage and Distribution:** Hydrogen necessitates advanced storage solutions such as high-pressure tanks or cryogenic systems, while the toxicity and corrosiveness of ammonia require specialized handling infrastructure.
- **Knock and Ignition Control:** The high reactivity of hydrogen necessitates meticulous regulation of ignition timing to avert pre-ignition and knocking in LTC engines.

10.2.3 Fuel Additives and Blending Strategies

Fuel additives and blending methods are crucial for enhancing LTC performance by altering fuel reactivity, minimizing emissions, and improving ignition control.

10.2.3.1 Oxygenated Additives

Oxygenated fuel additives, including ethers and esters, enhance combustion efficiency by raising oxygen levels and facilitating complete oxidation. Notable oxygenated additives comprise:

- **Dimethyl Ether:** A promising alternative fuel characterized by excellent ignition properties, diminished soot production, and improved combustion stability.
- **Ethanol and Butanol:** Alcohol-based additives that enhance knock resistance, optimize fuel atomization, and lower CO emissions.

10.2.3.2 Cetane and Octane Enhancers

LTC engines necessitate precise management of ignition delay to optimize efficiency and emissions. Cetane and octane enhancers contribute to achieving the desired auto-ignition properties:

- **Cetane Improvers:** Organic nitrates and peroxides are frequently utilized to boost cetane numbers in diesel-like LTC fuels.
- **Octane Boosters:** Aromatic compounds and metallic additives, such as manganese-based formulations, aid in stabilizing ignition in gasoline-like LTC fuels.

10.2.3.3 Nanoparticle-Based Additives

Recent developments in nanotechnology have led to the introduction of nanoparticle additives aimed at improving LTC fuel performance [12]. Metal and metal oxide nanoparticles, such as cerium oxide, aluminium oxide, and iron-based compounds, enhance combustion efficiency, decrease emissions, and improve thermal conductivity.

10.2.4 Fuel Property Optimization for LTC

Optimizing the properties of fuel is essential for achieving stable and efficient combustion in LTC. Important considerations regarding fuel properties include:

10.2.4.1 Ignition Delay and Reactivity Control

LTC fuels need to achieve a balance between auto-ignition delay and the controlled progression of combustion. Researchers adjust the chemical composition of fuels to refine ignition characteristics, thereby minimizing the risks of premature ignition while ensuring complete combustion.

10.2.4.2 Volatility and Evaporation Characteristics

Effective vaporization and mixing of fuel are vital for the preparation of a homogeneous charge in LTC. Fuel formulations that possess suitable volatility profiles improve the preparation of the mixture, resulting in enhanced combustion efficiency.

10.2.4.3 Lubricity and Wear Protection

Given that LTC functions at lower temperatures, the lubricity of fuel becomes a significant factor in mitigating engine wear. Additives like fatty acid methyl esters (FAME) and organic lubricants contribute to reducing friction and improving durability.

10.2.5 Future Trends in Fuel Development for LTC

10.2.5.1 Renewable and Carbon-Neutral Fuels

The future of LTC is contingent upon the adoption of sustainable fuel alternatives. Fuels that are carbon-neutral, sourced from renewable hydrogen, captured CO_2, and bio-based feedstocks, will be instrumental in achieving combustion processes that are free from emissions.

10.2.5.2 AI-Driven Fuel Formulation Optimization

The application of ML and AI-driven simulations is becoming increasingly prevalent in the development of optimal fuel formulations specifically designed for LTC. These AI models assess combustion kinetics, emissions behaviour, and thermodynamic performance to pinpoint the most effective fuel blends available.

10.2.5.3 Co-Optimization of Fuels and Engines

Future research in LTC fuels will prioritize the co-optimization of fuels alongside engine hardware. Joint initiatives among fuel chemists, engine manufacturers, and policymakers are expected to yield integrated solutions that enhance both efficiency and sustainability in fuel-engine systems.

Advancements in fuel formulations are vital for the effective execution of LTC strategies. The exploration of biofuels, synthetic fuels, hydrogen, and advanced additives presents promising avenues for improving LTC efficiency while minimizing environmental repercussions. Ongoing research will persist in investigating innovative fuel solutions, AI-driven optimization techniques, and sustainable alternatives to transform LTC technologies [13–15].

10.3 ADVANCED IGNITION AND INJECTION STRATEGIES

10.3.1 Importance of Ignition and Injection Strategies in LTC

The successful application of LTC technologies depends on effective injection and ignition techniques. To achieve lower emissions and greater efficiency than conventional HTC, LTC depends on carefully regulated ignition timing and fuel–air mixture preparation. Because of the unique combustion properties of LTC environments, traditional spark and compression ignition (CI) systems frequently require modifications to operate efficiently in these settings. To maximize combustion timing, minimize knocking, and enhance overall stability, novel ignition and injection techniques are being researched.

10.3.2 Ignition Strategies for LTC

LTC necessitates ignition strategies that guarantee dependable and uniform combustion across a range of operating conditions. The following ignition methodologies have been extensively researched:

10.3.2.1 Compression Ignition with Reactivity Control

Reactivity-Controlled Compression Ignition (RCCI), a sophisticated LTC technique, uses two fuels with varying degrees of reactivity. A low-reactivity fuel, such as gasoline or ethanol, is mixed with air to regulate ignition timing, while a high-reactivity fuel, such as diesel or biodiesel, is injected directly. A few benefits of RCCI include lower nitrogen oxide (NO_x) and PM emissions, increased fuel choice flexibility, and enhanced thermal efficiency due to better combustion timing.

However, RCCI needs sophisticated control systems and precise fuel blending to ensure steady operation.

10.3.2.2 Homogeneous Charge Compression Ignition (HCCI)

The basis of Homogeneous Charge Compression Ignition (HCCI) is the auto-ignition of a uniformly premixed fuel–air mixture under compression. The production of NO_x and soot is reduced because combustion occurs all at once inside the cylinder due to the lack of a direct spark or diesel pilot ignition. The following are the primary challenges facing HCCI:

- Limited operating range due to misfire and knock hazards.
- Inability to control combustion phasing.
- Need for real-time combustion monitoring and control.

10.3.2.3 Spark-Assisted Compression Ignition (SACI)

Spark-Assisted Compression Ignition (SACI) integrates the principles of SI and CI. The ignition process is initiated by a spark plug, while the auto-ignition driven by compression facilitates the propagation of combustion. SACI contributes to extending the operational range of LTC engines, reducing cycle-to-cycle variations, and providing improved ignition control in lean and dilute conditions.

10.3.2.4 Plasma and Laser Ignition

For LTC applications, cutting-edge ignition techniques like laser and plasma-assisted ignition are presently being investigated. Laser ignition uses concentrated laser beams to cause plasma breakdowns, while plasma-assisted ignition uses high-energy plasma discharges to start combustion. These methods offer the following benefits:

- Increased stability of ignition in extremely low conditions;
- Quicker development of flame kernels;
- Potentially lower ignition delays.

However, before they are widely adopted, issues with system complexity and cost must be addressed.

10.3.3 Fuel Injection Strategies for LTC

Optimizing fuel injection parameters is critical for achieving stable LTC combustion. Key injection strategies include:

10.3.3.1 Direct Injection (DI)

The process of injecting fuel straight into the combustion chamber is known as 'direct injection', and it improves combustion efficiency and allows for precise control of the fuel–air mixture. The creation of stratified charge, which reduces the possibility of premature ignition, better fuel use and thermal efficiency, and improved emissions control as a result of better air–fuel mixing are some benefits of DI.

10.3.3.2 Port Fuel Injection (PFI) and Dual Injection

Before the fuel enters the cylinder, it is premixed with intake air using port fuel injection (PFI). This technique is frequently combined with DI to produce dual-injection systems. Better control over fuel stratification, higher efficiency at low loads, and a decreased propensity for engine knock are some advantages of dual injection.

10.3.3.3 Multi-Pulse and Split Injection

Split injection and multi-pulse injection involve injecting fuel in multiple stages as opposed to a single injection event. This method enables precise timing control of ignition, a decrease in soot and NO_x emissions, and enhanced rates of pressure increase (Table 10.1).

10.3.4 Comparison of Ignition and Injec tion Strategies

TABLE 10.1
Comparison of Ignition and Injection Strategies

Strategy	Description	Advantages	Challenges
RCCI	Dual-fuel compression ignition (CI)	Low NO_x/particulate matter emissions, high efficiency	Requires precise fuel blending and control
HCCI	Auto-ignition of homogeneous charge	Near-zero NO_x and soot, high efficiency	Difficult combustion phasing control
SACI	Combination of spark ignition and CI	Extended operating range, better ignition control	Increased complexity
Plasma ignition	Plasma discharge to initiate combustion	Improved lean-burn stability	High system cost
Laser ignition	Laser-induced breakdown for ignition	Fast ignition, reduced misfires	Expensive and complex
Direct injection (DI)	Fuel injected directly into cylinder	Improved fuel–air mixing, reduced knock	Requires advanced injectors
Port fuel injection (PFI)	Fuel injected before intake valve	Uniform mixture formation	Less precise combustion control
Dual injection	Combination of DI and PFI	Improved combustion stability	Increased fuel system complexity
Multi-pulse injection	Fuel injected in multiple pulses	Optimized emissions, reduced pressure rise rates	Requires precise timing control

10.3.5 Future Trends in Ignition and Injection for LTC

10.3.5.1 Intelligent Control and AI Integration

AI and ML algorithms are being used in LTC engines to optimize ignition and injection timing in real time. AI-driven strategies can adapt to changing operational conditions, boost combustion stability and efficiency, and reduce variations from one cycle to the next.

10.3.5.2 Advanced Fuel Injectors

Next-generation fuel injectors, including piezoelectric and ultra-high-pressure injectors, will facilitate precise fuel metering and improved spray atomization, thereby further enhancing LTC performance.

10.3.5.3 Hybrid Combustion Approaches

The adoption of hybrid combustion strategies, which dynamically switch between different ignition and injection modes to maximize efficiency under varying load conditions, is anticipated for future LTC engines.

To realize LTC's full potential, improvements in injection and ignition techniques are necessary. Promising approaches to attaining low-emission, high-efficiency combustion include RCCI, HCCI, SACI, and multi-pulse injection. Ongoing research will continue to focus on next-generation fuel injection technologies, intelligent control systems, and AI-driven optimization to enhance LTC performance and dependability.

10.4 COMPUTATIONAL MODELLING AND AI IN LTC OPTIMIZATION

10.4.1 Introduction to Computational Modelling in LTC

Computational modelling is a crucial tool for advancing and refining LTC methods. Because LTC processes are complicated and involve intricate chemical kinetics, heat transfer, and turbulent flows, high-fidelity simulations are crucial for comprehending and forecasting combustion behaviour under a range of operating conditions [16].

Even though they are required, traditional experimental techniques are frequently costly and time-consuming. Computational techniques such as chemical kinetics modelling, computational fluid dynamics (CFD), and AI-based ML models allow researchers to design and enhance LTC engines more rapidly with minimal physical testing. These methodologies assist in:

- Predicting combustion characteristics, including ignition timing, heat release rates, and emissions.
- Understanding the influence of fuel properties on LTC performance.
- Optimizing injection timing, air–fuel mixture preparation, and ignition strategies.
- Minimizing design cycle duration and expenses.

10.4.2 Computational Fluid Dynamics (CFD) for LTC

CFD is widely used for simulating combustion phenomena in LTC systems. It involves solving the fundamental governing equations of fluid flow, heat transfer, and chemical reactions to predict combustion behaviour. CFD models solve the Navier–Stokes equations, which govern fluid dynamics, along with energy and species transport equations [17]. The governing equations (continuity, momentum, and energy) equations are solved numerically using discretization methods such as finite difference, finite volume, or finite element methods.

10.4.2.1 Turbulence and Combustion Modelling

Since LTC occurs in highly turbulent conditions, accurate turbulence and combustion models are required. Some commonly used models include:

- **Reynolds-Averaged Navier–Stokes Models:** Suitable for steady-state simulations but limited in resolving fine turbulence structures.
- **Large Eddy Simulation:** Provides better accuracy by resolving larger turbulent eddies but requires higher computational power.

- **Direct Numerical Simulation:** The most accurate approach, solving all turbulence scales, but impractical for engine-scale simulations due to computational costs.

For combustion, detailed chemical kinetics models (e.g., GRI-Mech and Chemkin) are often coupled with CFD solvers to capture complex reaction mechanisms governing LTC.

10.4.3 Chemical Kinetics and Reaction Mechanism Development

Given that it helps predict ignition delay, combustion timing, and pollutant formation, chemical kinetics modelling is crucial to LTC research. To model chemical kinetics in LTC environments, researchers use a variety of models, such as zero-dimensional, one-dimensional, and multi-dimensional.

10.4.3.1 Mechanism Reduction Techniques

Methods for mechanism reduction are employed to increase the efficiency of calculations because complete reaction mechanisms can contain thousands of species and reactions:

- Skeletal mechanisms preserve important combustion pathways while eliminating superfluous species and reactions.
- Sensitivity analysis aims to identify critical reactions that have a major influence on ignition and emissions.
- Artificial neural network (ANN)-based reductions: Instead of using complex kinetic equations, this approach uses AI techniques to estimate reaction rates.

10.4.4 Machine Learning and AI in LTC Optimization

LTC research is being revolutionized by AI and ML techniques that enable faster optimization and real-time adaptive control.

10.4.4.1 AI Applications in LTC

ML models are being used for:

- **Forecasting Ignition Delay and Heat Release Rates:** AI-powered models can substitute for costly CFD simulations.
- **Optimization of Fuel Formulation:** AI algorithms evaluate fuel property data to determine the best blends for LTC.
- **Control of Injection and Ignition:** AI-driven controllers modify injection timing and ignition settings in real time to enhance LTC performance (Table 10.2).

10.4.4.2 Common Machine Learning Techniques for LTC

TABLE 10.2
Common Machine Learning Techniques for LTC

Machine Learning Method	Application in LTC	Advantages	Challenges
Artificial neural networks	Predicting combustion characteristics	High accuracy, self-learning capability	Requires large training datasets
Support vector machines	Classifying combustion regimes	Works well with small datasets	Computationally expensive
Genetic algorithms	Optimizing engine parameters	Finds optimal solutions quickly	May converge to local optima
Deep reinforcement learning	Adaptive control of LTC	Learns from real-time data	Requires large computational resources
Principal component analysis	Reducing dimensionality in combustion data	Extracts meaningful trends	Loss of information in simplifications

10.4.5 AI-Based Real-Time Control for LTC

The incorporation of AI into engine management systems enables real-time control algorithms to dynamically enhance LTC performance. Advanced control strategies encompass:

- **Model Predictive Control:** This method forecasts and adjusts combustion parameters based on real-time engine data.
- **Reinforcement Learning-based Controllers:** AI agents learn the best control strategies by making mistakes.
- **Edge Computing for On-Board AI:** This technique enables incredibly quick decision-making by directly integrating ML models into engine control units.

10.4.6 Future Trends in Computational Modelling and AI for LTC

As computational power continues to grow, the future of LTC modelling and optimization is expected to witness:

10.4.6.1 High-Fidelity Hybrid Models

Hybrid models that combine physics-based CFD with data-driven AI approaches will enhance prediction accuracy while reducing computational time.

10.4.6.2 Digital Twins for LTC Engine Development

A digital twin serves as a virtual model of a real-world LTC engine, consistently refreshed with real-time information. The utilization of digital twins will facilitate:

- The virtual assessment of novel LTC strategies prior to their physical execution.
- Predictive maintenance and identification of faults in LTC engines.
- Real-time optimization of combustion parameters.

10.4.6.3 Quantum Computing for Combustion Simulations

The development of quantum computing could revolutionize LTC research by reducing the time it takes to complete complex combustion simulations, which currently take days, to a matter of minutes.

Combining AI and computational modelling has become crucial for optimizing LTC because it allows for faster insights and lowers experimental costs. Researchers can enhance LTC strategies with the help of CFD, chemical kinetics modelling, and AI-enhanced predictive models [18]. It is anticipated that developments in digital twins, hybrid modelling, and quantum computing will accelerate the development of high-efficiency, low-emission LTC engines.

10.5 HYBRIDIZATION AND RENEWABLE ENERGY INTEGRATION

10.5.1 Introduction

Hybrid powertrains, which combine LTC techniques with renewable energy sources, have attracted a lot of attention due to the growing global focus on sustainable energy solutions. Even with LTC, conventional ICEs struggle to reduce emissions and increase efficiency. Researchers hope to develop propulsion systems that are cleaner, more sustainable, and more efficient by combining hybridization and renewable energy.

To increase fuel efficiency and reduce emissions, hybridization entails combining multiple power sources, typically an ICE with an electric powertrain. Additional sustainability is promoted by the use of renewable fuels, such as hydrogen, biofuels, and synthetic fuels, which guarantees that LTC technologies meet international decarbonization goals.

10.5.2 Benefits of Hybridization and Renewable Energy Integration

Integrating LTC with hybrid powertrains and renewable energy offers several advantages:

1. Lower Emissions
2. Improved Fuel Efficiency
3. Flexibility in Fuel Choice
4. Optimized Power Management

10.5.3 Hybrid Powertrain Configurations for LTC

Hybridization in LTC-based engines is categorized into different powertrain architectures, each with unique advantages (Table 10.3):

TABLE 10.3
Hybrid Powertrain Configurations for LTC

Hybrid Type	Description	Advantages	Challenges
Mild hybrid	Small electric motor assists internal combustion engine (ICE), but cannot run independently	Improved fuel efficiency, regenerative braking	Limited electric-only driving capability
Full hybrid	ICE and electric motor can operate independently or together	Significant fuel savings, reduced emissions	Higher complexity and cost
Plug-in hybrid	Larger battery allows extended electric-only range	Can run without fossil fuels for short trips	Requires external charging infrastructure
Series hybrid	ICE acts as a generator; electric motor drives wheels	Maximum efficiency in stop-and-go traffic	Dependence on battery for propulsion
Parallel hybrid	ICE and electric motor both provide propulsion	Higher overall efficiency	More complex drivetrain management

For LTC applications, parallel and series hybrid configurations are particularly promising as they allow for optimized load distribution between the electric motor and LTC engine, reducing fuel consumption and emissions.

10.5.4 Renewable Fuels for LTC Hybrid Systems

The transition to renewable fuels is crucial for achieving zero-emission or carbon-neutral operations, even though hybridization increases energy efficiency. Investigations into different renewable fuels for LTC-integrated hybrid systems are presently underway.

10.5.4.1 Biofuels

Biomass such as plant oils, algae, and agricultural waste are the source of biofuels. These fuels may substantially decrease lifecycle carbon emissions and are compatible with current LTC engines.

- **FAME or Biodiesel:** This fuel has a high cetane number, but it may not work well in cold climates.
- **HVO:** This fuel has improved combustion properties and reduced emissions.
- **Alcohol Fuels (Ethanol, Methanol, and Butanol):** These fuels' high oxygen content increases LTC efficiency, but more work needs to be done to optimize their compatibility with blends.

10.5.4.2 Hydrogen and Ammonia

Hydrogen (H_2) and ammonia (NH_3) are fuels that do not contain carbon and can be utilized in LTC engines, either independently or in combination with conventional fuels.

1. **Hydrogen:** It offers a high energy density and quick combustion; however, issues related to storage and distribution persist.
2. **Ammonia:** This substance serves as a potential carrier for hydrogen, exhibiting improved storage stability, yet necessitates modifications in combustion because of its reduced flame speed.

10.5.4.3 Synthetic Fuels (E-Fuels)

A carbon-neutral substitute is provided by synthetic fuels, which are made using FT and PtL synthesis. They are extremely sustainable because they are produced using captured CO_2 and renewable hydrogen.

10.5.5 Renewable Energy Integration in LTC Systems

To further enhance sustainability, LTC engines can be integrated with renewable energy sources such as solar, wind, and waste heat recovery systems (WHRS).

10.5.5.1 Solar Energy Integration

Battery charging in hybrid cars can be facilitated by solar energy, which lessens dependency on the electrical grid. Plug-in hybrid batteries are charged by solar-assisted hybrid LTC systems using photovoltaic panels, which also help to preheat fuels for improved combustion efficiency and power auxiliary systems like lights, air conditioning, and sensors.

10.5.5.2 Wind Energy Integration

Off-grid charging of hybrid long-term battery vehicles or electrolysis of renewable hydrogen are the two main uses of wind energy. Green hydrogen can be produced by wind-powered hydrogen stations for LTC-hybrid uses.

10.5.5.3 Waste Heat Recovery Systems (WHRS)

Due to the operation of LTC engines at reduced temperatures, there are considerable losses of waste heat. The implementation of waste heat recovery can enhance overall efficiency.

Thermoelectric generators are capable of transforming exhaust heat into electrical energy, which can be used to recharge the batteries of hybrid vehicles. Organic Rankine Cycle Systems: These systems harness low-temperature waste heat to produce supplementary power.

10.5.6 Challenges and Future Directions

Despite the benefits of hybridization and renewable energy integration, several technical, economic, and infrastructure challenges remain.

10.5.6.1 Technical Challenges

- **Fuel Compatibility Issues:** To attain the best performance in LTC, some biofuels and synthetic fuels require engine modifications.
- **Battery Weight and Efficiency:** Increasing battery capacity increases battery efficiency, but it also adds weight, which reduces overall fuel savings.
- **Storage of Hydrogen and Ammonia:** The toxicity of ammonia and the necessity of high-pressure tanks for hydrogen pose major barriers to widespread adoption.

10.5.6.2 Economic and Infrastructure Barriers

- **High Initial Costs:** Despite the possibility of significant long-term savings, the deployment of hybrid and renewable LTC systems necessitates large initial investments.
- **Availability of Renewable Fuels:** The production of synthetic fuels and the lack of infrastructure for hydrogen refuelling stations slow down the rate of adoption.
- **Grid Integration:** Smart grids and distributed energy management systems must be developed to integrate renewable energy sources into hybrid LTC powertrains.

10.5.7 Future Outlook

The future of LTC hybridization and renewable integration will focus on:

1. **Advanced Battery Technologies:** Development of solid-state batteries with higher energy density and faster charging.
2. **Hydrogen Fuel Cell Hybrids:** Combining LTC with fuel cell technology for near-zero emissions.
3. **AI-Based Energy Management:** ML algorithms optimizing real-time power distribution among LTC, electric motors, and renewable energy sources.
4. **Decentralized Renewable Energy Systems:** On-site hydrogen production using excess solar/wind power for hybrid LTC applications.
5. **Government Policy and Incentives:** Regulations promoting green fuel adoption and carbon credits for LTC hybrid users.

10.5.8 Conclusion

The integration of hybridization with renewable energy sources offers a significant opportunity to enhance the efficiency and sustainability of LTC technologies. Through the utilization of sophisticated hybrid powertrains, renewable fuels, and the recovery of waste heat, LTC engines can reach near-zero emissions while still operating at high efficiency. To address the technical and economic hurdles, it is essential for researchers, industry participants, and policymakers to work together in advancing the development of the next generation of clean propulsion systems.

10.6 EMISSION REDUCTION AND REGULATORY CHALLENGES

LTC methods in ICEs are motivated by the necessity for emission regulation, providing considerable decreases in nitrogen oxide (NO_x) and PM while preserving high thermal efficiency; nevertheless, obstacles such as incomplete combustion resulting in elevated UHC and carbon monoxide (CO) emissions, cold-start difficulties due to reduced flame temperatures, and fuel-specific pollutants like aldehydes from biodiesel and alcohol-based fuels continue to exist. To tackle these issues and adhere to stringent global standards such as Euro 7, EPA Tier 4, and BS-VI, approaches including advanced after-treatment technologies—comprising three-way catalysts, diesel particulate filters, and selective catalytic reduction—as well as combustion optimization methods like EGR, oxygenated fuel additives, and multiple injection strategies are utilized to improve fuel–air mixing and minimize emissions [19]. Future research in LTC will concentrate on hybrid and electrified systems, AI-driven combustion control, and carbon-neutral fuels such as hydrogen and synthetic alternatives to meet changing regulations, ensuring cleaner and more efficient technologies despite persistent challenges with UHC, CO, and compliance with regulations.

10.7 POLICY AND MARKET TRENDS INFLUENCING LTC ADOPTION

LTC strategies in ICEs are propelled by stringent global policies such as Corporate Average Fuel Economy standards, Euro 7, and EPA Tier 4 regulations, which demand enhanced fuel efficiency and reduced CO_2, NO_x, and particulate emissions, while carbon neutrality goals and renewable fuel mandates, including biofuel blending and subsidies for hydrogen and e-fuels, bolster LTC's appeal due to its compatibility with low-carbon fuels; however, market challenges like competition from the expanding electric vehicle (EV) sector, limited hydrogen and synthetic fuel infrastructure, and high production costs of low-carbon fuels impede widespread adoption. Future trends, including co-optimization of fuels and LTC engines, integration with hybrid powertrains, and increased government funding for advanced combustion research, are poised to enhance LTC's role in sustainable transportation, provided policy support and fuel infrastructure continue to develop.

10.8 CONCLUSION AND FUTURE OUTLOOK

Driven by advances in fuel formulations (biofuels, hydrogen, ammonia, and oxygenated additives), advanced ignition and injection strategies (multiple injections and plasma-assisted ignition), computational modelling with AI and CFD, hybridization with renewable energy integration, and strict regulatory frameworks like Euro 7 and EPA Tier 4, LTC has emerged as a transformative approach for ICEs, balancing high efficiency with reduced emissions. However, obstacles like cold-start instability, limited fuel infrastructure for hydrogen and synthetic fuels, and competition from EVs and hydrogen fuel cells threaten its adoption, requiring large investments in fuel distribution and adaptive ignition technologies. Co-optimization of engines and fuels, integration with hybrid powertrains, and AI-driven real-time combustion control are some of the opportunities for LTC. These include co-optimization of fuels and engines, integration with hybrid powertrains, and AI-driven real-time combustion control, while future research should prioritize carbon-neutral fuels, multi-fuel LTC engines, automated control using AI and IoT, circular economy models like waste-to-fuel technologies, and advanced after-treatment solutions to address UHCs and carbon monoxide (CO), ensuring LTC's role in sustainable transportation amidst evolving market and policy landscapes.

LTC provides an exceptionally effective and low-emission route that complies with the changing fuel economy and emissions regulations as the global transportation sector strives for carbon neutrality. However, continued advancements in fuel science, especially in the creation of synthetic and renewable fuels, are necessary for the successful commercialization of LTC.

- Developments in real-time feedback systems and AI-based combustion control technologies.
- Infrastructure expenditures to support the distribution of LTC-compatible alternative fuels.
- Incentive programmes and policy frameworks that promote the use of electric and hybrid powertrains in combination with low-carbon combustion technologies.

10.8.1 The Road Ahead: The Role of LTC in the Clean Energy Transition

In applications (such as heavy-duty transportation, marine, and aviation) where electrification is impractical, LTC is still crucial for decarbonizing ICEs. Despite the market challenges brought on by EVs' growing dominance, the long-term viability of LTC in a sustainable transportation ecosystem will be determined by hybridized LTC solutions, AI-driven combustion modelling, and the expansion of renewable fuel supply chains.

By eliminating current barriers and leveraging interdisciplinary innovations, LTC can significantly contribute to a cleaner and more efficient future for ICEs.

REFERENCES

1. Agarwal, A. K., Gupta, J. G. and Dhar, A., 2017. Potential and challenges for large-scale application of biodiesel in automotive sector. *Progress in Energy and Combustion Science*, 61, pp. 113–149.
2. Bhattacharjee, S. and Zhou, M., 2020. Machine learning for combustion modeling in advanced internal combustion engines. *Energy and AI*, 2, p. 100021.
3. Dec, J. E., 2009. Advanced compression-ignition engines: Understanding the in-cylinder processes. *Proceedings of the Combustion Institute*, 32(2), pp. 2727–2742.
4. Johnson, T. V., 2015. Review of vehicular emissions trends. *SAE International Journal of Engines*, 8(3), pp. 1152–1167.
5. Kalghatgi, G. T., Hildingsson, L. and Johansson, B., 2011. Low NO_x and low smoke operation of a diesel engine using gasoline-like fuels. *Journal of Engineering for Gas Turbines and Power*, 133(9), p. 092803.

6. Kokjohn, S. L., Hanson, R. M., Splitter, D. A. and Reitz, R. D., 2011. Fuel reactivity controlled compression ignition (RCCI): A pathway to controlled high-efficiency clean combustion. *International Journal of Engine Research*, 12(3), pp. 209–226.
7. Lawler, B., Splitter, D., Szybist, J. and Kaul, B., 2017. Thermally stratified compression ignition: A new advanced low temperature combustion mode with load flexibility. *Applied Energy*, 189, pp. 122–132.
8. Lu, Y., Pan, J. and Liu, X., 2019. Hydrogen combustion in internal combustion engines: Challenges and opportunities. *Renewable and Sustainable Energy Reviews*, 111, pp. 250–262.
9. Manente, V., Johansson, B. and Tunestal, P., 2010. Partially premixed combustion at high load using gasoline and ethanol, its effects on HCCI combustion. *SAE Technical Paper*, 2010-01-2193.
10. Mueller, C. J., Boehman, A. L. and Martin, G. C., 2009. An experimental investigation of the origin of increased NO_x emissions when fueling a heavy-duty compression-ignition engine with soy biodiesel. *SAE Technical Paper*, 2009-01-1792.
11. Onishi, T., Tanaka, K. and Yamada, T., 2020. Development of advanced combustion control for low-temperature combustion engines. *Journal of Automobile Engineering*, 234(5), pp. 1345–1356.
12. Paykani, A., Garcia, A., Frouzakis, C. E. and Boulouchos, K., 2022. Machine learning for low-temperature combustion: A review. *Fuel*, 315, p. 123123.
13. Reitz, R. D. and Duraisamy, G., 2015. Review of high efficiency and clean reactivity controlled compression ignition (RCCI) combustion in internal combustion engines. *Progress in Energy and Combustion Science*, 46, pp. 12–71.
14. Sarathy, S. M., Oßwald, P., Hansen, N. and Kohse-Höinghaus, K., 2014. Alcohol combustion chemistry. *Progress in Energy and Combustion Science*, 44, pp. 40–102.
15. Splitter, D. A., Wissink, M. L., DelVescovo, D. and Reitz, R. D., 2016. RCCI engine operation towards 60% thermal efficiency. *SAE Technical Paper*, 2016-01-0777.
16. Turns, S. R., 2013. *An Introduction to Combustion: Concepts and Applications*, 3rd ed. Boston, MA: McGraw-Hill.
17. Zhang, Y., Li, H. and Zhao, H., 2018. Advanced ignition strategies for low-temperature combustion engines. *Applied Thermal Engineering*, 134, pp. 391–402.
18. Zhao, H., Li, Y. and Ma, X., 2017. After-treatment technologies for diesel engines: An overview. *Energy Conversion and Management*, 145, pp. 334–346.
19. Zhou, D., Yang, W. and Li, J., 2021. Artificial intelligence applications in low-temperature combustion modeling. *Energy Reports*, 7, pp. 342–356.

11 Case Studies and Applications in Automotive and Power Generation

Manas Ranjan Padhi, Saulesh Minazhova, Dillip Kumar Mohanta, and Surya Narayan Panda

11.1 INTRODUCTION

The automotive sector is a cornerstone of the global economy, encompassing the design, manufacturing, and distribution of motor vehicles. In recent years, its landscape has been shaped by rapid electrification, adoption of advanced driver-assistance systems, and a push toward automation in manufacturing. Manufacturers are increasingly leveraging lightweight materials, digital prototyping, and AI-driven design to enhance both efficiency and performance. The industry also faces ongoing challenges with supply chain resilience and regulatory pressures to lower vehicle emissions, accelerating the shift toward electric and hybrid drivetrains [1].

The power generation sector is responsible for producing electricity from a variety of sources like conventional fossil fuels, nuclear, and a growing share from renewables such as wind, solar, and hydroelectric energy. This sector is undergoing a significant transformation as it prioritizes lower-carbon energy solutions to address climate change and regulatory demands. Key advancements include improvements in grid stability, battery energy storage, and intelligent control systems that maximize efficiency and integrate intermittent renewable resources. Both sectors are critical for meeting global sustainability targets, and their convergence through trends like electric vehicles and decentralized power generation underscores the importance of innovation in mechanical engineering applications.

Case studies play a crucial role in applied research by providing an in-depth understanding of real-world phenomena within their specific contexts. Unlike purely theoretical research, case studies focus on practical applications, allowing researchers to explore complex issues, challenges, and solutions in detail. They serve as valuable tools for identifying best practices, testing theories, and generating insights that can be directly implemented to improve industry practices.

In sectors such as automotive and power generation, case studies help analyze the effectiveness of new technologies, safety protocols, and operational strategies. They enable researchers and practitioners to evaluate the impact of innovations in real operational environments, offering nuanced perspectives that broader quantitative

DOI: 10.1201/9781003633631-11

studies might overlook. Additionally, case studies facilitate learning from successes and failures, helping organizations adapt and optimize their processes.

Furthermore, case studies are instrumental in informing policy decisions and guiding strategic planning by illustrating tangible outcomes and lessons learned. They foster a deeper understanding of complex systems, enabling stakeholders to make evidence-based decisions. Overall, the depth, contextual relevance, and practical insights provided by case studies make them indispensable in applied research, driving continuous improvement and innovation across industries [2].

Energy efficiency, emissions, and performance are critical factors in advancing sustainable development in the automotive and power generation sectors. Improving energy efficiency reduces fuel consumption and operational costs, leading to economic savings and decreased resource depletion. It also lowers greenhouse gas emissions, mitigating climate change and environmental pollution. Emissions management is vital for complying with regulatory standards and enhancing air quality, which directly benefits public health [3]. Performance, encompassing reliability, safety, and operational effectiveness, ensures that vehicles and power systems meet user demands efficiently and sustainably. Balancing these factors is essential for reducing the carbon footprint of energy-intensive industries while maintaining economic viability. Innovations aimed at enhancing energy efficiency and reducing emissions contribute to a cleaner environment and promote the adoption of greener technologies. Ultimately, focusing on these aspects supports sustainable growth, environmental protection, and energy security, making them fundamental priorities for modern industry and policy development [4].

11.2 AUTOMOTIVE APPLICATIONS

11.2.1 Engine Simulation and Performance Analysis

Engine simulation and performance analysis involve using computer-based models to study and predict how an engine operates under various conditions. Engine simulation recreates the physical and chemical processes within an engine such as combustion, airflow, fuel injection, and heat transfer, allowing engineers to analyze performance, efficiency, and emissions without building physical prototypes. Performance analysis involves evaluating these simulated results to optimize engine parameters, improve power output, reduce fuel consumption, and minimize emissions. Together, they enable the design of more efficient, reliable, and environmentally friendly engines by providing valuable insights into how different configurations and operating conditions affect overall engine behavior [5].

Combustion modeling enables the optimization of energy efficiency, emissions, and engine performance across SI, diesel, and low-temperature combustion (LTC) modes, supporting both applied research and practical control/design in automotive and power generation sectors. The shift toward advanced models (multi-zone, kinetic, and data-driven) reflects the growing need for predictive accuracy and multifaceted system analysis.

Spark ignition (SI) combustion modeling involves using zero-dimensional or quasi-dimensional approaches like the two-zone model or multi-zone computational fluid dynamics (CFD). These models simulate flame propagation and combustion characteristics, accounting for turbulent flame propagation and eddy burning.

Advanced CFD models use advanced ignition kernels, turbulence models, and chemical kinetics to simulate cycle-to-cycle variability and wall quenching phenomena. These models are used to predict combustion stability, cyclic variation, and the effects of dilution strategies, with excellent agreement with experimental data under various spark timings and fueling strategies [6].

Diesel combustion models utilize heat release analysis and calibration methods to predict heat release and pressure across load/speed conditions. Modern approaches include physical models based on injection parameters and in-cylinder conditions, as well as machine-learning/data-driven methods. Dual-fuel and CFD approaches extend single- and multi-zone approaches to dual-fuel setups, often combining detailed kinetic mechanisms for multiple fuels and validating with in-cylinder pressure and emission measurements [7].

LTC modeling involves complex kinetics, high EGR rates, advanced fuel–air mixing, and precise control of combustion phasing. Multi-zone or multidimensional models are necessary to capture the distributed combustion characteristic of LTC. Thermo-kinetic multi-zone models divide the combustion chamber into distinct zones with distinct temperature and composition, accurately modeling LTC events and simulating nitrogen oxides (NO_x)/soot formation pathways. LTC models are used for stationary gas and dual-fuel engines to achieve lower emissions and better thermal efficiency [8].

11.2.2 LTC Technologies and Strategies

LTC strategies are characterized by their ability to minimize NO_x and particulate emissions, improve fuel economy, and operate efficiently at part loads. However, challenges such as combustion stability, controlling knock, and fuel requirements need ongoing research and advanced control systems. Table 11.1 provides

TABLE 11.1
LTC Strategies Comparison [9]

Strategy	Main Control	Fuel Type(s)	Key Advantages	Main Challenge
HCCI	Mixture, intake	Single (gasoline/diesel)	Ultra-low NO_x/soot	Ignition control, load range
PCCI	Injection timing	Single (diesel and advanced injectors)	Control and low emissions	Complexity and mixing timing
RCCI	Fuel reactivity	Dual (e.g., diesel/gasoline)	Flexible phasing	Dual-fuel hardware/cal.
Plasma-Assisted	Electrical plasma	Conventional or alternative fuels	Ignite at low temperature	Cost and system design
GCI	Injection and strategy	Gasoline, similar low-reactivity	Control and low emissions	Combustion stability
Fuel Reformation	Reformate ratio	Single with onboard reformation	Simpler fueling system	Reformer optimization

a comparative summary of various LTC strategies based on four key parameters. Modern LTC strategies: HCCI, PCCI, RCCI, plasma-assisted combustion, GCI, and fuel reformation provide the foundation for next-generation engines. Each strategy targets precise, low-temperature autoignition, ultra-low emissions, and high thermal efficiency, but with unique approaches to mixture preparation, ignition control, and practical implementation. Their deployment is enabling clean, efficient combustion for future transportation and power applications.

11.2.3 Control and Optimization Techniques

Control and optimization techniques are vital for the successful implementation of LTC in engines. Advanced sensors and actuators play a crucial role by providing real-time data on key parameters like pressure, temperature, and emissions, allowing precise adjustments during operation. Model-based control systems, including predictive algorithms such as model predictive control (MPC), help anticipate combustion behavior and optimize parameters like fuel injection timing, air–fuel ratio, and exhaust gas recirculation (EGR) to maintain stability and efficiency. Additionally, multiparameter optimization techniques simultaneously adjust variables such as boost pressure and fuel delivery to achieve desired combustion characteristics while minimizing NO_x and particulate emissions. Incorporating machine learning and artificial intelligence further enhances control by analyzing large datasets to develop adaptive strategies, improving stability and performance across different load conditions. These techniques collectively address LTC challenges such as combustion stability, knock, and operational flexibility, thus enabling cleaner, more efficient engine operation with reduced emissions. Table 11.2 outlines key technologies used to control and optimize LTC.

TABLE 11.2
Control and Optimization Techniques for LTC [10]

Technique	Purpose/Effect	Key Tools/Methods
Multiple Injection Strategies	Control mixture formation and phasing	Pilot/main/post injections, injection pressure/timing
Variable Valve Timing	Modulate exhaust gas recirculation (EGR) and compression ratio	LIVC, valve overlap, and variable intake/exhaust timing
EGR Control	Reduce NO_x and manage ignition timing	Variable cooled EGR and EGR rate tuning
Boost Control	Maintain oxygen availability and boost pressure	Turbocharger (VGT), and wastegate control
Real-time Combustion Sensing	Feedback for combustion phasing and stability	In-cylinder pressure sensors and lambda sensors
Knock Mitigation	Avoid engine damage, ensure smooth combustion	Fuel blending, phasing adjustment, and dilution optimization

11.2.4 Case Studies in Automotive Applications

Case studies in LTC focus on real-world implementations that demonstrate their benefits and challenges in automotive engines. One notable example is the development of Homogeneous Charge Compression Ignition (HCCI) engines, which aim to achieve combustion at lower temperatures to reduce NO_x and particulate emissions while maintaining high efficiency. A case study involving HCCI engines highlights the importance of precise control over air–fuel mixture, temperature, and compression ratio to ensure stable operation across various loads. Another study examines Reactivity Controlled Compression Ignition (RCCI), where two fuels with different reactivities are combined to optimize combustion timing and temperature, leading to significant emissions reduction. These case studies often utilize advanced simulation tools like GT-Power and experimental testing to evaluate performance, emissions, and operational stability. The insights gained from such applications help in overcoming challenges like knocking and combustion instability, paving the way for cleaner, more efficient automotive engines based on LTC principles.

a. Passenger car engines using HCCI/RCCI [11]

   Studies show HCCI can be implemented in both laboratory and production-intent engines. For example, Mercedes-Benz's "DiesOtto" concept engine and academic efforts have demonstrated full load ranges and high thermal efficiency in small-displacement, high-speed gasoline engines. HCCI strategies offer high efficiency (approaching or exceeding diesel engines at low loads) with near-zero NO_x and soot emissions under carefully controlled conditions.

   RCCI demonstrations (notably by the University of Wisconsin-Madison and GM) have moved into in-vehicle testing. RCCI utilizes dual-fuel operation: a low-reactivity fuel (like gasoline) is port injected, followed by direct injection of a high-reactivity fuel (like diesel), enabling precise phasing control. Real vehicle testing showed RCCI could achieve up to 20% higher fuel efficiency and meet stringent NO_x/particulate matter (PM) emission limits without aftertreatment systems, confirming its viability for passenger vehicles.

b. Commercial Diesel Engines with LTC Techniques [12]

   LTC modes such as HCCI, RCCI, and Modulated Kinetics combustion have been demonstrated in unmodified or slightly modified commercial diesel engines sometimes requiring only changes to injection strategy or utilization of EGR. Experiments on multicylinder, turbocharged diesels with optimized pistons and injection events have shown that LTC operation significantly reduces NO_x and soot, while maintaining or even improving power output and thermal efficiency compared to standard diesel operation.

   Studies found that LTC could be integrated into OEM diesel platforms without new hardware, and innovative piston/chamber designs further improved combustion efficiency and reduced hydrocarbon emissions. Some configurations allow compliance with Euro VI and future emission standards without NO_x/PM aftertreatment.

LTC (HCCI/RCCI) has potential for ultra-low NO_x and PM emissions, improving fuel economy by up to 20% compared to SI/diesel engines. Commercial viability is demonstrated in research projects and prototype testing for passenger and heavy-duty applications. Advanced control strategies are critical for practical implementation. LTC is seen as a leading technology for emissions legislation.

Research shows that LTC-enabled vehicles and engines can reduce NO_x and PM by over 90% compared to conventional diesel operation, with measured NO_x typically below 100 ppm and near-zero soot/PM across various loads. Biodiesel and fuel blends in LTC modes maintain low NO_x/PM and can further reduce unburned HCs and CO due to higher fuel-bound oxygen content.

Multimode LTC engines show a 19%–20% fuel economy improvement over conventional diesel engines, with brake thermal efficiency surpassing the best SI and diesel configurations. Real-world applications show that fuel economy benefits persist in drive-cycle simulations, with significant gains in combined city/highway usage. System integration, including EGR and boost control, enables LTC engines to maintain high efficiency while meeting strict emissions requirements (Table 11.3).

TABLE 11.3
LTC: Case Studies in Automotive

	Case Study 1: HCCI in Gasoline Passenger Car Engines
Project	Honda Experimental HCCI Engine
Objective	NO_x and PM emissions while improving fuel efficiency
Approach	Used a single-fuel HCCI strategy with controlled intake temperature and EGR; Employed VVT and advanced fuel injection timing
Results	NO_x emissions reduced by up to 90% compared to SI engines; Fuel economy improved by 15%–20%; Challenge: Narrow operating load range and cold-start limitations
	Case Study 2: PCCI in Light-Duty Diesel Engines
Project	Daimler PCCI Prototype for Euro 6 Compliance
Objective	Meet upcoming emission regulations without relying heavily on aftertreatment
Approach	Advanced multiple injection strategy; Optimized charge preparation and mixing time; Used late post-injections to control soot
Results	Significant NO_x and soot reduction without SCR; Smooth combustion achieved under various loads; Challenge: Precise control needed to avoid misfire or knock
	Case Study 3: RCCI in Dual-Fuel Heavy-Duty Engines
Project	University of Wisconsin & Cummins Collaboration
Engine	Dual-fuel diesel/gasoline engine retrofitted for RCCI
Approach	Used gasoline as low-reactivity fuel via port injection; Used diesel as high-reactivity fuel via direct injection; Adjusted injection timing and fuel ratios dynamically
Results	Up to 80% NO_x and 70% soot reduction; Thermal efficiency exceeded 50% at certain loads; Challenge: Dual-fuel complexity and real-time control systems
	Case Study 4: PCCI in Off-Road Construction Equipment
Project	John Deere Tier IV Compliant Diesel Engine
Objective	Comply with US EPA Tier IV off-road emission regulations

(Continued)

TABLE 11.3 (*Continued*)
LTC: Case Studies in Automotive

	Case Study 4: PCCI in Off-Road Construction Equipment
Approach	Used advanced diesel injectors with split injection; Delayed injection timing to promote premixing; Employed EGR and variable valve actuation for phasing control
Results	Significant reduction in NO_x and particulate matter; Maintained torque output with moderate fuel penalty; Challenge: Complexity in injector design and exhaust control integration
	Case Study 5: HCCI Using Ethanol in Urban Buses
Project	Swedish HCCI Bus Prototype (Scania + Lund Univ.)
Fuel	Ethanol (renewable)
Goal	Demonstrate renewable fuel usage with LTC benefits in public transport
Approach	HCCI enabled via intake air heating and high compression; Controlled combustion phasing through intake temperature and EGR; Operated successfully in stop-go urban cycles
Results	Reduced PM and NO_x nearly to zero; CO_2-neutral fuel use; Challenge: Cold-start and partial-load control

11.3 POWER GENERATION APPLICATIONS

LTC offers significant benefits for power generation by reducing emissions and improving efficiency in reciprocating engines and gas turbines. LTC technologies, such as HCCI and RCCI, enable combustion at lower temperatures, which markedly decreases NO_x and particulate matter emissions. This makes LTC particularly attractive for stationary power plants and microgrid systems aiming to meet stringent environmental regulations. LTC also enhances fuel efficiency, leading to reduced fuel consumption and operational costs. In gas turbines, research is ongoing to adapt LTC principles to achieve cleaner combustion with lower NO_x emissions, especially when using alternative fuels or biofuels. Overall, LTC in power generation applications contributes to sustainable energy production by minimizing environmental impact, increasing fuel economy, and enabling cleaner operation of power systems, aligning with global efforts to transition toward greener energy solutions [13].

a. Stationary Engines for Combined Heat and Power (CHP) [14]

   Stationary engines for CHP applications use reciprocating internal combustion engines, such as SI and compression ignition, for reliability, quick start-up, high part-load efficiency, and flexible operation. LTC strategies like HCCI, PCCI, and RCCI are increasingly used to achieve ultra-low NO_x and particulate emissions. These engines can use various fuels and are suitable for distributed power generation across industrial, commercial, and institutional settings. Research and field tests show that LTC, optimized piston geometry, and modern injection systems significantly improve thermal efficiency.

b. Gas Turbines with LTC-Inspired Strategies [15]

LTC-inspired strategies are used in gas turbines to reduce NO_x emissions. Lean-premixed or staged combustion, which moderates flame temperature and achieves more distributed heat release, is used in stationary gas turbines. Advanced low-NO_x gas turbines use dilution and lean premixing to operate below 1700 K, minimizing thermal NO_x formation without compromising efficiency. Special injector and combustor designs for alternative fuels, such as hydrogen and bio-syngas, also achieve low emissions and stable combustion.

c. Integration with Renewable Fuels (Biofuels and Hydrogen) [16]

LTC strategies have enabled the use of biofuels and hydrogen in stationary engines, resulting in lower-carbon and cleaner operation. Blending hydrogen with biofuels in LTC engines improves brake thermal efficiency and reduces key emissions, with up to 65% reduction in hydrocarbons, 37% in CO, 8% in CO_2, and 20% in smoke opacity compared to diesel. Studies on biodiesel and hydrogenation-derived renewable diesel show significant greenhouse gas reductions over fossil diesel while maintaining equipment compatibility and operational performance. LTC enables stationary CHP and power engines to use blended or single renewable fuels without major engine modifications, supporting decarbonization and compliance with renewable content regulations.

d. Engine Longevity and Maintenance Impacts

LTC can improve engine longevity by reducing thermal stress and component temperatures, extending service intervals. However, it may lead to higher levels of unburned HC and partially oxidized intermediates, increasing deposit formation and oil dilution risk. Maintenance schedules and oil monitoring are crucial. Reciprocating engines can be managed by in-house teams, while gas turbines using aggressive dilution or advanced LTC strategies may require careful material selection and corrosion management. LTC can reduce emissions aftertreatment and lower operational costs, but it requires optimization of combustion control and periodic inspection of intake, exhaust, and oil systems (Table 11.4).

TABLE 11.4
LTC in Power Generation

Application Area	LTC Strategy/ Advancement	Key Benefits	Maintenance/ Impact
CHP Reciprocating Engines	HCCI, RCCI, PCCI, optimized geometry	High efficiency, ultra-low NO_x/PM	Reduced thermal stress; HC/CO deposits
Gas Turbines	Lean-premixed, staged combustion, LDI	Ultra-low NO_x, stable emissions	Moisture/chemical management needed
Renewable Fuels	H_2/biofuel blends, HDRD, biodiesel	Decarbonization, broader fuel use	Mostly compatible, monitor deposits
Longevity & Maintenance	Cooler operation, careful controls	Potentially extended life	Monitor oil and combustion system

LTC technology enables stationary CHP engines and gas turbines to achieve high efficiency, low emissions, and broad renewable fuel compatibility. It also enhances environmental benefits and sustainability through hydrogen and biofuel blending. Maintenance in LTC-equipped power systems requires attention to increased HC/CO emissions and potential deposits.

These developments position LTC as a robust, flexible foundation for next-generation clean stationary power, ready to integrate with renewable fuel strategies and meet future regulatory and operational demands.

Table 11.5 is the comprehensive presentation of the LTC case studies in power generation.

TABLE 11.5
LTC Case Studies in Power Generation

	Case Study 1: Plasma-Assisted LTC in Micro Gas Turbines
Project	DOE/NREL Program on Plasma Ignition for Biofuels
Application	Stationary power generation using syngas
Approach	Plasma torches used to initiate combustion at lower temperatures; Enabled use of low-calorific fuels like biomass-derived syngas; Controlled via plasma generators and optical sensors
Results	Reliable ignition and combustion stability at low temperatures; Reduced NO_x by over 60% compared to standard combustors; Challenge: High cost and system integration in commercial units
	Case Study 2: LTC in Marine Engines using Ammonia
Project	MAN Energy Solutions—Dual-Fuel 2-Stroke Engine
Fuel	Ammonia with pilot diesel injection
Approach	Used Homogeneous Charge Compression Ignition (HCCI)-inspired combustion control to manage ammonia ignition; Engine operated at low temperatures to limit NO_x formation
Results	Drastic reduction in GHG emissions from marine vessels; Met IMO Tier III NO_x standards without SCR; Challenge: Ammonia toxicity, storage, and pilot ignition control
	Case Study 3: LTC with Natural Gas in Stationary Power Generators
Project	Caterpillar Gas Engine Development
Fuel	Lean-burn natural gasStrategy: Lean Premixed Compression Ignition (LPCI)
Application	Stationary power generation for industrial microgrids
Approach	Used turbocharged natural gas engines with lean premixing; Controlled combustion timing via mixture heating and spark support; Incorporated knock sensors and closed-loop fuel–air control
Results	Achieved ultra-low NO_x emissions without aftertreatment; Improved fuel conversion efficiency by ~43%; Challenge: Knock management and misfire sensitivity
	Case Study 4: LTC-Inspired Strategy in Aerospace Auxiliary Power Units (APUs)
Project	NASA-AFRL Clean Combustion Research
System	Small turbine APU combustor retrofitted for LTC
Fuel	Jet-A and alternative aviation biofuels

(*Continued*)

TABLE 11.5 (*Continued*)
LTC Case Studies in Power Generation

Case Study 4: LTC-Inspired Strategy in Aerospace Auxiliary Power Units (APUs)	
Approach	Employed plasma-assisted ignition for LTC stability; Used preheating and controlled swirl for homogenous mixing; Focused on flame stability at high altitude/low pressure
Results	Stable combustion at lower temperatures, reducing NO_x; Enabled altitude ignition and lean operation; Challenge: Electrical system complexity and power draw
Case Study 5: LTC-Based Dual-Fuel Retrofit of Diesel Gensets	
Project	Indian Institute of Science—Rural Electrification Project
Fuel	Diesel and ethanol blend
Application	Community microgrid for off-grid villages
Approach	Introduced ethanol via port injection, diesel for pilot ignition; Controlled combustion phasing using inlet throttling and injection timing; Utilized real-time sensing for knock and misfire prevention
Results	30%–40% reduction in diesel consumption; CO_2 emissions decreased by 25%, Challenge: Ethanol storage, blending, and cold-start issues
Case Study 6: LTC in Micro Gas Turbines with Hydrogen-Rich Fuels	
Project	EU HYFLEXPOWER (Siemens and Engie)
Fuel	Hydrogen/ammonia blends
Application	Low-carbon electricity in industrial parks
Approach	Modified combustors with lean premixed LTC strategy; Managed ignition stability with pilot burners; Used AI-based control for dynamic phasing
Results	NO_x emissions under 10 ppm with 30% hydrogen; Power generation flexibility under variable loads; Challenge: Flame stability and flashback risk with hydrogen
Case Study 7: Plasma-Assisted Combustion in Industrial Boilers	
Project	Russian Academy of Sciences—Boiler Modernization
Fuel	Low-calorific coal gas and biomass syngas
Application	Grid-connected thermal power plant
Approach	Plasma torches used to initiate combustion of low-reactivity fuels; Enabled stable flame at ~800°C–1000°C, avoiding flameout; Online diagnostics used for flame shape and emissions
Results	Increased combustion stability; CO and NO_x reduced by 40%–60%; Challenge: High cost and reliability of plasma equipment

11.4 SIMULATION AND MODELING APPROACHES

a. Combustion Modeling Tools [17]

Combustion modeling tools include CFD tools like ANSYS Fluent, CONVERGE CFD, and AVL FIRE, which are used to resolve in-cylinder processes like turbulence, spray breakup, vaporization, chemical kinetics, ignition, and emissions. KIVA-family Codes are used for LTC research, while Large Eddy Simulation (LES) captures turbulence-flame interaction and scalar mixing for accurate combustion and pollutant formation prediction in LTC engines. Chemkin is used for zero- and one-dimensional chemical kinetic modeling, and GT-Power is a 1-D engine simulation tool for system-level LTC studies, transient performance analysis, and calibration of injection/EGR strategies.

b. In-Cylinder Flow and Mixing Models [18]

In-cylinder flow and mixing models include Direct Numerical Simulation (DNS), LES, RANS, and multi-zone thermo-kinetic models. DNS provides high fidelity for resolving turbulence, mixing, and chemistry, but is limited to academic studies due to computational cost. LES captures cycle-to-cycle variability and mixture stratification effects, while RANS accounts for sub-grid turbulence, scalar mixing, and chemical kinetics.

c. Predictive Models for Ignition and Emission Control [19]

Linear/Nonlinear Parameter Varying models predict combustion phasing, IMEP, and pressure rise using real-time variables. Fast MPC is used for stable LTC operation. Fast models predict heat release rate and phasing prediction, essential for adaptive engine management. Emission models estimate real-time NO_x, PM, HC, and CO formation, enabling feedback and feedforward emission control for LTC engines.

d. Use of AI and Machine Learning in LTC Optimization [20]

Machine learning models, such as regression, random forest, support vector machines, and neural networks, can predict ignition delay, heat release, combustion mode, and key fuel properties using historical data and engine parameters. These models aid in automated control and can be used for optimization, such as generating surrogate models for rapid system exploration. Genetic algorithms and ensemble methods can optimize injection, EGR, and hardware settings for peak efficiency and ultra-low emissions. Hybrid approaches combine ML models with classic physics-based models for online adaptation and cycle-to-cycle control of LTC phasing and emissions.

Table 11.6 is a comparative overview of various computational and modeling tools used for analyzing and optimizing LTC systems. Advanced modeling, including high-fidelity CFD, predictive control-oriented models, and AI/ML, is crucial for

TABLE 11.6
Modeling Approaches in LTC

Approach/Tool	Application in LTC	Key Features/Notes
CFD (ANSYS, CONVERGE, and KIVA)	Detailed 3D in-cylinder modeling, flow, mixing, chemistry	High fidelity, resolves turbulence, and chemistry
Chemkin	Detailed chemical kinetics, ignition studies	Fast/accurate for 0D/1D, mechanism validation
GT-Power	System-level engine/transient simulations	Fast, widely adopted for complete engine studies
LES/DNS	In-cylinder turbulence and mixing analysis	Captures cycle variability, stratification
Predictive/MPC Models	Real-time ignition/phasing/emission control	Enables controller adaptation and feedback
AI/ML	Optimization, state prediction, and pattern recognition	Accelerates calibration, surrogate model creation

designing and operating LTC engines. This integration accelerates optimization, supports real-time emissions and phasing control, and accurately predicts cycle-to-cycle variability, combustion phasing, and emissions. These approaches are forming the backbone of research and commercial LTC engine development, bridging experimental results, rapid prototyping, and advanced control system design.

11.5 CONCLUSIONS

- LTC strategies (HCCI, PCCI, and RCCI) significantly reduce NO_x and PM by enabling autoignition of dilute, premixed or partially premixed mixtures at lower temperatures.
- Automotive applications demonstrate up to 90% reductions in NO_x and soot emissions with fuel economy improvements of 15%–20%, validating LTC's potential for clean, efficient passenger and commercial vehicles.
- Advanced multi-injection, variable valve timing, and precise EGR control are crucial for managing combustion phasing, stability, and emissions, especially under transient and cold-start conditions.
- Stationary power generation and heavy-duty sectors benefit from LTC through enhanced fuel flexibility—allowing integration of renewable fuels such as hydrogen and biofuels—while maintaining compliance with stringent emission standards.
- Challenges remain in extending LTC's stable operating range, controlling cold-start emissions, and ensuring long-term durability due to increased hydrocarbons and deposits.
- Continued innovations in combustion control, fuel blending, and hybrid system integration are framing LTC as a viable pathway for future ultra-clean and efficient engines in diverse applications.

REFERENCES

1. Hwang, J., Maharjan, K., & Cho, H. (2023). A review of hydrogen utilization in power generation and transportation sectors: Achievements and future challenges. *International Journal of Hydrogen Energy*, 48(74), 28629–28648. https://doi.org/10.1016/j.ijhydene.2023.04.024
2. Li, S., Wang, W., Diao, H., & Wang, L. (2022). Measuring the efficiency of energy and carbon emissions: A review of definitions, models, and input-output variables. *Energies*, 15(3), 962. https://doi.org/10.3390/en15030962
3. Juan, A. A., Ammouriova, M., Tsertsvadze, V., Osorio, C., Fuster, N., & Ahsini, Y. (2023). Promoting energy efficiency and emissions reduction in urban areas with key performance indicators and data analytics. *Energies*, 16(20), 7195. https://doi.org/10.3390/en16207195
4. Shu, C., Cao, Y., Sun, T., Zhang, T., & Wang, X. (2024). Energy recovery in automotive systems: A review of regenerative braking, flywheel, thermoelectric, Rankin cycle and electric turbo compound. *Applied and Computational Engineering*, 85(1), 44–60. https://doi.org/10.54254/2755-2721/85/20240623
5. Saikia, P. J., Dutta, P. P., & Kalita, P. (2021). Simulation and experimental performance studies of a computerized spark ignition engine. *Lecture Notes in Mechanical Engineering* (pp. 441–454). Springer, Cham. https://doi.org/10.1007/978-981-15-7711-6_45

6. Liang, L., & Reitz, R. D. (2006). Spark Ignition Engine Combustion Modeling Using a Level Set Method with Detailed Chemistry. *SAE Technical Papers on CD-ROM/SAE Technical Paper Series*. https://doi.org/10.4271/2006-01-0243
7. Hong, S., Wooldridge, M. S., Im, H. G., Assanis, D. N., & Kurtz, E. (2008). Modeling of diesel combustion, soot and NO emissions based on a modified Eddy dissipation concept. *Combustion Science and Technology*, 180(8), 1421–1448. https://doi.org/10.1080/00102200802119340
8. Gupta, S. K., & Krishnasamy, A. (2024). A relative comparison of HCCI, PCCI, and RCCI combustion strategies: An alternative fuels perspective. *International Journal of Engine Research*, 25(6), 1078–1092. https://doi.org/10.1177/14680874231216664
9. Maurya, R. K. (2017). Low temperature combustion engines. *Mechanical Engineering Series* (pp. 31–133). Springer, Cham. https://doi.org/10.1007/978-3-319-68508-3_2
10. Waschl, H., Kolmanovsky, I., Steinbuch, M., & Re, D. L. (2014). Optimization and optimal control in automotive systems. *Lecture Notes in Control and Information Sciences*. Springer, Cham. https://doi.org/10.1007/978-3-319-05371-4
11. Szamrej, G., Karczewski, M., & Chojnowski, J. (2021). A review of technical solutions for RCCI engines. *Silniki Spalinowe/Combustion Engines*, 189(2), 36–46. https://doi.org/10.19206/ce-142551
12. Beatrice, C., Avolio, G., Bertoli, C., Del Giacomo, N., Guido, C., & Migliaccio, M. (2007). Critical aspects on the control in the low temperature combustion systems for high performance DI diesel engines. *Oil & Gas Science and Technology – Revue D'IFP Energies Nouvelles*, 62(4), 471–482. https://doi.org/10.2516/ogst:2007058
13. Jhalani, A., Sharma, D., Sharma, P. K., Singh, D., Jhalani, S., & Dubey, N. J. P. (2021). A comprehensive review on low-temperature combustion technologies for emission reduction in diesel engines. *International Journal of Automotive and Mechanical Engineering*, 18(4), 710.
14. Chaudhari, V., & Deshmukh, D. (2020). Diesel and diesel-gasoline fuelled premixed low temperature combustion (LTC) engine mode for clean combustion. *Fuel*, 266, 116982. https://doi.org/10.1016/j.fuel.2019.116982
15. Singh, A. P., & Agarwal, A. K. (2018). Low temperature combustion engines and mode switching strategies – A review. *Journal of Energy and Environmental Sustainability*, 6, 42–47. https://doi.org/10.47469/jees.2018.v06.100066
16. No, S. (2019). *Application of Liquid Biofuels to Internal Combustion Engines. Green Energy and Technology*. Springer, Singapore. https://doi.org/10.1007/978-981-13-6737-3
17. Pelosin, M., Novella, R., Bracho, G., Fernandes, C., Lucchini, T., Marmorini, L., & Zhou, Q. (2022). Combustion modeling approach for the optimization of a temperature controlled reactivity compression ignition engine fueled with iso-octane. *Energies*, 15(21), 8216. https://doi.org/10.3390/en15218216
18. Vasudev, A., Mikulski, M., Balakrishnan, P. R., Storm, X., & Hunicz, J. (2022). Thermo-kinetic multi-zone modelling of low temperature combustion engines. *Progress in Energy and Combustion Science*, 91, 100998. https://doi.org/10.1016/j.pecs.2022.100998
19. Knop, V., Kircher, H., Jay, S., Béard, P., Da Cruz, A. P., & Colin, O. (2008). Quantitative pollutant modelling: An essential prerequisite for diesel HCCI and LTC engine design. *Oil & Gas Science and Technology – Revue D'IFP Energies Nouvelles*, 63(4), 495–515. https://doi.org/10.2516/ogst:2008022
20. Batool, S., Naber, J. D., & Shahbakhti, M. (2024). Machine learning approaches for identification of heat release shapes in a low temperature combustion engine for control applications. *Control Engineering Practice*, 144, 105838. https://doi.org/10.1016/j.conengprac.2023.105838

12 Exploring the Depth and Navigating the Complexities of Low-Temperature Combustion Process

A Review

Masilamani Gowtham, Ramakrishnan Prakash, and Yerlan Sarsenbayev

12.1 INTRODUCTION

The diesel engine (DE) is the most preferred prime mover for producing high thermal efficiency and good operating conditions. It plays a crucial role by supporting the fuel economy of the transport sector [1]. Increasing number of automobiles along with an increase in the world's population leads to an increase in the usage of petroleum fuels [2]. The DE releases harmful gases like nitrogen oxide carbon dioxide, CO, unburned hydrocarbon (UHC), and particulate matter (PM) [3]. India is ready to implement the Bharat stage VI (BS VI) emission level norms to maintain the sulphur content at maximum of 10 ppm, PM at 0.012 g/kWh, and nitrogen oxides at 0.4 g/kWh levels. It is very difficult to maintain this emission level by use of a conventional DE [4]. Engineers, scientists, and researchers, and also academicians, are working together to develop resolutions that help maintain BS VI emission standards while ensuring efficient energy utilization [5–8].

High concentrations of reduction fuels are used in the low-temperature combustion (LTC) process to reduce the total combustion temperature and prolong the ignition delay (ID), which allows the injected fuel to evaporate and minimizes mixture inhomogeneities while lowering NO_x and PM emissions and adhering to BS VI emission standards. Some drawbacks of these methods are that the maximum levels of CO and hydrocarbon emissions are discharged at low load conditions. Some advanced techniques like achieving a homogeneous charge, limited operating conditions, and lack of combustion phase control are used. Some other extensive research works include high exhaust gas recirculation (EGR), boosting techniques, high pressure injection, and multiple and variable injection methods. This paper presents an overview of the

DOI: 10.1201/9781003633631-12

LTC technique, its methods, and methods to reach BS VI emission norms on DE. Different challenges in tackling the advanced LTC process and its basic chemical kinetics involved in each combustion stage are observed [9].

12.1.1 Chemical Factors Influencing Combustion

Chemical kinetics is fundamental to understand and examine the advanced combustion characteristics. Diesel fuel consists of compounds like alkenes, aromatics, chain alkanes, olefins, formaldehyde, and OH. These components present challenges in the assessment of chemical kinetics [10]. The major combustion process is categorized into four types: primary pre-ignition, primary ignition, secondary pre-ignition, and secondary ignition. In pre-ignition, the oxidation of fuel molecules generates keto-hydro peroxide (KHP). In the primary ignition stage, a group of inlets of radicals oxidize the fuel, increasing the temperature. In the primary pre-ignition stage, the resulting final products are formaldehyde, hydrogen peroxide, and carbon monoxide. In the secondary pre-ignition stage, or simply continuation of the primary stage, the final products are hydrogen peroxide and carbon monoxide, which react with methanol and oxidize the hydrocarbons to form hydrogen peroxide and hydroperoxy radicals. The temperature also increases to nearly 1000–1200 K, which is critical to start the decomposition of hydrogen peroxide. The final products of the additional stage are hydrogen peroxide, UHCs, and carbon monoxide [11–14].

12.1.2 Equivalence Ratio with Temperature

Equivalence ratio and temperature are necessary to understand the normal combustion and LTC processes. It is designed by estimating the soot and NO_x development at a certain equivalence ratio and temperature for a given time [15].

12.1.3 Heat Release Rate

Three separate phases define conventional diesel combustion: premixed or fumigation, diffusion or mixing-controlled, and ID. In contrast, LTC includes primary pre-ignition, primary ignition, secondary pre-ignition, and secondary ignition phases [16,17]. Paul et al. conducted experiments on the LTC process using three different injection systems: conventional, early direct injection (DI), and late DI. The heat release and cylinder pressure curves reveal differences between normal diesel combustion and the LTC process. A similar initial incline in the heat release curve due to fuel vaporization is observed, but the overall combustion patterns differ considerably between them. In LTC, the heat release rate is reduced by around three times. The experiment by Paul et al. shows that considerable heat release takes place during conventional, diffusion-controlled combustion but not during early LTC. Furthermore, the increase in cylinder pressure rise rate is more pronounced during low-temperature ignition compared to standard diesel combustion [16]. The reactants of the LTC process have an important role and aid in the production of the mixture, and at the end, they deactivate the ignition of hydrocarbon fuels. This is completely diverse from the typical diesel combustion process, where fewer, rapid low-temperature reactions

(LTRs) are observed. The role of low-temperature flame reactants is understandable in the heat release of the LTC process.

12.2 LTC COMBUSTION AND EMISSION ANALYSIS

One of the most effective internal combustion (IC) engines on the market is a sophisticated DE, which offers remarkable fuel efficiency and reduces greenhouse gas emissions. But these engines also release nitrogen oxides and PM, which can seriously harm both public health and the environment. As a result, numerous technologies and techniques have been developed to reduce the emissions of these dangerous pollutants. Three key approaches include use of alternative fuels like biodiesel, modification of the combustion process, and implementation of exhaust after-treatment systems [18]. One recently developed technique, known as LTC, has shown promise in reducing these undesirable emissions. This approach is widely applicable, spanning light-duty to heavy-duty engine applications. Low-temperature combustion strategies are advanced processes that use a low combustion temperature that favourably modifies the chemistry of nitrogen oxides and soot development [19]. The three primary categories of LTCs are reactivity-regulated compression ignition (CI), premixed charge compression ignition (PCCI), and HCCI, as shown in Figure 12.1.

12.2.1 HCCI Combustion and Emission Strategies

In contrast to spark ignition (SI) or CI engines using diesel fuel, HCCI includes a combination of SI(homogeneous charge SI) and CI (stratified charge compressive ignition) processes [20,21]. In the HCCI process, the charge particles are mixed prior to the start of combustion (SOC), and the premixture particles are automatically ignited due to the high temperature and pressure present in the compression stroke [22]. This homogeneous mixture recruits the combustion procedure instantaneously at altered positions of the combustion chamber [23].

In the HCCI process, the heat release can occur in two phases. In the first stage, due to kinetic chemical reactions in the LTR, less heat is released, as compared to the

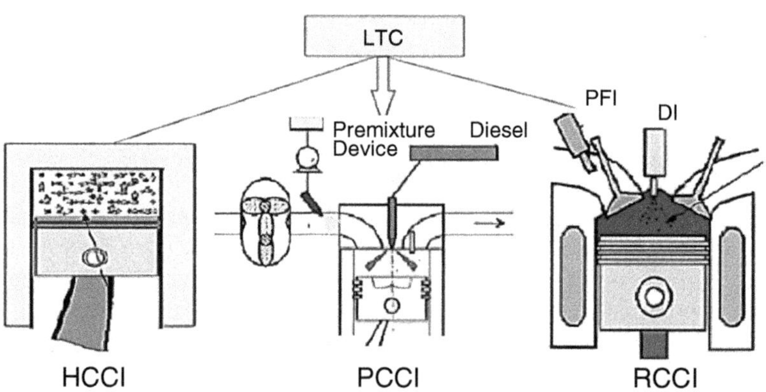

FIGURE 12.1 Types of LTC combustion processes.

second stage. The approximate temperature range is 1000 K. Approximately 7%–10% of the energy is released in the LTR. In the second stage or high-temperature reaction (HTR), substantial heat release and generation of considerable power are observed [23]. In the isomerization process, the hydrocarbon molecule is converted into a hydroperoxy alkyl particle. This molecule is then transformed by addition of another oxygen molecule, undergoing an addition reaction to form an oxo hydroperoxide particle. The oxo hydroperoxide is further decomposed and isomerized into a KHP and hydroxide particle [24]. The Negative Temperature Coefficient (NTC) region in combustion refers to temperature zone (typically 500–800 K) where the overall oxidation reaction rate decreases despite an increase in temperature. The hydrogen and oxygen reactions during the HTR are carried out under controlled conditions [25]. The overall process includes oxidation of carbon monoxide into carbon dioxide. Jiang et al. used a HCCI combustion process, in which gasoline was added into the cylinder through a port injector to produce a homogeneous air–fuel mixture. The fuel–air charge is powered by the self-ignition of a tiny amount of diesel. The HCCI procedure is used to diminish emissions of nitrogen oxide and PM [26]. Because of the increased weakening of the charge by the presence of air or residual gases, the gas temperature for the duration of HCCI combustion is lesser than 1800–1900 K (Table 12.1).

12.2.2 PCCI Combustion and Emission Strategies

During the full load condition, the calculation and control of the combustion process is very difficult in both PCCI and HCCI when compared with a normal DE because chemical kinetics alone controls the overall rate of heat release, which can result in unmanageable noise and an unusual rise in pressure. Current research on CI engines mainly focuses on developing better fuel management systems to overcome these problems [19]. Challenges with combustion control, regulated auto-ignition, limited operating range, cold start, and attaining a homogeneous air–fuel mixture are some of the limitations of the HCCI combustion process [43–45]. The PCCI engine method was developed to lessen these shortcomings. A controlled ID interval to finish the combustion process is regulated by PCCI, a variation of the HCCI mode. This is accomplished by using EGR, raising the injection pressure, and decreasing the compression ratio. Three primary fuel injection techniques are used in the PCCI combustion process: incomplete fuel evaporation, advance direct injection, and port fuel injection (PFI). Nevertheless, these techniques may also result in higher CO and UHC emissions. Advanced combustion methods, like narrow fuel injection and EGR, can assist in reducing wall impingement and enhancing emissions to overcome the inherent trade-off between nitrogen oxides and particulate matter (PM)/smoke emissions in diesel engines, as shown in Figure 12.2 (Table 12.2).

12.2.3 RCCI Combustion and Emission Strategies

The RCCI technique is one of the best methods in the LTC process. In this technique, two different fuels are used. Each has its own fuel properties; the primary fuel is called high-reactivity fuel (HRF), and low-reactivity fuel (LRF) is used as the

TABLE 12.1
Performance and Emission Parameters in the HCCI Engine

Key Parameters	Test Engine	Fuel Used	Performance (BTE and BSFC)	Emission	References
Injection pressure	1-C4-S direct injection (DI)	Diesel	HT 0.8% ↑	NO_x ↓ Soot ↓	[27]
	1-C, 4-S DI Variable injection 400–700 bar	Rapeseed methyl ester	-Na-	NO_x 33% ↓ HC 43% ↓ Particulate matter (PM) and soot tend to zero level	[28]
Piston bowl geometry	1C 4S	*n*-Dodecane	-Na-	NO_x ↓	[29]
	4-C DI-DE	Gasoline – diesel fuel.	Brake thermal efficiency (BTE) ↑ brake specific fuel consumption (BSFC) ↓	NO_x ↓ HC, CO ↓ PM and soot ↑	[30]
Compression ratio	1-C, 4-SDE	Grease and tallow biodiesel	-Na-	NO_x ↓ HC ↓ PM and soot	[31]
	1-C, 4-S DE	Yellow grease biodiesel	BTE 32% ↑	NO_x 16% ↓ CO same PM 33% ↑	[32]
Intake charge temperature	1-C, 4-S DE	Ethanol	BTE 43% ↑	NO_x 11 Ppm Smoke 0.1% ↓ HC and CO ↑	[33]
	1-C, 4-S DE	Hydrogen	BTE 46% ↑	NO_x ↑ and ↓	[34]
Exhaust gas recirculation (EGR)	1-C, 4-S DE	Soyabean	Na EGR up to 32%	NO_x 12% ↓ CO 22% ↓ PM 33% ↑	[35]
	1-C, 4-S DE, and CR 16.1:1	Mixed biodiesel	BTE 5.5 % ↓ EGR up to 60%	NO_x ↓ as EGR ↑ CO and HC ↑ PM ↓ as EGR ↑	[36]

(Continued)

TABLE 12.1 (*Continued*)
Performance and Emission Parameters in the HCCI Engine

Key Parameters	Test Engine	Fuel Used	Performance (BTE and BSFC)	Emission	References
Super or turbo charging	1-C, 4-S DE	Diesel	BTE ↑ BSFC ↓	NO_x ↑,CO ↑ HC ↑	[37]
	1-C, 4- SDE	Diesel	BTE ↑	NO_x ↑, CO ↑ HC ↑	[38]
Injector characteristics	Naturally aspirated diesel engine	Hydrogen	BTE 24.2% ↑	NO_x 8.5 g/kWh CO_2 ↑	[39]
	1-C, 4-S diesel engine MP injector	Diesel	BTE 15% ↑	HC 56% ↓ Smoke 24% ↓ NO_x 6% ↓	[40]
	1-C, 4-S MP injector	71 % n-heptane 21% iso octane	BTE 11.2% ↑	CO and HC lower at 152° injection angle, PM lower at 124°	[41]
	1-C, 4-S diesel engine dual injection	Diesel	-Na-	NO_x 47% ↓, CO 36% ↑ HC 40% ↑	[28]
	1-C,4-S CR:17.5:1, port fuel injection with fuel vaporiser	Biodiesel	BTE 5.5% ↓ BSFC 3%–5% ↑	NO_x 87% ↓ CO 20% ↓ HC and PM ↑	[42]

FIGURE 12.2 Basic layout of the PCCI engine.

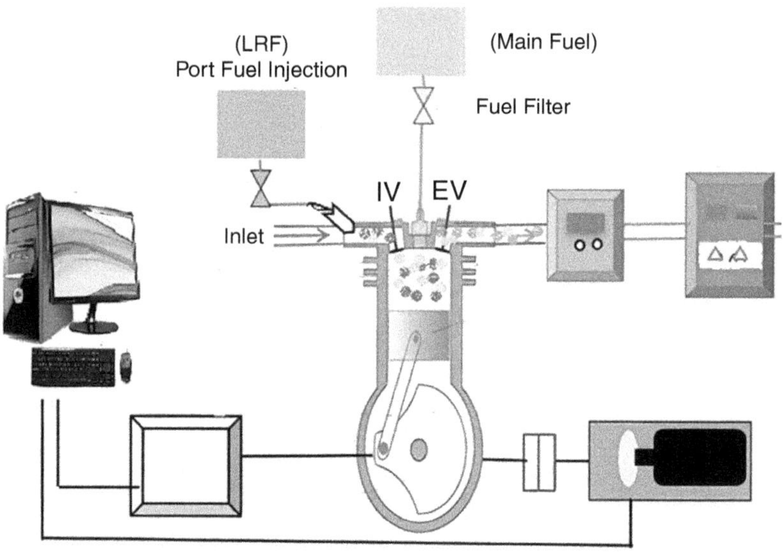

FIGURE 12.3 Pictorial view of the RCCI engine with its working model.

secondary fuel. Both fuels acknowledge in-cylinder fuel mixing and embrace various injection techniques and fittings [59–61], as shown in Figure 12.3.

The EGR method is used to regulate the in-cylinder reactivity which was manipulated to improve the combustion staging and reduce emissions, thus prompting maximum engine efficiency and minimum NO_x and soot discharges [62]. In RCCI

TABLE 12.2
Performance and Emission Parameters in the PCCI Engine

Key Parameters	Test Engine	Fuel Used	Performance	Emission	Reference
Injection pressure	4-C, direct injection (DI), diesel engine, and (DE)	Diesel	Brake thermal efficiency (BTE) 12% ↑	NO_x and particulate matter (PM) ↓ CO ↑ HC ↑	[46]
	4-C, DI, and DE	Soy-based methyl ester	–	HC 64% ↓, NO*x* ↑, and PM 94% ↑	[25]
PBG	1-C, DI, and CR: 16.1:1	Methanol	BTE ↑, brake specific fuel consumption (BSFC) ↓	NO_x 32% and PM 12% ↓ CO ↑ HC ↑	[47]
	1-C, DI, and CR: 16.1:1	Diesel	BTE ↑, BSFC ↓	NO_x 39% and PM 10% ↓ CO ↑ HC ↑	[48]
CR	1-C, DI DE, and CR: 18.4:1	Ultra-low sulphur diesel	BTE 15% ↑	NO_x ↓ at full load. PM ↑ at full load.	[49]
	4-C, DI DE, and CR: 17.5:1	*n*-butanol O_2 up to 19.5%	BTE ↑, BSFC ↓	NO_x 20% ↓ HC ↑ *n*-butanol ↑ PM ↓	[50]
ICT	4-C, DI And CR: 16.5:1	Gasoline with diesel	BTE ↑, BSFC ↓	NO_x ↓ CO and HC ↑ PM ↓	[51]
	4C, DI, and CR: 16.5:1	diesel	BTE ↑, BSFC ↓	NO_x 26%↓ CO and HC ↑ PM ↓	[52]

(Continued)

TABLE 12.2 (*Continued*)
Performance and Emission Parameters in the PCCI Engine

Key Parameters	Test Engine	Fuel Used	Performance	Emission	Reference
EGR	4C, DI, and DE	Low sulphur diesel	BTE 12% ↑ Up to 40% EGR	NO_x 0 g/kwh, same PM, CO 60% and HC 46%↑	[53]
	1-C and CI	Gasoline and diesel	BTE 24% ↑ Up to 35% EGR	NO_x 42% ↓ CO and UHC ↑ PM ↓	[54]
Super charging	4C and DI	Soybean methyl ester.	BTE ↑ BSFC ↑	NO_x ↓at final ↑, CO ↓, HC 43% ↓	[55]
	4C, DI, and CR: 17.5:1	*n*-butanol O_2 up to 19.5%	BTE ↑, BSFC ↓	NO_x 20% ↓ UHC ↑ *n*-butanol ↑ PM ↓	[56]
Injector characteristics	4C and DI	Diesel dual injection	BTE 24% ↑	NO_x 71% ↓ CO ↓ and PM ↑ at full load, HC 64% ↓.	[57]
	4C and DI	Biodiesel	BTE 16% ↑	NO_x ↓ 26–35 ppm, CO 17% ↓, PM ↑	[58]

devices, a PFI is attached to the inlet section or manifold section. The gasoline or light fuels, LRF, is injected through the port fuel injector [63]. In the cylindrical section, during the intake stroke, the injected LRF and the air are mixed thoroughly. The diesel or biodiesel, HRF, is passed through the DI into the cylinder through single, binary, or three-way fuel addition for the duration of the compression stroke [64]. The reactivity-regulated compression ignition engine differs from the traditional dual-fuel engine in the combustion mode. Decoupling the completion of the directly injected fuel's injection from the beginning of combustion is essential for RCCI engines [65,66]. This promotes thorough mixing and a diluted charge, which offers extremely low nitrogen oxide concentrations and increases the efficiency of collecting discharged soot throughout an extensive range of engine loads and speeds. Because of the limited ability to modify fuel reactivity based on engine operating circumstances, the RCCI technique allows for greater control over the combustion process than single-fuel LTC systems [67,68] (Table 12.3).

12.3 CHALLENGES IN COMBUSTION AND EMISSION CONTROL TECHNIQUES IN LTC

12.3.1 Challenges of LTC Combustion

One of the most effective combustion techniques for burning biodiesel is LTC, a recently developed ignition process. However, the LTC system reduces NO_x emissions while simultaneously addressing a number of challenges that aid its implementation in real-world application cars [81,82], as shown in Figure 12.4.

We recognized a few disadvantages of the LTC system and approaches to overcome the weaknesses of LTC, which has been quickly examined below. The LTC method is a recently used method for effective combustion of both diesel and biodiesel fuels [19,83]. Even though NO_x emissions were greatly decreased under LTC combustion, there are still many issues with its implementation in actual automobiles [84–87]. A few limitations of the LTC approach were identified by this study, and solutions to these issues are briefly discussed below.

12.3.1.1 Combustion Control Rate and Ignition Timing

Developing controlled activities of LTC engines over an extensive variety of quickness and loads is difficult. If combustion occurs too earlier, power decrease will happen along with substantial damages to the engine. If the combustion process occurs too late, there is more chance for a misfire. In LTC engines, no immediate technique for controlling SOC (like start or fuel infusion) is accessible. The SOC in LTC is mostly administered by chemical kinetics in this manner [88]. Chemical kinetics is controlled by the temperature time-history of the air–fuel ratio (AFR) as well as the fuel's physical and chemical qualities. Some factors affecting the combustion phasing in LTC combustion are self-ignition characteristics of the fuel, AFR, flow amount and reactivities of the exhaust gas, combustion duration, residual rate, homogeneous fuel–air mixture ratio, thermodynamic properties, heat transfer from the engine, compression ratio, etc. The starting charge composition, the initial pressure and temperature form, the level of compression efforts, and the extent of heat

TABLE 12.3
Performance and Emission Parameters in the RCCI Engine

Key Parameters	Test Engine	Fuel Used	Performance	Emission	References
Injection pressure	1-C, 4-S, and direct injection (DI)	Gasoline + DTBP	BTE ↑	NO_x ↓ CO ↑, PM ↓ and HC =	[69]
	1-C, 4-S, and DI	Biodiesel and gasoline	BTE ↑	5° CA, NOx ↓ CO and Soot ↓ HC↑	[70]
Piston bowl geometry	1-C, DI, diesel engine (DE) CR: 16.1:1	Natural gas	BTE ↑	NO_x ↓, HC and CO ↑,	[71]
	1-C, DI, and DE CR: 16.1:1	Natural gas/diesel	BTE↑	At high load conditions NO_x ↓, HC and CO ↑,	[72]
Compression ratio	1C, DI, and DE	Gasoline + diesel, methanol + diesel	At 400K BTE↑	–	[73]
	6C, DI, and DE	Gasoline	BTE 47% ↑	High load, low speed. NO_x ↓ PM ↓ and HC ↑	[45]
Intake charge temperature	1-C, 4-S DI	Gasoline + ethanol	BTE ↑	NO_x ↓ CO ↑, PM ↓ and HC ↑	[74]

(Continued)

TABLE 12.3 (*Continued*)
Performance and Emission Parameters in the RCCI Engine

Key Parameters	Test Engine	Fuel Used	Performance	Emission	References
Exhaust gas recirculation	4c, turbocharged HD, common rail direct injection (CRDI) DE with an intercooler	Alcohol	BTE ↑	NO_x ↓ At low load CO ↑, PM ↓ and HC ↓	[75]
	4c, turbo with an intercooler	Diesel/natural gas	BTE ↑	NO_x ↓ (high load) At low load CO ↑, PM ↓ and HC ↓	[76]
Super or turbo charging	6c, RP 163 kw, turbocharged HD, DI, and DE	Diesel and Iso-butanol	Butanol ↑- -BTE ↑ and BSFC ↓	Butanol ↑-- NO_x ↓ CO ↓, PM ↓ and UHC ↑	[77]
	6c, turbocharged HD, CRDI, and DE	Butanol and diesel	BSFC=	At 75% load at 1500 rpm, NO_x ↓ CO ↓, PM ↓ and UHC ↓	[78]
Injector characteristics	1C, 4S diesel engine	Hydrous ethanol	BTE ↑ and BSFC ↓	CO ↑, PM ↓ and HC ↑ NO_x ↓	[79]
	4C, 4S, turbo, diesel engine	2-butanol/diesel	BTE ↑ and BSFC ↓	At high load, NO_x ↓ CO ↑, PM ↓, and HC ↑	[80]

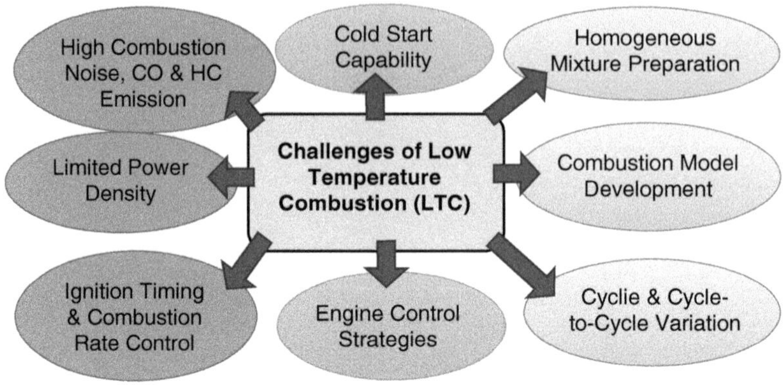

FIGURE 12.4 Challenges of LTC combustion.

transfer distribution from the engine are the four key components that are necessary to achieve improved combustion efficiency and lower emissions. Some other methods introduced by scientists and researchers are changing the temperature time-history in addition to moderating inlet air temperature [89,90]; different EGR [91]; variable compression ratio (VCR) [92,71]; variable valve timing (VVT) and exhaust gas tramping [93–95], a variation of injection timing [96,97]; changeable coolant temperature [98]; and injection of water [99]. Future approaches for varying mixture reactivities like modifying two or more fuel [100,101] additives used in LTC need to be advanced to enhance combustion in an extensive variety of load speed mixtures of the engines [102,103].

12.3.1.2 Limited Power Densities

LTC engines are expected to work at heavy loads and speeds like diesel and petrol engines. In any case, the LTC engine works acceptably from zero load to intermediate load conditions and slightly encounters challenges at maximum load conditions. In higher load conditions, the combustion rate is high, resulting in a substantial increase in pressure, leading to overcoming the pressure fluctuations. It produces heavy noise, engine damage, and high level of exhaust emissions. This additionally constrains the output power/thickness and lays a heavy risk on LTC engine. Research in this field is mandatory to generate innovative ideas and comprehension to control chemical kinetics of the air–fuel mixture. Extended combustion times at high engine loads would be the outcome, and engine noise and external damage would be avoided. One methodology to broaden the working scope of the LTC engine is to support the admission of air either by supercharging [104,105] or turbo charging [106–108]. Also, another system for conquering the problem of the lesser power density of the LTC engine is dual-mode activities like SI-HCCI [109,110] and CI-HCCI [88,111] hybrid combustion processes.

12.3.1.3 Combustion Noise and Regulated Emissions

In the LTC process, ignition happens all over the combustion chamber. Henceforth, at higher load conditions, a more richer fuel–air mixture is utilized, prompting higher RoPR and high-recurrence heavy noise, in contrast with normal direct compression ignition or DISI engines [89]. The CO and HC emissions are formed due to the combustion of either lean mixture or rich mixture of the stoichiometric ratio in an IC engine. The lean mixture leads to inflammability and rich mixture effects owing to a lack of oxidant. The LTC engine emits a fundamentally lesser amount of NO_x and PM; however, a significantly lower amount of NO_x and PM is generated. The necessary quantity of charge particles is formed in the cracks during the compression stroke and escapes during the combustion stroke without being fully combusted in the homogeneous combustion process. The initial gas temperature is very low to combust a high amount of this unburned species, when it re-enters into the CC at exhaust stroke. The results indicate an insignificant increase in the CO and UHC emissions when compared with the normal combustion process [112]. A minimum combustion temperature range of less than 1400 or 1500 K is needed for the CO-to-CO_2 reaction process at low load conditions, with an increase in combustion efficiency [90]. Also, at low load conditions, the direct in-cylinder injection technique can better control the CO and UHC emissions. At intermediate and peak load conditions, the after-treatment process is best for reducing the UHC and CO emissions.

12.3.1.4 Cold Start Ability

Ignition in the LTC combustion process is extremely sensitive to the temperature of the intake charge. Depending on the engine's operating circumstances and the fuel's chemical properties, different initial temperatures are necessity for self-ignition. At the time of cold-start conditions, the lowermost temperature of the compressed charge particle restricts the LTC engines from initial combustion. Several techniques are adopted to improve the cold start during LTC, like lighting plug systems, some of fuel additives, and cumulative crusting in VCR and VVT systems [113]. Nevertheless, the most useful method is the dual-mode approach, whereby the engine is sparked via the SI or CI approach and then warmed-up sufficiently, followed by the LTC process.

12.3.1.5 Homogeneous Mixture Preparation

In the LTC combustion process, ignition is considerably delicate to the intake charge temperature.

The initial temperature required for self-ignition varies based on the fuel properties and engine operating conditions. At the time of the cold-start condition, the lowest temperature of the compressed charge particle restricts the LTC engine from initial combustion [114], as shown in Figure 12.5. Several techniques are adopted to improve the cold-start condition during LTC, like using lighting plugs, fuel additives, and increasing the crusting in VCR and VVT systems. The dual-mode approach, which starts the engine in the dual mode and switches to the LTC method once the engine has reached a suitable temperature, is the most practicable approach, nevertheless. The injection timing was determined at the compression stroke in the early DI method, and it is set after the TDC in the late DI process. High injection pressure techniques combined with many tiny holes optimize spray disintegration, resulting in uniform preparation. Lowering HC and PM emissions, preventing oil weakening in LTC engines, and achieving greater brake thermal efficiency (BTE) all depend on homogeneous preparation and separation from fuel separator operations [115–117].

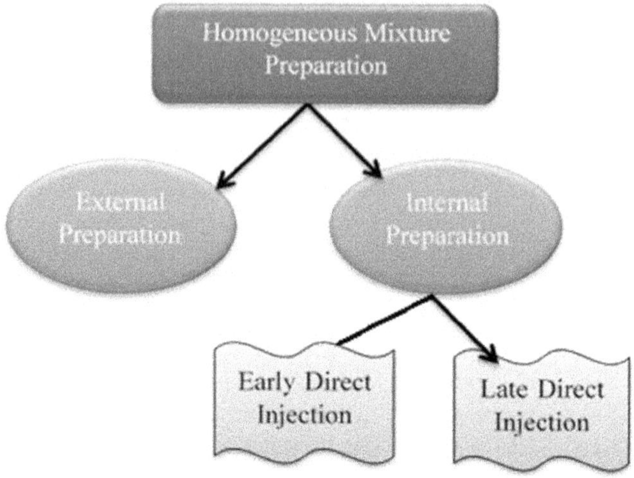

FIGURE 12.5 Strategies of mixture preparation.

12.3.1.6 Cyclic Process and Cylinder-to-Cylinder Deviations

The cyclic varieties in the LTC engine are regularly minimum in contrast with the normal SI engine. Anyway, these varieties may be substantially superior under certain working conditions (e.g., very lean engine operation and lower inlet temperature) [118,119]. Cyclic varieties straightforwardly impact two notable components of the LTC engine, that is, restricted lower load boundaries and testing combustion stage control. In the LTC process, ignition timing, spark timing, and combustion are not directly controlled, as in the case of normal engines [120]. Hasty mode exchanging amongst LTC and conventional combustion (diesel/petrol) is a decent technique to enhance engines' working extent. Most of the experiments were done on a single cylinder engine, nevertheless. Because of its geometric section and the different lengths of its individual manifolds, the multi-cylinder engine with an attached LTC displayed cylinder-to-cylinder changes. These little variations result in a defect in ignition timings, thus causing the combustion rates to fluctuate incredibly [121].

12.3.1.7 Exact Chemical Reaction Device and Combustion Model

Numerous design plans and research methodologies in the field of research and development rely on numerical methods and computational fluid dynamics (CFD) simulations. To precisely anticipate particular highlights and developments from the LTC engine, it is important to create multidimensional CFD ignition models that can be combined with point-by-point substance energy models to conquer the challenge of quickly and extravagantly assessing combustion framework plans.

12.3.2 Control Strategies of LTC Combustion and Emission

The steady and proficient process of LTC requires accurately controllable combustion timings. Major experiments with regard to the LTC process on combustion control focus on the start of ignition (SOI), which relies upon the internal cylinder temperature, internal cylinder pressure, and AFR in the cylindrical section. No additional device is used to begin the SOC. The combustion should be done in sufficient timing, and if too advanced or too retarded in the engine, it causes unacceptable yield increase pressure, heavy noise, and engine damage.

Additionally, the advanced ignition leads to release of a high amount of NO_x, late combustion causes incomplete combustion, leading to significant release of UHC and CO. Too late combustion can lead to misfire and extend the cycle by cycle, which can cause another misfire, and the engine may stall [122]. The closed-loop combustion control is important to make sure to measure the combustion timings. In the LTC combustion process, uniquely, the ones working in HCCI condition don't have a straight actuator for the combustion timing. Different techniques are adopted to estimate the combustion timings, such as 10%–50% of heat release and peak pressure rise rate [123], and different pressure sensors are applied for combustion point sensing. In the laboratories, piezo-electric transducers are used to measure the interior cylinder pressure. Because of the high CO levels and less lifespan of the device, it is not used in the industries. The ion current sensor is also used, but it has some drawbacks like it gives only local information, it is suitable for use in a homogeneous mixture, and it depends on fuel type, its chemical properties, and working conditions. Researchers are suggested to activate the combustion stages in LTC engines by VVA, VCR, dual fuel, and thermal administration [124–126]. All techniques have their own advantages and disadvantages.

12.3.2.1 Dual Fuel Control Strategy

In the dual fuel control system, dualistic oils (primary and secondary fuels) that have different chemical properties are used. The primary fuel used as an extraordinary octane number (ON), and the secondary fuel will act as the lesser octane fuel [127]. Since blends of two distinct fuels in varying amounts alter self-ignition properties, the ignition stage in LTC was controlled by the dissimilar self-ignition properties of both fuels. Numerous studies have examined the regular fuels, the mixing of on its own factor fuels, and PRFs [128]. For instance, studies have documented the benefits and varying degrees of feasibility of fuel mixtures such as ethanol, butanol, and n-heptane [129], diesel and n-heptane [130], diesel and gasoline [131,124], and natural gas and gasoline [125]. Researchers conducted PCCI-based experiments with diesel and ethanol as fuel for DEs. The ethanol fuel was fumigated and passed into the inlet manifold section along with air. The experiment was performed at different compression ratios; it was suggested that OH radicals can be controlled by the compression ratio [132].

12.3.2.2 VVT and VCR

In the VVT process, EGR and compression control play major roles. VVT is a rapid method for optimizing engine consumption. Exhaust gas is recirculated from the exit section to escalate the initial temperature. The second way to regulate the initial temperature is to advance or pause the inlet valve closing (IVC) timings. As a result, the engine's compression ratio is less active. There is growing interest in using VVA to control the HCCI process, and researchers are considering it using both design and analysis [133,134].

12.3.2.3 EGR Strategy

This is an important technique to attain low soot particles and NO_x oxides liberated from the engine without a restrictively high level of fuel consumption, due to the less ignition phasing. The most popular approach to change the cylinder temperature is to adjust the EGR percentage, which also regulates the SOC. The study [135] thoroughly examined the HCCI combustion and emission parameters by using the Ti and EGR with different ratios. The four important constituents which highly impact the combustion and emission parameters are as follows: first, the preheating effect: the engine inlet air temperature is maximized when the hot EGR is combined with the air/fuel mixtures. The second stage is the dilution effect, where increasing the EGR level minimizes the oxygen content in the inlet section. Third, heat capacity impact: the complete heat ability of the mixed form of air, fuel, and EGR. It leads to higher heat release, CO_2, and water vapour, at the end of the compression stroke, resulting in temperature reduction. Fourth, chemical effect: all the unburned particles undergo chemical reactions with CO_2, CO, NO_x, HC, NO, etc.

In the present investigation, EGR introduction acts plays a major role in the complete combustion and emission process and release of its particles. All these particles have high effects in regulating SOC and CD, and EGR is a well-organized control approach in HCCI–DI combustion [136–138]. The pre-advanced ignition causes a knocking in the normal combustion process. The EGR process can be classified into interior and exterior modes. The interior EGR can be changed by varying the connection period. The EGR controller and exhaust back-pressure controller affect the external EGR rate. A higher EGR ratio reduces the combustion temperature, and its results show the pyrolysis of fuel and soot particles. Kimura et al. [139] suggested that reducing the inlet gas temperature from 117°C to 68°C leads to minimization of soot emissions to nearly 91% and NO_x level by 35%–40%. Ishiikawa et al. [140] studied the impact of heat exchanger proficiency with application of the EGR cooling reaction for heat rejection from 2.8 to 7.5 kW. Automatically, the inlet temperature decreases from 110°C to 35°C, and the charge density and oxygen level increase. Their result shows that reduction in NO_x level slightly increases the HC emissions. Idicheria et al. [141] explained that temperature increase simultaneously increases high-level soot formation with high percentage of EGR to without EGR in the PCCI combustion. He concluded that EGR cooling reduced soot formation. Hargy et al. [142] reported that the maximum equivalence ratio can be obtained at a higher percentage of EGR, and they successfully proved NO_x level reduction and soot formation.

12.3.2.4 Different Fuel Additives

The fuel organization has a minor effect on the reaction rate and ignition timing once reactions have been started; in any event, it characterizes the self-ignition temperature, consequently regulating the SOC in the LTC process. The fuel additive compounds largely control the LTR process, which thus influences the initiation of the primary reaction process. The primary fuel components, such as density, latent heat of evaporation, and heating value, have an important influence on the physical delay component of the ID. Additionally, the fuel characteristics and self-ignition are impacted by the chemical delay component. All of the outcomes include an extended ID (SOI or SOC), including maximum fuel density, volatilization problems, and high latent heat of vaporization due to a lower fuel–air combination temperature. Normal approaches to classify the fuels are based on the simplicity of self-ignition called as cetane number (CN) and to work against to the self-ignition called as ON [143]. A higher CN shows lesser resistance of self-ignition, called chain paraffin. The higher ON shows opposition against self-ignition, which incorporates spread chain paraffin. Petroleum fuels have high ON, so they have little no or little LTC at 950 K, whereas conventional diesel indicates an LTC reaction requiring an initial temperature at 750 K [144]. Stark et al. [145] performed an experiment on the effect of the fuel properties on the HCCI technique and reported that lower CN fuel is better for HCCI fuel engines. The lesser CN fuel, with better combustion, enhances the HCCI combustion peak pressure because of the lower combustion timing, which is highly suitable for a better fuel air mixture in the HCCI process. Mack et al. [112] explained the impact of using additive di-tertiary butyl peroxide (DTBP) on HCCI ignition utilizing engine analysis and numerical demonstration. A minor amount of DTBP is mixed with 100% pure ethanol with diethyl ether (DEE) for performing DE testing to change the fuel injection timings with different engine loads. In every load condition increase, the DTBP fuel is used. The numerical model examination was additionally done to approve the test results. In spite of the fact that additive substances enhanced the combustion timings for both pure ethanol fuel using different techniques like blend and fumigation, the added substances brought about a considerable increase in ethanol levels. It affirms both warm and better impacts on the expansion of the DTBP.

12.3.2.5 Split-and-Spray Injection

By utilizing the split injection technique and better positioning the fuel spray point on the geometrical portion of the cylinder bowl, problems from the LTC process like uncontrolled combustion can be reduced. Kook et al. have done an experiment on the split injection method and concluded that LTC ignition provided good control combustion. Before the compression stroke, that is (100° bTDC), a high level of fuel can be injected. It leads to premixing, and a less quantity of fuel can be injected nearer to TDC. This result shows the NO_x emission is reduced and better engine efficiency is obtained. Lee and Reitz [146] performed an experiment on the influence of spray pointing on the piston bowl section and its emission control. He suggested that emission levels are minimized when the spray is located at the piston edge (close to the squish area). It leads to an escalation in the premixture of fuel and air, which causes complete combustion and reduces soot and NO_x emissions. According to certain other scientists, modest load conditions result in good thermal efficiency and minimal NO_x emissions.

12.3.2.6 Closed-Loop Controller

Different combustion drawbacks and control approaches are implemented previously, and several strategies were used in experimental work. In the HCCI process, a closed-loop system is the first method implemented for dual fuel control [147].

Dual-fuel, single-cylinder experiments require injection of two different fuels. The proportional, integral, and derivative (PID) device is used to regulate the combustion phase in the HCCI process. Strandh et al. [148] performed the experiment to control the combustion drawbacks using DE, by varying the IVC timings. Three distinct approaches such as the model predictive control (MPC), Linear-Quadratic-Gaussian, and PID techniques are used. MPC is the most effective of these three approaches.

12.3.2.7 Dual-Mode Operation

The enhancement of an LTC process, particularly at maximum load conditions, is a lasting objective; although in the long term, the dual combustion system looks for a practical execution of LTC in instead of IC engines, a combination of both normal CI and SI engines. The main difficulty of the double combustion process is the price and complication associated with the growth of the system. Many scientists and researchers suggested that it is the best one to execute; it is an intermediate between the LTC and SI process. It is highly used at high load conditions, operated under stoichiometric conditions. Numerous different control procedures are attempted for HCCI control systems. Three methods are the physical model-based control system, experimental method-based control system, and physically adjusted regulators. Physical-based control systems are highly used in the HCCI process. In closed-loop systems, other equipment like PID controllers, AVT, HVT, and water-powered valve timing are used.

12.3.3 Emissions Control Methods

Normally, reduction in NO_x emission can be distinguished into two streams a. pre-combustion b. after or post-combustion treatment.

12.3.3.1 Different Additives

Diesel and biodiesel are mixed with various additives, including metal-based additives, oxygenated additives, stability additives, ignition enhancers, and CN enhancers [149]. Some antioxidants and oxygenated additives used are dimethyl ester, ethanol, and methanol [150], and CI [151] are used to reduce the NO_x level.

12.3.3.2 Exhaust Gas Recirculation Process

This system, which regulates the oxygen density and combustion temperature, is very helpful in reducing the NO_x emissions released from the DE [152,153]. This process reduces the combustion gas temperature, without any change in the dilution and inert gas effect with high specific heat ratios of CO_2 and water [154,155]. In the above two processes, superior heat capacity compared to that of air is achieved [156]. The amount of oxygen content in the suction air is low due to the exhaust gas being recirculated into the suction or intake valve. The AFR becomes very low due to the decreased quantity of oxygen for complete combustion, finally affecting the engine emission level. In the normal combustion process, maximum energy is required to preheat the suction air and fuel mixture. Because of these actions, the suction valve will have less oxygen, the flame temperature will decrease, and fewer nitrogen oxides will form.

12.3.3.3 Injection Method and Emulsion Method

Two techniques are employed in this process: direct water droplet injection into the combustion chamber [157]. In the experiment to ascertain the performance and emission in the DI turbocharged DE, water is injected into the inlet manifold section. It should not induce any variations in the cylinder pressure and HRR. However, when applied in the PCCI engine, it plays a major role in reducing the NO_x emission by up to 50% by use of diesel and biodiesel fuel. The main drawback of the experiment was reduced CO and increased brake specific fuel consumption (BSFC) level; the BTE was also reduced. The water and diesel combined emulsion process is highly useful to minimize the emission levels of NO_x, SO_x, carbon monoxide, and PM without a decline in the engine's performance. This method minimizes the local Soupçon by water evaporation, to absorb heat from the combustion chamber; it influences the reduction in nitrogen oxide emissions [158,159]. Tran and Ghojel [160] had done the experiment on both models via water–diesel emulsion and direct water injection to reduce NO_x emissions. Water is directly injected into the engine cylinder. These water droplets get converted into steam, leading to extraordinary high temperature and pressure; higher temperature and pressure are present inside the engine cylinder at the time of compression stroke. It is used to reduce the engine temperature and also decrease the NO_x emission level nearly 10–60% compared to the normal process.

Some salient points are as follows.

- The reduction in nitrogen oxides in HCCI, PCCI, and RCCI modes depends on a lengthier ID and a lesser rate of combustion, which decreases the in-cylinder temperature and causes the pressure to rise proportionately.
- Air–fuel proportion and adjacent oxygen fixation are more important indicators of nitrogen oxide emissions under a given combustion temperature than in-cylinder temperature. In terms of nitrogen oxide emissions, late injection premixed LTC is better than early injection premixed LTC.
- Propelled injection helped the HCCI mode achieve higher EGR to control oxides of nitrogen emission as cutting-edge injection causes improved in-cylinder temperature.
- With the help of the fuel reactivity grade inside the cylinder, the reactivity-controlled combustion mode has succeeded in maintaining the nitrogen oxides level lesser at higher loads even with a low EGR rate.
- Minimizing the nitrogen oxide level by the higher percentage of EGR, ID, and increased late injection time.
- ID is highly dependent upon the chemical kinetics process of biodiesel than temperature reduced by using EGR.
- The delayed injection is better than early ignition for reducing the nitrogen oxide emission in biodiesel usage. The maximum level of oxygen and high CN of biodiesels can maintain a greater EGR and late SOI-caused LTC. Along these lines, reduced NO_x emissions were observed.
- Adding additives in biodiesel fuel can decrease NO_x level and increase hydrocarbon and emission levels.
- By using the EGR method, the NO_x level can be reduced by 88.5%, increase BSFC level, and minimize engine efficiency. Some advantages like low cost, less space, and easy control exist.

- Water injection and ET method can reduce NO_x level up to 39%, reduce smoke and PM also, and increase hydrocarbon and carbon monoxide level.

12.3.4 Control of Soot and Particulate Matter

The primary component of diesel and soot particulates is carbonaceous material (soot) produced by combustion, onto which some organic and inorganic chemicals have been adsorbed. The PM ranges are from 20 nm to 10 mm and are categorized as soluble and insoluble organic fractions. The dry fraction is normally called as soot. Soot is developed from incomplete combustion of fuels, it is nucleated from vapour to solid state in a rich fuel region with higher temperature [161]. By using the water/diesel emulsion process, soot and PM are reduced nearly 81%–89%. Microemulsion is another process in which fuel is dispersed into smaller sized water droplets compared to normal water/diesel emulsion fuel. It has well-collaborating properties, and greater atomization causes lesser emission [162].

Jacobs and Asanis [49] carried out an experiment on AFR and found out that it is lower than the stoichiometric ratio, and when injected at 25° bTDC, it leads to complete combustion. It showed a very low soot level that is 0.03 FSN. Parks Li et al. [162] also performed an experiment using a PCCI engine using advanced injection with EGR and obtained the same result. PM level is decreased up to 51% when compared to normal combustion [163]. Singh et al. [164] performed the experiment on a dual injection system. The first-stage injection starts at 22° bTDC at low load conditions. In the second stage, combustion occurs at 15° aTDC, which causes enhanced mixing and reduced soot particle level.

- PM/soot discharge increases with the expansion of the EGR rate, headway, or difficulty of injection timing to accomplish the LTC process.
- Under soot development temperature, soot emission does not rely upon the nearby equality proportion.
- The LTC combustion process with a narrow angle of injection yields a less amount of soot emission.
- Very advanced injection and late injection in the LTC combustion process cause maximum release of soot/PM level.

12.3.5 Control of Carbon Monoxide and Unburnt Hydrocarbon

CO is primarily produced because of the uneven combustion of the fuel. Incomplete combustion happens owing to a decrease in the ignition temperature, and the oxidation to carbon monoxide is maximum. In the HCCI combustion process, the CO level is very high compared with the commercial DE; it is a crucial task in the HCCI combustion process [26,165]. Ganesh et al. [106] reported that measurements of carbon emissions depend on the performance of combustion, while the combustion efficiency can be mentioned as the ratio of carbon dioxide to the total amount of fuel carbon introduced in the exhaust, which includes hydrocarbons, carbon dioxide, and CO. The cylindrical section's fissures, which are extremely cold for full consumption, produce CO, although at a lesser concentration. In the complete conversion of carbon emissions, the temperature of 1500 K is maintained. Tormos et al. [166] reported that in the LTC combustion process using biodiesel, the CO and UHC are reduced up to 27% and 45% compared with diesel fuel, respectively. Use of biodiesel fuel is better than that of diesel in the LTC process, yielding low combustion emission levels. Considering the use of biodiesel in the LTC process, some parameters play an important role, namely, injection timing [167], injection pressure, load condition, and injection model [83], in a cylinder, and intake air temperature [168].

In traditional diesel fuel ignition, the primary explanations behind hydrocarbon discharge are catching of fuel in the cleft area of the combustion chamber, the less temperature mass extinguishing of the oxidation responses, locally finished lean or over-rich blend, liquid wall films for exorbitant spray impingement, and incomplete combustion of the fuel [169]. Opat et al. [170] reported that the temperature effect on the engine is not completely controlled by the hydrocarbon and carbon monoxide emission levels. The important parameter is fuel impingement in the cylindrical bowl. Much research has been carried out to get less UHC and CO emissions. Changes in piston bowl and fuel film formation and spray injection are the major reasons for higher hydrocarbon emission in premixed LTC combustion [114].

There are two major reasons for UHC release in the exhaust section: fuel that escapes combustion in the chamber or halfway hydrocarbons that are formed at the time of combustion and incompletely combusted. This is caused due to misfire or local flame extinction and presence of gaps on the cylindrical section. UHC and CO are increased in a homogeneous charge CI engine. It is the major challenge in the HCCI engine.

Some of the important points from reference papers suggest on emission levels using diesel fuel.

- In the LTC process, higher levels of HC and CO are emitted because of a decrease in cylinder combustion temperature and oxygen impact ratio.
- In LTC modes, higher ID with overmixing of fuel makes it too lean for a complete combustion process, so hydrocarbon and carbon monoxide emission levels are increased.
- Higher levels of HC and CO are produced by spray impingement on piston bowl section variation in PCCI and HCCI processes.
- In the LTC combustion process, CO emission highly relies on the equivalence ratio.
- In the LTC combustion process, hydrocarbon emission is highly dependent on ID.
- In the LTC process, using biodiesel reduces the UHC and CO, but total emission is higher than that of a DE.
- Early and late injection of biodiesel yields maximum emission levels; in the PCCI process, using a short premix charge in LTC gives good emission results.
- In PCCI and HCCI process, using biodiesel with a higher injection pressure, multiple injection processes, and varying intake pressure gives low hydrocarbon and CO emission levels compared with the DE.

12.4 CONCLUSION

The LTC is the best combustion method for producing better thermal efficiency in petrol engine fuels and ultra-less NO_x and soot emissions in a DE. Some drawbacks are found in this process. For commercial adaptation, it can be produced at large amounts, but has several additional problems such as operational range, the need for a closed-loop control of SOC, and higher emissions levels of HC and CO.

The LTC combustion technique has been developed recently in light of the requirement for enhanced thermal efficiency of DEs and less NO_x, ash soot, and PM emissions from the engines. Some remarkable operations can be made in the LTC process; however, several difficulties are encountered in many LTC engines; for example, control of SOC and higher emission levels of HC and CO are the major difficulties faced in LTC method implementation. Fuel adaptability is another essential method of LTC, wherein fuel properties are the main factor to have an impact on the chemical kinetics, it helps monitor the LTC operation. In light of engine design, working conditions, and control techniques, different alternative fuels like biodiesel, compressed natural gas, di-ethyl methyl ester, DEE, alcohol, kerosene, and some other fuels are used in the LTC engine. Self-ignition was well organized by alteration in the fuel properties or fuel so as to make it all the more chemically reactive by addition of promoters. Based on this concept, researchers have developed a new concept called RCCI combustion. In this process, two different fuels with dissimilar chemical kinetics are used. Here, combustion regulation is the main task, which might be resolute owing to use of different methods like VCR, VVT, EGR, and dual fuel injection, although these methods have their own advantages and disadvantages. Many researchers have suggested using dual fuel combustion, which appears to be suitable for practical use as this process is close to medium tenancy. In this methodology, a normal system (like CI or SI) involves the combination of LTC processes. By including this process, the cost of the vehicle is increased. Because complexity still remains the main problem, it prevents commercialization as of now. Consequently, for a long period of time, the implementation of the LTC process, flexible fuel injection technique, and EGR system for high mixture preparation is carried out, giving maximum boosting power at high load conditions, and some added parameters are investigated and used so that commercially feasible LTC vehicles are manufactured and used.

NOMENCLATURE

ADI	Advance direct injection
BSFC	Brake specific fuel consumption
BS-VI	Bharat standard
BTE	Brake thermal efficiency
CA	Degree of Crank angle
CAI	Control auto-ignition
CFD	Computational fluid dynamics
CI	Compression ignition
CN	Cetane number

CNG	Compressed natural gas
CO	Carbon monoxide
CO_2	Carbon dioxide
CRDI	Common rail direct injection
DE	Diesel engine
DEE	Diethyl ether
DI	Direct injection
DICI	Direct compression ignition
DME	Di-ethyl methyl ester
DTBP	Di-tertiary butyl peroxide
EGR	Exhaust gas recirculation
HCCI	Homogeneous charge compression ignition
HTR	High-temperature reaction
IC	Internal combustion
IVC	Inlet valve close
KHP	Ketohydro peroxide
LQG	Linear-Quadratic-Gaussian
LTC	Low-temperature combustion
LTR	Low-temperature reaction
MPC	Model predictive control
NO_x	Oxides of nitrogen
NTC	Negative temperature coefficient
ON	Octane number
PAH	Polycyclic aromatic hydrocarbons
PCCI	Premixed charge compression ignition
PFI	Port fuel injection
PID	Proportional, integral, and derivative
PM	Particulate matter
ppm	Parts per million
RCCI	Reactivity-controlled compression ignition
SI	Spark ignition
SOC	Start of combustion
SOI	Start of ignition
UHC	Unburned hydrocarbon
VCR	Variable compression ratio
VVT	Variable valve timing

REFERENCES

1. Kumar P, Rehman A. Bio-diesel in homogeneous charge compression ignition (HCCI) combustion. *Renew Sustain Energy Rev*, 2016, 56:536–50.
2. Francisco P, Sarah C, Kate B. *Cost of Emission Reduction Technologies for Heavy Duty Vehicles.* International Council for Clean Transportation, Washington, DC, 2016.
3. Pandey RK, Rehman A, Sarviya RM. Impact of alternative fuel properties on fuel spray behavior and atomization. *Renew Sustain Energy Rev*, 2012, 16:1762–78.
4. Kumar S, Chaube A, Jain SK. Sustainability issues for promotion of Jatropha biodiesel in Indian scenario: a review. *Renew Sustain Energy Rev*, 2012, 16:1089–98.

5. Kumar A, Sharma S. Potential non-edible oil resources as biodiesel feedstock: an Indian perspective. *Renew Sustain Energy Rev*, 2011, 15(4):1791–800.
6. Fazal MA, Haseeb ASMA, Masjuki HH. Biodiesel feasibility study: an evaluation of material compatibility; performance; emission and engine durability. *Renew Sustain Energy Rev*, 2011, 15(2):1314–24.
7. Koh MY, Mohd Ghazi TI. A review of biodiesel production from Jatropha curcas L. oil. *Renew Sustain Energy Rev*, 2011, 15(5):2240–51.
8. Bhuiya MMK, Rasul MG, Khan MMK, Ashwath N, Azad AK, Hazrat MA. Prospects of 2nd generation biodiesel as a sustainable fuel – part 2: properties, performance and emission characteristics. *Renew Sustain Energy Rev*, 55:1129–46.
9. Jain S, Sharma M. Stability of biodiesel and its blends: a review. *Renew Sustain Energy Rev*, 2010, 14(2):667–78.
10. Naik C, Westbrook C. Kinetic Modeling of Combustion Characteristics of Real Biodiesel Fuels. In: Proceedings of the 6th US National Combustion Meeting, Ann Arbor, MI, 2009: pp. 1–5.
11. Lai JY, Lin KC, Violi A. Biodiesel combustion: advances in chemical kinetic modeling. *Prog Energy Combust Sci*, 2011, 37(1):1–14.
12. Mofijur M, Rasul M, Hyde J, Azad A, Mamat R, Bhuiya M. Role of biofuel and their binary (diesel–biodiesel) and ternary (ethanol–biodiesel–diesel) blends on internal combustion engines emission reduction. *Renew Sustain Energy Rev*, 2016, 53:265–78.
13. Mofijur M, Rasul M, Hyde J. Recent developments on internal combustion engine performance and emissions fuelled with biodiesel-diesel-ethanol blends. *Procedia Eng*, 2015, 105:658–64.
14. Brown NJ, Bastien LA, Price PN. Transport properties for combustion modeling. *Prog Energy Combust Sci*, 2011, 37(5):565–82.
15. Khair MK. Jaaskelainen Hannu "Combustion in Diesel Engines" DieselNet.com, 2020.
16. Genzale C, Reitz R, Musculus M. Effects of piston bowl geometry on mixture development and late injection lowtemperature combustion in a heavy-duty diesel engine. *SAE Int J Engines*, 2009, 1(1):913–37. doi:10.4271/2008-01-1330
17. Derry LD, Dodds EM, Evans EB, Royle D. The effects of auxiliary fuels on smoke limited power output of diesel engine. *Proc Inst Mech Eng*, 1954, 168:280–6.
18. Banapurmath N, Tewari P, Gaitonde V. Experimental investigations on performance and emission characteristics of Honge oil biodiesel (HOME) operated compression ignition engine. *Renew Energy*, 2012, 48:193–201.
19. Lu X, Han D, Huang Z. Fuel design and management for the control of advanced compression-ignition combustion modes. *Prog Energy Combust Sci*, 2011, 37(6):741–83.
20. Su W, Lin T, Pei Y. A compound technology for HCCI combustion in a DI diesel engine based on the multi-pulse injection and the BUMP Combustion Chamber. SAE Paper No, 2003-01-0741, 2003.
21. Su W, Wang H, Liu B. Injection mode modulation for HCCI diesel combustion. SAE Paper No, 2005-01-0117, 2005.
22. Bessonette P, Schleyer C, Duffy K, Hardy W, Liechty M. Effects of fuel property changes on heavy-duty HCCI combustion. *Polymer*, 2007, 2013:4–8.
23. Zheng J, Yang W, Miller L, Nicholas PC. Prediction of pre-ignition reactivity and ignition delay for HCCI using a reduced chemical kinetic model. SAE Paper No, 2001-01-1025, 2001.
24. Kusaka J, Yamamoto T, Daisho Y. Simulating the homogeneous charge compression ignition process using a detailed kinetics model for *n*-heptane mixture. *Int J Engine Res*, 2000, 1(3):281–9.
25. Lapuerta M, Armas O, Ballesteros R, Fernández J. Diesel emissions from biofuels derived from Spanish potential vegetable oils. *Fuel*, 2005, 84:773–80.
26. Kanda T, Hakozaki T, Uchimoto T, Hatano J, Kitayama N, Sono H. PCCI operationwith early injection of conventional diesel fuel. *SAE Trans*, 2005, 114:584–93.

27. James Gunasekaran, E., Gunasekar, N., Gopi Kannan, K., Dhivagar, R., & Sathyanarayanan, S. (2024). Exploring the Depth and Navigating the Complexities of Low-Temperature Combustion process-A Review.
28. Mancaruso E, Vaglieco B. Optical investigation of the combustion behaviour inside the engine operating in HCCI mode and using alternative diesel fuel. *Exp Therm Fluid Sci*, 2010, 34(3):346–51.
29. Karthikeya Sharma T, Amba Prasad Rao G, Madhu Murthy K. Effective reduction of NO_x emissions of a HCCI (Homogeneous charge compression ignition) engine by enhanced rate of heat transfer under varying conditions of operation. *Energy*, 2015, 93:2102–15.
30. Lee S, Sungwook Park. Optimization of the piston bowl geometry and the operating conditions of a gasoline-diesel dual-fuel engine based on a compression ignition engine. *Energy*, 2017, 121:433–8.
31. Zheng M, Wang M, Reader GT, Mulenga MC, Tjong J. An improvement on low temperature combustion in neat biodiesel engine cycles. *SAE Int J Fuels Lubr*, 2009, 1(1):1120–32.
32. Manente V, Tunestål P, Johansson B. Influence of the Wall Temperature and Combustion Chamber Geometry on the Performance and Emissions of a Mini HCCI Engine Fueled with Diethyl Ether. SAE Technical Paper No. 2008-01-04-08, 2008.
33. Harisankar Bendu BB, Deepak VL, Murugan S. Application of GRNN for the prediction of performance and exhaust emissions in HCCI engine using ethanol. *Energy Conv Manag*, 2016, 122:165–73.
34. Maurya RK, Akhil N. Development of a new reduced hydrogen combustion mechanism with NO_x and parametric study of hydrogen HCCI combustion using stochastic reactor model. *Energy Conv Manag*, 2017, 132:65–81.
35. Veltman MK, Karra PK, Kong S-C. Effects of Biodiesel Blends on Emissions in Low Temperature Diesel Combustion. SAE Technical Paper No. 2009-01-0485, 2009.
36. Tsolakis A, Megaritis A, Yap D. Application of exhaust gas fuel reforming in diesel and homogeneous charge compression ignition (HCCI) engines fuelled with biofuels. *Energy*, 2008, 33(3):462–70.
37. Aulin H, Johansson T, Tunestal P, Johansson B. Control of a turbo charged NVO HCCI engine using a model based approach. *IFAC Proceed Vol*, 2009, 42:79–86.
38. Yang F, Gao G, Ouyang M, Chen L, Yang Y. Research on a diesel HCCI engine assisted by an ISG motor. *Appl Energy*, 2013, 101:718–29.
39. Mohamed Ibrahim M, Ramesh A. Investigations on the effects of intake temperature and charge dilution in a hydrogen fueled HCCI engine. *Int J Hydrogen Energy*, 2014, 39:14097–108.
40. Mathivanan K, Mallikarjuna JM, Ramesh A. Influence of multiple fuel injection strategies on performance and combustion characteristics of a diesel fuelled HCCI engine–An experimental investigation. *Exp Therm Fluid Sci*, 2016, 77:337–46.
41. Ganesan D, Nagarajan G, Ganesan S. Performance and Emission Analysis on Mixed-Mode Homogeneous Charge Compression Ignition (HCCI) Combustion of Biodiesel Fuel with External Mixture Formation. SAE Technical Paper No. 2011-01-2450, 2012.
42. Abu-Qudais M, Haddad O, Qudaisat M. The effect of alcohol fumigation on diesel engine performance and emissions. *Energy Convers Manage*, 2000, 41:389–99.
43. Kavuri C, Paz J, Kokjohn SL. A comparison of reactivity controlled compression ignition (RCCI) and gasoline compression ignition (GCI) strategies at high load, low speed conditions. *Energy Convers Manage*, 2016, 127:324–41.
44. Inagaki K, Fuyuto T, Nishikawa K, Nakakita K, Sakata I. Dual-Fuel PCI Combustion Controlled by in-Cylinder Stratification of Ignitability, SAE Technical Paper No. 2006-01-0028, 2006.

45. Epping K, Aceves S, Bechtold R, Dec J. The potential of HCCI combustion for high efficiency and low emissions. *Training*, 2002, 2013:11–4.
46. Jacobs TJ, Assanis DN. The attainment of premixed compression ignition lowtemperature combustion in a compression ignition direct injection engine. *Proc Combust Inst*, 2007, 31:2913–20.
47. Lava KR, Sannagoudraa J, Ganesh DB. Experimental exploration on the influence of different piston geometry and injection timing by using bio-diesel. *Mater Today Proceed*, 2017, 4:10879–85.
48. Zehni A, Saray RK, Poorghasemi K. Numerical comparison of PCCI combustion and emission of diesel and biodiesel fuels at low load conditions using 3D-CFD models coupled with chemical kinetics. *Appl Therm Eng*, 2017, 110:1483–99.
49. Laguitton O, Crua C, Cowell T, Heikal MR, Gold MR. The effect of compression ratio on exhaust emissions from a PCCI diesel engine. *Energy Convers Manage*, 2007, 48:2918–24.
50. Valentino G, Corcione FE, Iannuzzi SE, Serra S. Experimental study on performance and emissions of a high speed diesel engine fuelled with *n*-butanol diesel blends under premixed low temperature combustion. *Fuel*, 2012, 92:295–307.
51. Dempsey A, Walker N, Reitz R. Effect of piston bowl geometry on dual fuel reactivity controlled compression ignition (RCCI) in a light-duty engine operated with gasoline/diesel and methanol/diesel. *SAE Int J Engines*, 2013, 6(1):78–100. doi.org/10.4271/2013-01-0264
52. Torregrosa AJ, Broatch A, García A, Mónico LF. Sensitivity of combustion noise and NO_x and soot emissions to pilot injection in PCCI diesel engines. *Appl Energy*, 2013, 104:149–57.
53. Kiplimo R, Tomita E, Kawahara N, Yokobe S. Effects of spray impingement, injection parameters, and EGR on the combustion and emission characteristics of a PCCI diesel engine. *Appl Therm Eng*, 2012, 37:165–75.
54. Jain A, Singh AP, Agarwal AK. Effect of split fuel injection and EGR on NO_x and PM emission reduction in a low temperature combustion (LTC) mode diesel engine. *Energy*, 2017, 122:249–64.
55. Perini F, Miles PC, Reitz RD. A comprehensive modeling study of in – cylinder fluid flows in a high-swirl, light-duty optical diesel engine. *Comput Fluids*, 2014, 105:113–24.
56. Abd Alla GH, Soliman HA, Badr OA, Abd Rabbo MF. Effect of injection timing on the performance of a dual fuel engine. *Energy Convers Manag*, 2002, 43:269–77.
57. Mohammadi A, Kee S-S, Ishiyama T, Kakuta T, Matsumoto T. Implementation of ethanol diesel blend fuels in PCCI combustion. SAE Technical Paper No. 2005-01-3712, 2005.
58. McCormick RL, Tennant C, Hayes RR, Black S, Ireland J, McDaniel T, et al. Emission Standards. SAE Technical Paper No. 2005-01-2200, 2004.
59. Hanson RM, Kokjohn SL, Splitter DA, Reitz RD. An experimental investigation of fuel reactivity controlled PCCI combustion in a heavy-duty engine. *SAE Int J Engine*, 2010, 3:700–16.
60. Hanson R, Curran S, Wagner R, Kokjohn S, Splitter D, Reitz R. Piston bowl optimization for RCCI combustion in a light-duty multi-cylinder engine. *SAE Int J Engine*, 2012, 5:286–99.
61. Kokjohn SL, Reitz RD. Characterization of Dual-Fuel PCCI Combustion in a Lightduty Engine. In: *Proceedings of the International Multi-Dimensional Engine Modeling User's Group Meeting*, Detroit, 2010.
62. Benajes J, Molina S, García A, Belarte E, Vanvolsem M. An investigation on RCCI combustion in a heavy duty diesel engine using in-cylinder blending of diesel and gasoline fuels. *Appl Therm Eng*, 2014, 63:66–76.

63. Li Y, Jia M, Chang Y, Liu Y, Xie M, Wang T, et al. Parametric study and optimization of a RCCI (reactivity charge controlled compression ignition) engine fuelled with methanol and diesel. *Energy*, 2014, 65:319–32.
64. Li Y, Jia M, Chang Y, Fan W, Xie M, Wang T. Evaluation of the necessity ofexhaust gas recirculation employment for a methanol/diesel reactivity charge controlled compression ignition engine operated at medium loads. *Energy Convers Manag*, 2015, 101:40–51.
65. Zang R, Yao C. Numerical study of combustion and emission characteristics of a diesel methanol dual fuel (DMDF) engine. *Energy Fuel*, 2015, 29:3963–71.
66. Dempsey AB, Das Adhikary B, Viswanathan S, Reitz RD. Reactivity controlled compression ignition using premixed hydrated ethanol and direct injection diesel. *J Eng Gas Turbines Power*, 2012, 134:82806.
67. Han X, Divekar P, Reader G, Zheng M, Tjong J. Active injection control for enabling clean combustion in ethanol-diesel dual-fuel mode. *SAE Int J Engine*, 2015, 8:890–902.
68. Fang W, Fang J, Kittelson DB, Northrop WF. An experimental investigation of reactivity-controlled compression ignition combustion in a single-cylinder diesel engine using hydrous ethanol. *J Energy Resour Technol*, 2014, 137:311–21.
69. Splitter D, Reitz R, Hanson R. High efficiency, low emissions RCCI combustion by use of a fuel additive. SAE Technical Paper No. 2010-01-2167, 2010.
70. Lia J, Linga X, Liua D, Yangb W, Zhou D. Numerical study on double injection techniques in a gasoline and biodiesel fueled RCCI (reactivity controlled compression ignition) engine. *Appl Energy*, 2018, 211:382–92.
71. HyvAonen J, Haraldsson G, Johansson B. Super Charging HCCI to Extend the Operating Range in a Multi-Cylinder VCR HCCI Engine. SAE Technical Paper No. 2003-01-3214, 2003.
72. Kakaee A-H, Nasiri-Toosi A, Partovi B, Paykani A. Effects of piston bowl geometry on combustion and emissions characteristics of a natural gas/diesel RCCI engine. *Appl Therm Eng*, 2016, 102:1462–72.
73. Wu F, Wang J, Chen W, Shuai S. A study on emission performance of a diesel engine fueled with five typical methyl ester biodiesels. *Atmos Environ*, 2009, 43:1481–5.
74. Splitter D, Hanson R, Kokjohn S, Reitz R. Reactivity Controlled Compression Ignition (RCCI) Heavy-Duty Engine Operation at Mid-and High-Loads with Conventional and Alternative Fuels. SAE Technical Paper No. 2011-01-0363, 2011.
75. Ullman TL, Spreen KB, Mason RL. Effects of Cetane Number, Cetane Improver, Aromatics, and Oxygenates on 1994 Heavy-Duty Diesel Engine Emissions. SAE International Engineering Society for Advancing Mobility, Land Sea Air Andspace, 1994.
76. Salahi MM, Esfahanian V, Gharehghani A, Mirsalim M. Investigating the reactivity controlled compression ignition (RCCI) combustion strategy in a natural gas/diesel fueled engine with a pre-chamber. *Energy Convers Manag*, 2017, 132:40–53.
77. Kimura S, Aoki O, Kitahara Y, Aiyoshizawa E. Ultra-Clean Combustion Technology Combining a Low-Temperature and Premixed Combustion Concept for Meeting Future Emission Standards. SAE Technical Paper No. 2001-01-0200, 2001.
78. Pidol L, Lecointe B, Starck L, Jeuland N. Ethanol–biodiesel–diesel fuel blends: performances and emissions in conventional diesel and advanced low temperature combustions. *Fuel*, 2012, 93:329–38.
79. Liu H, Ma G, Hu B, Zheng Z, Yao M. Effects of port injection of hydrous ethanol on combustion and emission characteristics in dual-fuel reactivity controlled compression ignition (RCCI) mode. *Energy*, 2018, 145:592–602.
80. Pan S, Li X, Han W, Huang Y. An experimental investigation on multi-cylinder RCCI engine fueled with 2-butanol/diesel. *Energy Convers Manag*, 2017, 154:92–101.

81. Stanglmaier R, Roberts C. Homogeneous Charge Compression Ignition (HCCI): Benefits, Compromises, and Future Engine Applications. SAE Technical Paper No. 1999-01-3682, 1999.
82. Epping K, Aceves S, Bechtold R, Dec J. The Potential of HCCI Combustion for High Efficiency and Low Emissions. SAE Technical Paper No. 2002-01-1923, 2002.
83. Kitamura T, Ito T, Kitamura Y, Ueda M, Senda J, Fujimoto H. Soot Kinetic Modelling and Empirical Validation on Smokeless Diesel Combustion with Oxygenated Fuels. SAE Technical Paper No. 2003-01-1789, 2003.
84. Ramesh N, Mallikarjuna JM. Evaluation of incylinder mixture homogeneity in a diesel HCClengine ACFD analysis. *Eng Sci Technol Int J*, 2016, 19(2):917–25.
85. Ryan TW, Matheaus C. Fuel Requirements for HCCI Engine Operation. SAE Technical Paper No. 2003-01-1813, 2003.
86. Aceves SM, Flowers D, Frias JM, Loza FE, Pitz WJ, Dibble R. Fuel and Additive Characterization for HCCI Combustion. SAE Technical Paper No. 2003-01-1814, 2003.
87. Maurya RK, Agarwal AK. Combustion and Emission Behaviour of Ethanol Fuelled Homogeneous Charge Compression Ignition (HCCI) Engine. SAE Technical Paper No. 2008-28-0064, 2008.
88. Dev S, Chaudhary HB, Gothekar S, Juttu S, Walke NH, Marathe NV. Advanced Low Temperature Combustion Approach for BSVI. SAE Technical Paper No. 2017-26-0042, 2017.
89. Maurya RK, Agarwal AK. Effect of intake air temperature and air fuel ratio on particulating gasoline and *n*-butanol fuelled HCCI engine. *Int J Engine Res*, 2014, 15(7):789–804.
90. Haraldsson G, Tunestal P, Johansson B, Hy Aonen J. HCCI Closed Loop Combustion Control Using Fast Thermal Management. SAE Technical Paper No. 2004-01-0943, 2004.
91. Yamaoka S, Kakuya H, Nakagawa S, Okada T, Shimada A, Kihara Y. HCCI Operation Control in a Multi Cylinder Gasoline Engine. SAE Technical Paper No. 2005-01-0120, 2005.
92. Araldsson G, Tunestal P, Johansson B, Hy Aonen J. HCCI Combustion Phasing with Closed-Loop Combustion Control Using Variable Compression Ratio in a Multi-Cylinder Engine. SAE Technical Paper No. 2003-01-1830, 2003.
93. Caton PA, Song HH, Kaahaaina NB, Ed Wards CF. Residual effected homogeneous charge compression ignition with delayed intake valve closing at elevated compression ratio. *Int J Engine Res*, 2005, 6(4):399–419.
94. Law D, Kemp D, Allen J, Kirkpatrick G, Copland T. Controlled Combustion in an IC Engine with a Fully Variable Valve Train. SAE Technical Paper No. 2000-01-0251, 2000.
95. Agrell F, Angstr€Aom HE, Eriksson B, Wikander J, Linderyd J. Integrated Simulation and Engine Test of Closed-Loop HCCI Control by Aid of Variable Valve Timings. SAE Technical Paper No. 2003-01-0748, 2003.
96. Handford DI, Checkel MD. Extending the Load Range of a Natural Gas HCCI Engine Using Direct Injected Pilot Charge and External EGR. SAE Technical Paper No. 2009-01-1884, 2009.
97. Nakagome K, Shimazaki N, Niimura K, Kobayashi S. Combustion and Emission Characteristics of Premixed Lean Diesel Combustion Engine. SAE Technical Paper No. 970898, 1997.
98. Milovanovic T, Blundell D, Pearson R, Turner J, Chen R. Enlarging the Operational Range of a Gasoline HCCI Engine by Controlling the Coolant Temperature. SAE Technical Paper No. 2005-01-0157, 2005.
99. Christensen M, Johansson B. Homogeneous Charge Compression Ignition with Water Injection. SAE Technical Paper No. 1999-01-0182, 1999.

100. Akagawa H, Miyamoto T, Harada A, Sasaki S, Shimazaki N, Hashizume T, et al. Approaches to Solve Problems of the Premixed Lean Diesel Combustion. SAE Technical Paper No. 1999-01-0183, 1999.
101. Tunestal P, Olsson JO, Johansson B. Closed loop control of HCCI engines. In: Johansson R, Rantzer A, editors. *Non Linear and Hybrid Systems in Automotive Control.* Springer-Verlag, New York, 2003.
102. Yap D, Megaritis A, Peucheret S, Wyszynski ML, Xu H. Effect of Hydrogen Addition on Natural Gas HCCI Combustion. SAE Technical Paper No. 2004-01-1972, 2004.
103. Maurya RK, Agarwal AK. Experimental investigations of particulate size and number distribution in an ethanol and methanol fuelled HCCI engine. *ASMEJ Energy Resour Technol*, 2015, 137(1):012201–10.
104. Iwabuchi Y, Kawai K, Shoji T, Takeda Y. Trial of New Concept Diesel Combustion System Premixed Compression Ignition Combustion. SAE Technical Paper No. 1999-01-0185, 1999.
105. Sun R, Thomas R, Gray CL. An HCCI Engine: Power Plant for a Hybrid Vehicle. SAE Technical Paper No. 2004-01-0933, 2004.
106. Olsson JO, Tunestal P, Johansson B. Boosting for High Load HCCI. SAE Technical Paper No. 2004-01-0940, 2004.
107. Olsson JO, Tunestal P, Haraldsson G, Johansson B. A Turbo Charged Dual Fuel HCCI Engine. SAE Technical Paper No. 2001-01-1896, 2000.
108. Koopmans L, Strom H, Lundgren S, Backlund O, Denbratt I. Demonstrating a SI-HCCI-SI Mode Change on a Volvo5-Cylinder Electronic Valve Control Engine. SAE Technical Paper No. 2003-01-0753, 2003.
109. Santoso H, Matthews J, Cheng WK. Managing SI-HCCI Dual Mode Engine Operation. SAE Technical Paper No. 2005-01-0162, 2005.
110. Guezennec Y, Musardo C, Staccia B, Midlam-Mohler S, Calo E, Pisu P, et al. Supervisory Control for NO_x Reduction of an HEV with a Mixed-Mode HCCI/DI Engine. SAE Technical Paper No. 2004-05-0123, 2004.
111. US Department of Energy. *A Report to the U.S. Congress: Homogeneous Charge Compression Ignition (HCCI) Technology*. US Department of Energy, New York, 2001.
112. Mack JH, Flowers DL, Buchholz BA, Dibble RW. Investigation of HCCI combustion of diethyl ether and ethanol mixtures using carbon 14 tracing and numerical simulations. *Proc Combust Inst*, 2005, 30(2):2693–700.
113. Kawano D, Naito H, Suzuki H, Ishii H, Hori S, Goto Y, et al. Effects of Fuel Properties on Combustion and Exhaust Emissions of Homogeneous Charge Compression Ignition (HCCI) Engine. SAE Technical Paper No. 2004-01-1966, 2004.
114. Harada A, Shimazaki N, Sator S, Miyamoto T, Akagawa H, Tsujimura K. The Effects of Mixture Formation on Premixed Lean Diesel Combustion. SAE Technical Paper No. 980533, 1998.
115. Christensen M, Johansson B, Einewall P. Homogeneous Charge Compression Ignition (HCCI) Using Isooctane, Ethanol and Natural Gas a Comparison with Spark Ignition Operation. SAE Technical Paper No. 972874, 1997.
116. Aoyama T, Hattori Y, Mizuta J, Sato Y. An Experimental Study on Premixed Charge Compression Ignition Gasoline Engine. SAE Technical Paper No. 960081, 1996.
117. Kalghatgi GT, Head RA. Combustion limits and efficiency in a homogeneous charge compression ignition engine. *Int J Engine Res*, 2006, 7(3):215–36.
118. Koopmans L, Backlund O, Denbratt I. Cycle to Cycle Variations: Their Influence on Cycle Resolved Gas Temperature and Unburned Hydrocarbons from a Camless Gasoline Compression Ignition Engine. SAE Technical Paper No. 2002-01-0110, 2002.
119. Shaver GM, Roelle MJ, Gerdes JC. Modeling cycle to cycle dynamics and mode transition in HCCI engines with variable valve actuation. *J Control Eng Pract*, 2006, 14:213–22.

120. Shahbakhti M, Koch CR. Characterizing the cyclic variability of ignition timing in a homogeneous charge compression ignition engine fuelled with *n*-heptane/iso-octane blend fuels. *Int J Engine Res*, 2008, 9:361–97.
121. Olsson J, Tunest LP, Johansson B, Fiveland S, Agama R,Willi M, etal. Compression ratio influence on maximum load of a natural gas fuelled HCCI engine. SAE Technical Paper No. 2002-01-0111, 2002.
122. Tunestal P, Johansson B. HCCI control. In: Zhao H, editor. *HCCI and CAI Engines for the Automotive Industry*. Wood head Publishing Limited, England, 2007.
123. Bahng GW, Jang D, Kim Y, Shin M. Anew technology to over come the limits of HCCI engine through fuel modification. *Appl Therm Eng*, 2016, 98:810–5.
124. Gowthaman S, Sathiyagnanam AP. Effects of charge temperature and fuel injection pressure on HCCI engine. *Alex Eng J*, 2016, 55(1):119–25.
125. Kawano D, Suzuki H, Ishii H, Goto Y. Ignition and combustion control of diesel HCCI. SAE Technical Paper No. 2005-01-2132, 2005.
126. Aldawood A, Mosbach S, Kraft M. HCCI Combustion control using dual fuel approach: experimental and modelling investigations. SAE Technical Paper No. 2012-01-1117, 2012.
127. Xingcai L, Yuchun H, Linlin Z, Zhen H. Experimental study on the auto ignition and combustion characteristics in the homogeneous charge compression ignition (HCCI) combustion operation with ethanol/n-heptane blend fuels by port injection. *Fuel*, 2006, 85:2622–31.
128. Ma J, Lu X, Ji L, Huang Z. An experimental study of HCCI-DI combustion and emissions in a diesel engine with dual fuel. *Int J Therm Sci*, 2008, 47:1235–42.
129. Kokjohn S, Hanson R, Splitter D, Reitz R. Experiments and modelling of dual fuel HCCI and PCCI combustion using in-cylinder fuel blending. SAE Technical Paper No. 2009-01-2647, 2009.
130. Singh AP, Agarwal AK. Diesoline, diesohol and diesosene fuelled HCCI engine development. *J Energy Res Technol*, 2016, 138:052212.
131. Genchi G, Pipitone E. Preliminary experimental study on double fuel HCCI com bustion. *Energy Procedia*, 2015, 81:784–93.
132. Shaver GM, Gerdes JC, Roelle M. Physics Based Closed Loop Control of Phasing, Peak Pressure and Workout Put in HCCI Engine Sutilizing Variable Valve Actuation. In: *Proceedings of the American Control Conference*, Stanford, CA, 2004, pp. 150–5.
133. Milovanovic N, Chen R, Turner J. Influence of the variable valve timing strategy on the control of a homogeneous charge compression (HCCI) engine. SAE Technical Paper No. 2004-01-1899, 2004.
134. Jia M, Li Y, Xie M, Wang T, Wang H, Reitz RD. The potential of high-load extension by using late intake valve closing for a diesel premixed charge compression ignition (PCCI) engine. *Energy Proceedia*, 2015, 66:33–6.
135. Babaji Mopoulos A, Lavoie GA, Assanis DN. Modelling HCCI combustion with high levels of residual gas fraction- acomparison of two VVA strategies. SAE Technical Paper No. 2003-01-3220, 2003.
136. Singh AP, Agarwal AK. Combustion characteristics of diesel HCCI engine: An experimental investigation using external mixture formation technique. *Appl Energy*, 2012, 99:116–25.
137. D'Ambrosio S, Ferrari A. Effects of exhaust gas recirculation in diesel engines featuring late PCCI type combustion strategies. *Energy Convers Manage*, 2015, 105:1269–80.
138. Cınar C, Uyumaz A, Polat S, Yılmaz E, Can O, Solmaz H. Combustion and performance characteristics of an HCCI engine utilizing trapped residual gas via reduced valve lift. *Appl Therm Eng*, 2016, 100:586–94.

139. Kimura S, Ogawa H, Matsui Y, Enomoto Y. An Experimental analysis of low temperature and premixed combustion for simultaneous reduction of NO_x and particulate emissions indirect injection diesel engines. *Int J Engine Res*, 2002, 3(4):249–59.
140. Ishikawa N, Ohkubo Y, Kudou K. Study on the Effects of EGR Cooler Performance on Combustion Properties of the Premixed Compression Ignition Combustion by Multi Cylinder DI Diesel Engine. SAE Technical Paper No. 2007-01-1881, 2007.
141. Aldicheria C, Pickett LM. Soot formation in diesel combustion under high EGR conditions. Sandia National Laboratories (No. SAND2005-2879C), 2005.
142. Hardy WL, Reitz RD. A Study on the Effects of High EGR, High Equivalence Ratio,and Mixing Time One Missions Levels in a Heavy Duty Diesel Engine for PCCI Combustion. SAE Technical Paper No. 2006-01-0026, 2006.
143. Niemeyer KE, Daly SR, Cannella WJ, Hagen CL. An ovel fuel performance index for low temperature combustion engines based on operating envelopes in light duty driving cycle simulations. *J Eng Gas Turbines Power*, 2015, 137(10):101–10.
144. Kalghatgi GT. Fuel effects in CAI gasoline engines. In: Zhao H, editor. *HCCI and CAI Engines for the Automotive Industry*. Wood head Publishing Limited, England, 2007.
145. Starck L, Lecointe B, Forti L, Jeuland N. Impact of fuel characteristics on HCCI combustion: performances and emissions. *Fuel*, 2010, 89:3069–77.
146. Lee S, Reitz RD. Spray Targeting to Minimizesoot and CO Formation in Premixed Charge Compression Ignition (PCCI) Combustion with a HSDI Diesel Engine. SAE Technical Paper No. 2006-01-0918, 2006.
147. Olsson J, Johansson B. Closed Loop Control of an HCCI Engine. SAE Technical Paper No. 2001-01-1031, 2001.
148. Strandh P, Bengtsson J, Johansson R, Tunestl P, Johansson B. Variable Valve Actuation for Timing Control of a Homogeneous Charge Compression Ignition Engine. SAE Technical Paper No. 2005-01-0147, 2005.
149. Midlam-Mohler S, Haas S, Guezennec Y, Bargende M, Rizzoni G. Mixed Mode Diesel HCCI/DI with External Mixture Preparation. SAE Technical Paper No. 2004-05-0446, 2004.
150. Hou JX, Qiao XQ. Characterization of knocking combustion in HCCI DME engine using wavelet packet transform. *Appl Energy*, 2010, 87:1239–46.
151. Dec JE. A Computational Study of the Effects of Low Fuel Loading. SAE Technical Paper No. 2002-01-1309, 2002.
152. Gomaa M, Alimin A, Kamarudin K. The effect of EGR rates on NO_x and smoke emissions of an IDI diesel engine fuelled with Jatropha biodiesel blends. *Int J Energy Environ Eng*, 2011, 2:477–90.
153. Agarwal D, Singh SK, Agarwal AK. Effect of exhaust gas recirculation (EGR) on performance, emissions, deposits and durability of a constant speed compression ignition engine. *Appl Energy*, 2011, 88:2900–7.
154. Shin B, Cho Y, Han D, Song S, Chun KM. Hydrogen effects on NO_x emissions and brake thermal efficiency in a diesel engine under low-temperature and heavy-EGR conditions. *Int J Hydrogen Energy*, 2011, 36:6281–91.
155. Turns SR. *An Introduction to Combustion: Concepts and Applications*. McGraw-Hill, New York, 1996.
156. Saleh HE. Effect of exhaust gas recirculation on diesel engine nitrogen oxide reduction operating with jojoba methyl ester. *Renew Energy*, 2009;34:2178–86.
157. Tauzia X, Maiboom A, Shah SR. Experimental study of inlet manifold water injection on combustion and emissions of an automotive direct injection diesel engine. *Energy*, 2010;35:3628–39.
158. Adnan R, Masjuki HH, Mahlia TMI. Performance and emission analysis of hydrogen fueled compression ignition engine with variable water injection timing. *Energy*, 2012, 43:416–26.

159. Badrana O, Emeishb S, Abu-Zaidc M, Abu-Rahmaa T, Al-Hasana M, AlRagheba M. Impact of emulsified water/diesel mixture on engine performance and environment. *Int J Thermal Environ Eng*, 2011, 3:1–7.
160. Tran X, Ghojel J. Impact of Introducing Water into the Combustion Chamber of Diesel Engines on Emissions-An Overview. In: *Proceedings of the 5th Asia-Pacific Conference on Combustion*, Adelaide SA, 2005, pp. 17–20.
161. Tree DR, Svensson KI. Soot processes in compression ignition engines. *Prog Energy Combust Sci*, 2007, 33:272–309.
162. Fu WB, Hou LY, Wang L, Ma FH. A unified model for the micro-explosion of emulsified droplets of oil and water. *Fuel Process Technol*, 2002, 79:107–19.
163. Parks Ii JE, Prikhodko V, Storey JME, Barone TL, Lewis Sr SA, Kass MD, et al. Emissions from premixed charge compression ignition (PCCI) combustion and affect on emission control devices. *Catal Today*, 2010,151:278–84.
164. Singh S, Reitz RD, Musculus MPB, Lachaux T. Simultaneous optical diagnostic imaging of low-temperature, double injection combustion in a heavy-duty DI diesel engine. *Combust Sci Technol*, 2007, 179:2381–414.
165. Shropshire GJ, Bashford LL. Comparison of ethanol fumigation systems for a diesel engine. *Agric Eng*, 1987, 65(5):17–24.
166. Tormos B, Novella R, García A, Gargar K. Comprehensive study of biodiesel fuel for HSDI engines in conventional and low temperature combustion conditions. *Renew Energy*, 2010, 35:368–78.
167. Zheng M, Mulenga MC, Reader GT, Wang M, Ting DSK. Influence of Biodiesel Fuel on Diesel Engine Performance and Emissions in Low Temperature Combustion. SAE Technical Paper No. 2006-01-3281, 2006.
168. Peterson BR, Ekoto IW, Miles PC. An Investigation into the Effects of Fuel Properties and Engine Load on UHC and CO Emissions from a Light-Duty Optical Diesel Engine Operating in a Partially Premixed Combustion Regime. SAE Technical Paper No. 2010-01-1470, 2010.
169. Swami Nathan S, Mallikarjuna J, Ramesh A. Effects of charge temperature and exhaust gas re-circulation on combustion and emission characteristics of an acetylene fuelled HCCI engine. *Fuel*, 2010, 89:515–21.
170. Opat R, Ra Y, Manuel A, Krieger R, Reitz RD, Foster DE, et al. Investigation of Mixing and Temperature Effects on HC/CO Emissions for Highly Dilute Low Temperature Combustion in a Light Duty Diesel Engine. SAE Technical Paper No. 2007-01-0193, 2007.

Index

For Product Safety Concerns and Information please contact our EU
representative GPSR@taylorandfrancis.com
Taylor & Francis Verlag GmbH, Kaufingerstraße 24, 80331 München, Germany

www.ingramcontent.com/pod-product-compliance
Ingram Content Group UK Ltd.
Pitfield, Milton Keynes, MK11 3LW, UK
UKHW022220190926
475702UK00003B/472
* 9 7 8 1 0 3 2 9 8 3 2 4 0 *